# Reforesting Landscapes

Landscape Series

Volume 10



**Aims and Scope**

Springer's innovative Landscape Series is committed to publishing high-quality manuscripts that approach the concept of landscape from a broad range of perspectives. Encouraging contributions on theory development, as well as more applied studies, the series attracts outstanding research from the natural and social sciences, and from the humanities and the arts. It also provides a leading forum for publications from interdisciplinary and transdisciplinary teams.

Drawing on, and synthesising, this integrative approach the Springer Landscape Series aims to add new and innovative insights into the multidimensional nature of landscapes. Landscapes provide homes and livelihoods to diverse peoples; they house historic – and prehistoric – artefacts; and they comprise complex physical, chemical and biological systems. They are also shaped and governed by human societies who base their existence on the use of the natural resources; people enjoy the aesthetic qualities and recreational facilities of landscapes, and people design new landscapes.

As interested in identifying best practice as it is in progressing landscape theory, the Landscape Series particularly welcomes problem-solving approaches and contributions to landscape management and planning. The ultimate goal is to facilitate both the application of landscape research to practice, and the feedback from practice into research.

For other titles published in this series, go to
www.springer.com/series/6211

Harini Nagendra • Jane Southworth
Editors

# Reforesting Landscapes

## Linking Pattern and Process

Springer

*Editors*
Harini Nagendra
Indiana University
Bloomington, IN
USA
and
Ashoka Trust for Research
in Ecology and Environment (ATREE)
Bangalore
India
nagendra@indiana.edu

Jane Southworth
University of Florida
Gainesville, FL
USA
jsouthwo@geog.ufl.edu

ISBN 978-1-4020-9655-6 e-ISBN 978-1-4020-9656-3
DOI 10.1007/978-1-4020-9656-3
Springer Dordrecht Heidelberg London New York

Library of Congress Control Number: 2009929348



Printed on acid-free paper

Springer is part of Springer Science+Business Media (www.springer.com)

# Foreword by the Series Editors

With the Springer Landscape Series we want to provide a much-needed forum for dealing with the complexity of landscape types that occur, and are studied, globally. It is crucial that the series highlights the richness of global landscape diversity – both in the landscapes themselves and in the approaches used in their study. Moreover, while the multiplicity of relevant academic disciplines and approaches is characteristic of landscape research, we also aim to provide a place where the synthesis and integration of different knowledge cultures is common practice.

This book, Reforesting Landscapes, the tenth in the series, marks a shift in the research perspective from focusing on deforestation to a broader view on regrowth, reforestation, and afforestation. It is particularly timely, as a growing body of literature gives evidence for forest regrowth in developed as well as developing countries. This trend has important implications for biodiversity, carbon sequestration, soil maintenance and reduction of greenhouse gases.

We are proud that this book is part of the Landscape Series, which focuses on integrative aspects in landscape research. The way the authors approach the subject is integrative in a number of ways. Regrowth, reforestation and afforestation are not discussed isolated, but within their biophysical, geographic, ecological, socio-economic, and institutional contexts. The authors have integrated experiences from multiple research locations across the world, thus giving a unique picture of similarities and differences of forestation processes in various physical and social environments. Data gathering reflects an interdisciplinary approach, in which methods and techniques from natural sciences and social sciences complement each other. Field-based examinations ensure practical applicability of the results. Integration was also reached across geographical scales, with some chapters giving a broader international overview and others exemplifying projects on a regional and local scale. Both volume editors are distinguished scholars, with broad experiences in landscape research, landscape ecology, and geography from a variety of different international institutions. The book gives a unique overview for everybody interested in processes of regrowth and reforestation. It will be warmly welcome by professionals as well as researchers and students across the world.

Munich and Toulouse, March 2009

Bärbel Tress
Gunther Tress
Henri Décamps

# Contents

# Chapter 1
# Reforestation: Challenges and Themes in Reforestation Research

**Jane Southworth and Harini Nagendra**

J. Southworth (✉)
Department of Geography and Land Use and Environmental Change Institute (LUECI), University of Florida, FL, USA
e-mail: jsouthwo@geog.ufl.edu

H. Nagendra
Indiana University, Bloomington, IN, USA and
Ashoka Trust for Research in Ecology and the Environment (ATREE), Bangalore, India

H. Nagendra and J. Southworth (eds.), *Reforesting Landscapes: Linking Pattern and Process*, Landscape Series 10,
DOI 10.1007/978-1-4020-9656-3_1, © Springer Science+Business Media B.V. 2010

## 1.1 Introduction

Tropical forest habitat continues to decline globally, with serious negative consequences for environmental sustainability (Rudel 2005). Perhaps as a consequence, studies of land cover change have long been dominated by discussions of deforestation. Most studies on land cover change have been focused on deforestation occurring in different countries, monitored by national level databases such as FAO, regional studies of deforestation in hotspots of deforestation such as the Amazon and Southeast Asia, national level studies and even smaller case studies of specific landscapes (Cropper et al. 2001 ; Seidl et al. 2001; Messina et al. 2006; Kao and Iida 2006). Yet in recent times there has been a growing awareness in the land use/land cover change research community, and amongst landscape ecologists, of the need to move away from the dominant focus on deforestation to examine the patterns and processes associated with reforesting landscapes (Rudel 2005).

There is an increasing body of literature which suggests a recent trend towards forest regrowth in regions across the world. Such forest transitions have been documented in economically developed countries in the temperate world, with the majority of these transitions having occurred towards the last half of the twentieth century. In the past couple of decades, there has been growing evidence of large scale forest regrowth also taking place in tropical and sub-tropical forests, across multiple continents. Even in landscapes which exhibit deforestation, a number of recent studies have increasingly focused on regrowth, reforestation and afforestation, often hand in hand with deforestation and degradation processes (Moon and Park 2004; Munroe et al. 2004; Nagendra et al. 2008; Southworth and Tucker 2001).

The dual and simultaneous focus on regrowth/regeneration and reforestation/afforestation is a welcome change and has serious implications for global biodiversity, carbon sequestration, soil maintenance and reduction of greenhouse gases that contribute to global climate change (Grainger 2008). Reforestation is often patchy, and rates of forest recovery are typically slower than initial rates of clearing. These emerging forests often do not contain the same species or supply the same range of ecosystem goods and services provided by old growth forests (Bentley 1989; Lugo 1992; Rudel et al. 2000). Nevertheless, secondary forests provide important environmental services that assist efforts towards sustainable development, increase carbon sequestration, assist in soil conservation and the stabilization of hydrological cycles, and increase overall biodiversity levels. Developing a more comprehensive understanding of the range of proximate and underlying factors that can help to promote reforestation is therefore critical, if we are to develop useful policy interventions to arrest or reverse deforestation, and encourage forest regrowth (Rudel et al. 2005). Yet it is important to recognize that forests are embedded within larger-level ecological, socio-economic and political settings, which have the capacity to significantly influence outcomes. Thus, discussions of context – biophysical, geographic, ecological, socio-economic and institutional – are essential to the development of our understanding of this area of study.

Despite the increase in case studies examining the patterns and processes of reforestation, though (Moon and Park 2004; Munroe et al. 2004; Nagendra et al.

2008; Southworth and Tucker 2001), there have been few efforts to integrate these findings across multiple research locations. If we are to identify and encourage such processes where they are occurring in different parts of the globe, there is a pressing need to establish broader frameworks to guide our understanding of the drivers associated with reforesting landscapes. A book addressing the issues relating to reforestation, regeneration and regrowth of forest cover from around the world is thus long overdue within the landscape research community.

## 1.2 Description of the Book

The idea for this edited volume emerged from such an awareness of the critical need for cross-site empirical studies examining the patterns and processes impacting reforestation in a variety of field contexts. In this book, we have integrated research findings from scientists working in a range of contexts and continents and utilizing a variety of integrated, inter-disciplinary approaches to examine reforestation. The cross-site examinations conducted here by scientists working in a variety of different field settings can help us narrow down the larger set of potential variables to identify specific factors that are important in a given context.

Reforestation and regrowth issues are addressed from multiple dimensions of ecosystem services, protected areas, social institutions, economic transitions, remediation of environmental problems, conservation, land abandonment, and both micro and macro level drivers of forest regrowth. This volume sets out to address these issues on a global scale, incorporating research from North America, South America, Central America, Africa, Asia and Europe. Consequently, a diversity of issues can be addressed, common threads discussed and compiled, and different drivers and patterns established.

This book is targeted towards an interdisciplinary research community working on issues relevant to the biophysical, geographic, socioeconomic and institutional processes associated with reforestation. Methods used to study these patterns and processes range from the use of techniques of satellite remote sensing, aerial photography, historical maps and GIS, to the collection of intensive field data, economic and social datasets, and the study of political and social institutions. These data are used in concert, which also leads the often interdisciplinary teams to directly address the issues of scale (spatial, spectral and temporal), the organizational and functional level of analysis, the timing of data acquisition and the linkage of the social and biophysical datasets, all necessary to answer their research questions.

To fully understand both the social and ecological dimensions of regional land cover change, requires both fine scale data and in depth field studies. On the other hand, placing specific case studies within the larger body of literature and linking multiple case studies is a pre-requisite to synthesis and theoretical progress (Rudel et al. 2005). Thus, critically, such field based examinations are also closely integrated with theoretically motivated examinations of literature, to distil the complex set of potential driving factors and narrow in on variables that are the most important in different contexts, achieving clarity without sacrificing relevant detail.

The book is organized in a hierarchical fashion, beginning with chapters that lay out broader frameworks for the study of reforesting landscapes, then moving to regional studies of reforestation, and from there, finally, to case studies of reforestation in specific landscapes. The geographic scope is vast and varied, covering countries as diverse as Bhutan, Madagascar, Peru, Poland, USA and Vietnam. While all authors address issues of specific importance to their landscapes of focus, there are some common themes that link these discussions.

We begin with a set of three chapters which discuss issues relevant to reforestation research across all landscapes. In Chapter 2, Grainger discusses the challenges of monitoring long term forest change. When deforestation and reforestation simultaneously take place in an area, problems of aggregation ensue, leading to substantial uncertainties about the nature and extent of forest change. Chapter 3 (Rudel) then uses a comparative historical approach to outline the human drivers of forest expansion, describing three pathways that give rise to reforestation under different social circumstances, and describing policy initiatives that would help encourage such change. Chapter 4 (Perz and Almeyda) presents an alternate tri-partite framework, drawing on concepts of hierarchy, heterarchy and panarchy to develop multi-scale perspectives of reforestation that reconcile short term, medium term and long term forest dynamics, and explore drivers of forest change at scales ranging from the local to the global.

Latin America, Eastern Europe and South Asia form three major regions of the world that have experienced large scale reforestation in recent times, and Chapters 5–7 discuss reforestation in these different regional contexts. Chapter 5 (Bray) focuses on the dynamics of forest transition in Mexico and Central America, while Chapter 6 (Taff et al.) examines reforestation in Central and Eastern Europe, and Chapter 7 (Nagendra) looks at the drivers of reforestation in South Asia.

Chapters 8–15 develop themes relevant to reforestation research in specific landscapes located in a variety of social, institutional, biophysical, economic and historical settings. These landscapes range from the Midwestern USA (Chapter 8, Evans et al.) to the Peruvian Amazon (Chapter 9, Crews and Moffett), northwestern Costa Rica (Chapter 10, Daniels), the Polish Carpathian mountains (Chapter 11, Kozak), Kibale National Park in Uganda (Chapter 12, Hartter et al.), southern Madagascar (Chapter 13, Elmqvist el al), Vietnam and Bhutan (Chapter 14, Meyfroidt and Lambin) and China (Chapter 15, Song and Zhang). A diversity of issues critical to understanding the social and ecological aspects of reforestation are addressed in this range of landscapes and socio-ecological contexts.

## 1.3 Challenges for Reforestation Research

Through this book, our endeavor is to map our current state of knowledge on reforestation, to outline the gaps in our understanding, and to identify the major challenges for reforestation research. Based on discussions with all authors contributing to this volume, we have identified major challenges critical to reforestation studies, which are further addressed in the chapters in this volume. These main themes or challenges within this field of study are discussed in Table 1.1, and linked to relevant chapters in the book which address these challenges or themes in greater detail.

These dominant themes and challenges will be addressed within this volume but their importance also deserves some brief attention here in the introduction, in terms of what we mean by each of these themes and how these are also challenges to the future research needs within these fields. The following section will briefly address these themes.

### *1.3.1 Definitions of Reforestation*

Definitions for the terms of reforestation, afforestation, regeneration, etc. are used differently in different disciplines and areas of research. This range of terms reflects the range of contexts and meanings that these terms represent, and as such, this diversity mirrors the interdisciplinarity of thought represented in reforestation research. This volume consequently reflects the same diversity of meaning. All authors clearly describe the terms they use and their meaning in each chapter, providing clarity. In a broader sense though, the use of different terms interchangeably, across different disciplines and methodologies, is a major problem not just within this arena, but also within many fields of research that similarly cross disciplinary and methodological boundaries. Unfortunately, however, the field of reforestation research is a nascent one and does not appear ready as yet for the emergence of standardized terminologies that researchers from different disciplines will be willing to adopt. The most we can do at this point is to clearly define each author's use of such terminology and to take care that the uses of such terms are appropriate for the specific subject matter under discussion.

### *1.3.2 Interdisciplinarity*

Human–Environment interactions and their study, especially under changing conditions, pose challenges to research requiring not only a *multidisciplinary approach*, with contributions from several social and biophysical science disciplines, but also an *interdisciplinary conceptual framework* that integrates social and ecological changes and their iterative feedbacks. Due to the complexity of social, economic, institutional, biophysical, ecological and policy drivers, reforestation requires an inherently interdisciplinary approach to a much greater degree than many other land use conversions. While reforestation may be a biophysical process and as such much in the realm of ecologists and other physical scientists, in order to understand the drivers of these processes and to better explain these systems we

**Table 1.1** Dominant themes within the study of reforestation and regeneration – compiled by Southworth and Nagendra with inputs from all authors of this volume

| Theme/challenge | Description | Chapters in this volume which address this topic |
|---|---|---|
| 1. Definitions of reforestation | The multitude of co-existing definitions of reforestation is a major issue. | All chapters address this issue, with many of the chapters using their own definitions |
| 2. Interdiscplinarity | Reforesting landscapes are linked social-ecological systems, and to understand the process of reforestation an interdisciplinary approach is required which links both social and ecological research. In addition, we need multiscalar studies, over varying temporal and spatial scales. | Chapters 2–8, 10–15 |
| 3. Multiplicity of spatial and temporal scales | Reforestation occurs over a range of time periods and spatial scales – thus, drivers of reforestation need to be studied using an approach that incorporates an awareness of spatial and temporal scale. | Chapters 2–8, 10, 11, 14, 15 |
| 4. New methodological approaches | Better methodological approaches are required in addition to traditional classification analyses, to get at intermediate steps of change and to evaluate modification processes that occur within forest classes in addition to conversion from forest to non-forest areas or vice versa. Further, satellite remote sensing is a relatively young area of research, and approaches such as aerial photography, continuous analyses of land cover change, land cover modeling and historical comparative research are essential for a more complete understanding of landscape processes. | Chapters 2, 3, 8, 9, 11, 12, 14 |
| 5. Reforestation as a Process | The drivers of reforestation are not simply the inverse of drivers of deforestation, and are often distinct. In addition to processes which lead to reforestation we also need to understand processes by which forest has been maintained on the landscape, focusing less on the deforestation process and more on reforestation or forest maintenance. | All chapters |
| 6. Global focus | We need to move away from an exclusive focus on hotspots of deforestation and on tropical forests, towards more diverse studies of a range of ecosystems, including wetlands, coniferous forests, dry tropical forests and woodland savanna. We must also acknowledge and attempt to understand the global differences in the processes of reforestation. | Chapters 2–4, 6, 10, 11, 13–15 |

| | | |
|---|---|---|
| 7. Urbanization | This is becoming a dominant process in many regions around the globe. While urbanization appears to be linked to reforestation in surrounding rural areas, the long term consequences of urbanization on reforestation are unclear. We need to have stronger linkages with forest transition theory and urban growth, and to also study this process over time and see if there are trends which can be discerned over time. | Chapters 3, 5, 6, 8, 11, 14 |
| 8. Forest transition theory | Can FTT ever relate to a global process of reforestation, and if not what are the implications of this?<br>How useful is FTT for developing versus developed and what is the role of protected areas or parks within this work? | Chapters 2–8, 10, 12–14 |
| 9. Cultural and ecosystem processes and services | Current socio-cultural and ecological understanding of reforestation is limited and we need much more research in this area, including an understanding of the ecological processes associated with, and the cultural and ecosystem services offered by reforestation. | Chapters 6, 8, 10, 11, 14, 15 |
| 10. Future expansion of plantations | Where do plantations fit within a reforestation dialogue and what is their future role? | Chapters 2, 3, 7, 10, 14, 15 |

must look much more closely at the social component. People are an integral part of the dynamic, be it in the form of a land abandonment which has left an area to regenerate and ultimately return to forest cover, or a human led replanting effort. Thus, in order to understand the causes and consequences of these systems and their implications for reforestation, we need to model both the social and physical components, and to better understand the social and economic determinants of different management strategies. We need a much better understanding of the interplay between top-down processes such as policies, and bottom-up responses of the local land managers and local communities. People play a key role which is currently not well understood, and we must acknowledge that most of these landscapes are 'working' landscapes, both socially and ecologically. Thus, there is a need for the development of new approaches and frameworks that integrate across disciplines, and integrate different methodological approaches for the study of reforestation.

### *1.3.3 Multiplicity of Spatial and Temporal Scales*

Changes in climate, population and land use are occurring and interacting simultaneously at different time and space scales (Milly et al. 2008; Lettenmaier et al. 2008). Nonlinearities and differences in timescales and characteristic response times across key interfaces between land cover, the atmosphere, and the surface complicate efforts to monitor and model environmental processes. Up-scaling and down-scaling in space and time is a challenging problem (Bloschl and Sivapalan 1995). Incorporation of both spatial and spectral information into land-cover change analyses greatly improves the amount of information available to modeling studies (Southworth et al. 2006). For example, Lambin and Strahler (1994) found that changes in the spatial extent of land cover patches across the landscape and its arrangement or pattern were more likely to reveal longer lasting and longer-term land-cover changes, while spectral differences and within class changes are more sensitive to shorter-term fluctuations, for example, inter-annual variability in climatic conditions.

Anthropogenic, ecological and land-surface processes interact in reforesting landscapes at multiple spatial and temporal scales to create characteristic patterns (O'Neill et al. 1996). The relationships between temporally and spatially varying processes and patterns are poorly understood because of the lack of spatio-temporal observations of real landscapes over significant stretches of time (Southworth et al. 2004). Interacting anthropogenic, ecological and land-surface processes occur in landscapes at multiple scales. If we are to understand and manage the causes and consequences of anthropogenic effects on reforesting landscapes, it is imperative that we develop approaches to understanding spatial and temporal variation, the processes that produce the patterns that we observe, and the ways in which pattern-process relationships change with scale. Remote sensing has traditionally been considered an ideal tool for providing data to describe landscape patterns and dynamics. However, our understanding of the scale dependency of landscape pattern-process interactions is limited (Moody and Woodcock 1995). Understanding scaling effects is critical to our ability to better understand, model and/or predict landscape dynamics of reforestation, and specifically for understanding the roles of spatial and temporal heterogeneity and the hierarchical arrangement of landscape elements (Qi and Wu 1996).

### 1.3.4 New Methodological Approaches

While remote sensing has helped to advance our studies of land cover change and their drivers, the techniques currently utilized are often quite limited and very static in their approaches. The most commonly used technique for studying land cover change is that of discrete land cover classification. This only enables the study of changes in land cover, and does not allow us to view the extent of modification within a land cover category. Thus, for instance, such an approach would enable us to understand the extent of reforestation (conversion from a non-forest category to a forest category), but not enable us to study increases or decreases in density within a forest category, which are also critical for issues such as biodiversity, carbon sequestration, soil conservation and water management, and hence may be of much more import to forest managers, planners and policy makers. As such, we need to develop and test the use of more advanced remote sensing techniques, focusing more on the creation of continuous datasets, and different approaches to land cover modeling. We also need robust and detailed strategies to monitor and map reforestation by remote sensing from local to regional, and up to global scales where we differentiate degraded forests from secondary growth, plantations from natural forest, and separate out processes of regeneration and degradation, as this is a precondition for any understanding of the causes and effects. Tied in with this goal, is the need to incorporate multiple data sources, across different spatial, spectral, temporal and radiometric resolutions, to enable a more complete answering of our questions. Less reliance on Landsat will be beneficial in the long run, and is currently being dictated due to the failure of the Landsat 7 ETM sensor, and the lack of Landsat 8 readiness. Such a data gap will have massive repercussions on the land change science community and hence the evaluation of the integration of different sensors and technologies is now essential.

In addition, an added problem or current limitation relates to the fact that deforestation is a much easier process to "see" in terms of remotely sensed analyses, as this is itself a quick process when compared to that of reforestation due to the time it takes for trees to grow and to be of a size where the process of regeneration and forest expansion is visible to the satellite. To study forest expansion, satellite remote sensing seems too 'young' a discipline as compared to the time needed in many landscapes for a tree to grow to maturity (i.e. over thirty years in some landscapes).

The use of aerial photography, and approaches such as historical comparative analysis become critical in this regard.

### 1.3.5 Reforestation as a Process

Within this theme are a number of issues. A major one is that we need to focus less on unidirectional change within the field of land change science overall. Within the arena of reforestation we need to look at it not as a one-way process, it is not simply the reverse of deforestation. Rather it is a separate process, usually with separate and different drivers and we must understand these issues in order to fully understand the process, and to therefore help increase the occurrence of reforestation over deforestation (Rudel et al. 2005). We cannot start to do this until we have actually understood, monitored, mapped and modeled these occurrences, from individual case studies to global analyses, as we have started to do for deforestation. Our understanding of reforestation is much more limited and we have a lot of catching up to do within this area. For example, in FTT what factors determine where (at what percent forest cover) the reforestation phase stabilizes or plateaus? How does this pattern and trend vary across the local landscape, regional, national and international? Secondly, just as we know little on the drivers of reforestation we know even less about those that actually maintain forest cover to begin with, that is, drivers that are maintaining existing forests are little acknowledged, for example, coffee agroforestry and sustainable forest management. We need more landscape or regional scale studies of the ecological effects of reforestation.

### 1.3.6 Global Focus

A broader focus on understudied areas, latitudes and ecosystems, beyond the traditionally popular ‘tropical hotspots’ of deforestation, specifically extending into regions of less studied but critical and endangered ecosystems such as wetlands, dry tropical forest, coniferous forest and woodland/savanna is essential, as in these locations the dynamics of reforestation are very different. More of a focus is also required on the intermediate, human dominated and fragmented land cover types, for example, pastures with trees, suburban subdivisions with trees – that is to say, those landscapes within which many of us reside. We must also acknowledge and attempt to understand the global differences in the processes of reforestation. Currently while some forests (e.g., in the Amazon) are being encroached upon by migrating populations, in some parts of the world (e.g., much of Eastern Europe), populations are migrating to cities, abandoning croplands, many of which then undergo reforestation. Interestingly, such reforestation on abandoned agricultural lands has its own set of concerns or perceived risks, with the associated loss of some cultural landscapes in those regions.

### 1.3.7 Urbanization

The processes of urbanization and reforestation are tightly coupled within many systems. It is the consolidation of people within such concentrated regions that allows for land abandonment and hence, often, reforestation to occur. Such reforestation however takes place in areas away from the city, and thus, the impacts of urbanization need to be viewed at a landscape scale in order to perceive these outcomes of reforestation. Given the projected trends of increasing urbanization anticipated globally, we must better understand this relationship between urbanization and reforestation, and take advantage of possible opportunities for land abandonment and consequent natural reforestation. We also need to better understand the limits to this relationship and to predict under which conditions such processes may not result in increased forest cover. At a time when the pace and extent of reforestation in near-urban areas of both developed and developing countries is significant, this leads us to question whether reforestation can be expected to continuously proceed alongside the process of urban sprawl.

### 1.3.8 Forest Transition Theory

Forest Transition Theory has been successfully applied to understand reforestation in a range of countries, particularly in North America and Europe. Yet, despite the popularity of this theory, one issue that often arises is whether we can expect a return to deforestation in the future in places that have thus far transitioned from deforestation to reforestation (Rudel 2005). The time frame within which our current studies are located are often limited to the last 100 years, if not the last 35 years, and tightly linked to the advent of satellite remote sensing technologies. However, we may simply be in a transition phase when reforestation occurs, rather than this being an end point. Our time frame may not be the appropriate one for study, and we may start to see these areas of reforestation once again follow new trajectories which may lead them back to deforested landscapes. As such, is this forest transition theory really useful? Along these same lines, can we expect developing countries that have not yet transitioned from deforestation to reforestation to follow the same trajectory of developed countries that created the FTT? What relevance does reforestation in developed countries have for reforestation potential in developing countries?

We need to separate out drivers of reforestation in terms of their temporal and spatial differentiation. Much of what seemed to be critical drivers of reforestation from the late nineteenth century to before World War two (economic growth etc.) are different from current drivers of reforestation; and reforestation in different developing economies, in different parts of the world takes very different trajectories, for example, Eastern Europe with land abandonment versus South Asia with community institutions and plantation forestry.

It currently appears that there is something like a 'spatial diffusion' of forest transitions from Western Europe and North America, to some developing and transition economies including Eastern Europe, Asia, and Latin America. If forest transition becomes a global process then forest area globally will increase, but this is not really feasible to expect. Rather, we raise the question that forest transition can possibly never be global in scope. Instead, perhaps only some regions may experience it, while other countries may again be clearing forests. As such is forest transitions theory much more cyclical than currently believed?

### *1.3.9 Cultural and Ecosystem Processes and Services*

While ecologists are coming to better understand the processes associated with reforestation across landscapes (Bentley 1989; Lugo 1992) in terms of their ecology, we are still quite limited in our studies related to the social, economic and institutional roles in such conversions. In addition, the ecological understanding in some instances is also still limited. For instance, even if we observe an increased species richness in a regenerated or replanted forest area, this does not necessarily have positive implications for the landscape. An increase in the number of species can come at the cost of ecological integrity or ecosystem function, if the landscape is invaded by exotic species. Thus, we still lack knowledge on restoring important ecological functions and ecosystem services of importance for the production landscape. This can only be achieved through increased field research on restoration ecology, with field measurements in vegetation plots and transects to increase our knowledge of the ecosystem functions and processes associated with reforestation in different landscapes, associated with different species assemblages.

Additional questions, linked more tightly to theoretical constructs such as system resilience, are also critical to understand. What is the importance of reforestation in building resilience to large-scale disturbances? The fact that multiple directions of forest increase and decrease coexist in a landscape, often reversibly, needs more consideration in forestry studies. Quantification of ecosystem services provided by regrowing forests, such as the impact of reforestation on soil fertility, carbon sequestration, biodiversity protection, or hydrological cycles, is also essential for managers. Cultural services also play a key role. For instance, while in many parts of South Asia, forests are viewed as sacred and reforestation is a desirable process, reforestation in large parts of Eastern Europe is socially viewed as undesirable, leading to the disappearance of the traditional agricultural landscape that culturally defined large parts of the region.

Understanding spatial and temporal scale dependencies in reforestating landscapes is critical here. How does resolution of our data affect, or even define, our understanding of the spatial patterns and temporal pace of forest expansion? How big must a patch of forest have to be to count as "forest"? What stem density, successional stage, canopy closure should characterize "forest" for different regions? This will be context and site-dependent – thus, are there empirical scaling relationships that can be used to develop relationships across sites? These are key questions in need of resolution.

### *1.3.10 Future Expansion of Plantations*

We need a better understanding of the role of plantations, as well as timber trades and trade policies regarding natural forest preservation. Do the increases in area for plantations worldwide alleviate the pressure on natural forests, or at the contrary lead to further deforestation by reducing incentives to manage natural forests? Do timber trade and policies lead to an adjustment to the optimal natural conditions of tree growth, or at the contrary to leakage and a "moving wall" of forest exploitation? This also links in with the issue of carbon sequestration, as we need to better understand the carbon sequestration potential of planned reforestation, natural reforestation and plantations. Further research also needs to be conducted into the effectiveness or failure of many programs currently widespread, that compensate local populations for facilitating reforestation on their lands, through approaches such as the purchase of carbon credits.

## 1.4 Concluding Remarks

In conclusion, this book represents one of the first large scale efforts to provide an interdisciplinary, cross-country, multi-scalar perspective on reforestation that includes an integration of theoretical perspectives with empirical analyses from many parts of the world. In this introductory chapter, we outline a number of themes and challenges that are particularly relevant for reforestation research. The rest of the chapters go on to address these and other issues relevant to reforestation at a variety of scales, using a range of diverse methodological and theoretical perspectives that provide an overarching, innovative, interdisciplinary approach to studying the patterns and processes (or the "whats", "hows" and "whys") of reforesting landscapes. In the concluding chapter, we will return to the themes outlaid here, use these to provide an assessment of thesis findings, and to highlight some of the major challenges for future research in this field. This volume, as presented here, will thus provide one of the first global assessments of reforestation process, a knowledge which is critical to understanding the future of forests and biodiversity in an increasingly human impacted world.

**Acknowledgments** The idea of this book grew out of many discussions at the Center for the Study of Institutions, Population and Environmental Change (CIPEC), at Indiana University, and we would like to thank our colleagues at CIPEC for the many helpful exchanges we have had with them relating to reforestation research. We are also indebted to our families - Jane Southworth would like to thank Romer Kohlhaas and Parker Kohlhaas, and Harini Nagendra would like to thank Venkatachalam Suri and Dhwani Suri for their patience with us as we worked our way through this book. Harini also gratefully acknowledges financial support from the Branco Weiss Society in Science fellowship.

## References


Bentley J (1989) Bread forests and new fields: the ecology of reforestation and forest clearing among small woodland owners in Portugal. J For Hist 17:188–195

Bloschl G, Sivapalan M (1995) Scale issues in hydrologic modeling – A review. Hydrol Process 9:251–290

Cropper M, Puri J, Griffiths C (2001) Predicting the location of deforestation: the role of roads and protected areas in North Thailand. Land Econ 77(2):172–186

Food and Agriculture Organization (FAO) (2006) Global Forest Resources Assessment 2005. Food and Agriculture Organization of the United Nations, Rome

Grainger A (2008) Difficulties in tracking the long-term global trend in tropical forest area. Proc Natl Acad Sci U S A 105:818–823

Kao D, Iida S (2006) Structural characteristics of logged evergreen forests in Preah Vihear, Cambodia, 3 years after logging. For Ecol Manag 225:62–73

Lambin EF, Strahler A (1994) Remotely-sensed indicators of land-cover change for multitemporal change-vector analysis. Int J Remote Sens 15:2099–2119

Lettenmaier D, Hayhoe P, Major D, Poff NL, Running S (2008) Water resources, in The effects of climate change on agriculture, land resources, water resources, and biodiversity. Climate Change Science Program Synthesis and Assessment Report 4.3

Lugo A (1992) Comparison of tropical tree plantations with secondary forests of similar age. Ecol Monogr 62:1–41

Messina JP, Walsh SJ, Mena CF, Delamater PL (2006) Land tenure and deforestation patterns in the Ecuadorian Amazon: conflicts in land conservation in frontier settings. Appl Geogr 26:113–128

Milly PCD, Betancourt J, Falkenmark M, Hirsch RM, Kundzewicz ZW, Lettenmaier DP, Stouffer RJ (2008) Stationarity is dead: whither water management? Science 319(5863):573–574

Moody A, Woodcock CE (1995) The influence of scale and the spatial characteristics of landscapes on land-cover mapping using remote sensing. Landscape Ecol 10:363–379

Moon KH, Park D (2004) The role and activities of NGOs in reforestation in the northeast Asian region. For Ecol Manag 201:75–81

Munroe DK, Southworth J, Tucker CM (2004) Modeling spatially and temporally complex land-cover change: the case of western Honduras. Prof Geogr 56(4):544–559

Nagendra H, Pareeth S, Sharma B, Schweik CM, Adhikari KR (2008) Forest fragmentation and regrowth in an institutional mosaic of community, government and private ownership in Nepal. Landscape Ecol 23:41–54

O'Neill RV, Hunsaker CT, Timmins SP, Jackson BL, Jones KB, Riitters KH, Wickham JD (1996) Scale problems in reporting landscape pattern at the regional scale. Landscape Ecol 11:169–180

Qi Y, Wu J (1996) Effects of changing spatial resolution on the results of landscape pattern analysis using spatial autocorrelation indices. Landscape Ecol 11:39–49

Rudel T, Perez-Lugo M, Zichal H (2000) When fields revert to forests: development and spontaneous reforestation in post-war Puerto Rico. Prof Geogr 52:386–397

Rudel T, Coomes O, Moran E, Achard F, Angelsen A, Xu J, Lambin E (2005) Forest transitions: towards a global understanding of land use change. Glob Environ Change 15:23–31

Rudel T (2005) Tropical forests: regional paths of destruction and regeneration in the late twentieth century. Columbia University Press, New York

Seidl AF, dos SV da Silva J (2001) Cattle ranching and deforestation in the Brazilian Pantanal. Ecol Econ 36:413–425

Southworth J, Tucker CM (2001) The influence of accessibility, local institutions, and socio-economic factors on forest cover change in the mountains of western Honduras. Mountain Res Dev 21(3):276–283

Southworth J, Munroe DK, Nagendra H (2004) Land cover change and landscape fragmentation – comparing the utility of continuous and discrete analyses for a Western Honduras region. Agric Ecosyst Environ 101(2–3):185–205

Southworth J, Cumming GS, Marsik M, Binford MW (2006) Linking spatial and temporal variation at multiple scales in a heterogenous landscape. Prof Geog 58(4):406–420

# Chapter 2
# The Bigger Picture – Tropical Forest Change in Context, Concept and Practice

**Alan Grainger**

## 2.1 Introduction

The trajectory of the long-term trend in global forest area is still very uncertain (Mather 2005). This is also apparent in inconsistencies between tropical area trends published by the UN Food and Agriculture Organization (FAO) (Fig. 2.1). Moreover, evidence for deforestation in the humid tropics based on FAO statistics or sampling by independent analysis of satellite images cannot be substantiated by evidence for a long-term decline in forest area obtained from independent surveys (Fig. 2.2) (Grainger 2008).

A. Grainger (✉)
School of Geography, University of Leeds, Leeds, UK
email: a.grainger@leeds.ac.uk

H. Nagendra and J. Southworth (eds.), *Reforesting Landscapes: Linking Pattern and Process,* Landscape Series 10,
DOI 10.1007/978-1-4020-9656-3_2, 

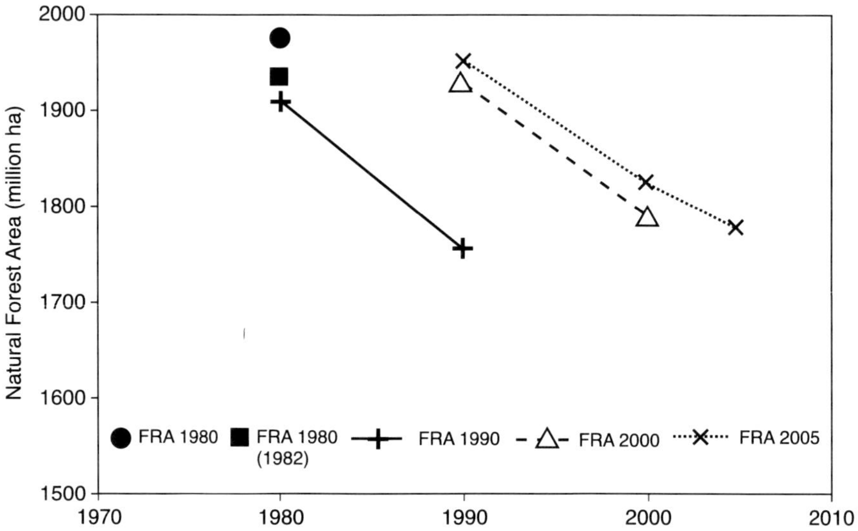


**Fig. 2.1** Trends in Natural Forest area in 90 tropical countries, 1980–2005, from data in FAO Forest Resources Assessments (FRAs) 1980, 1990, 2000, and 2005 (million hectares) (Grainger 2008)

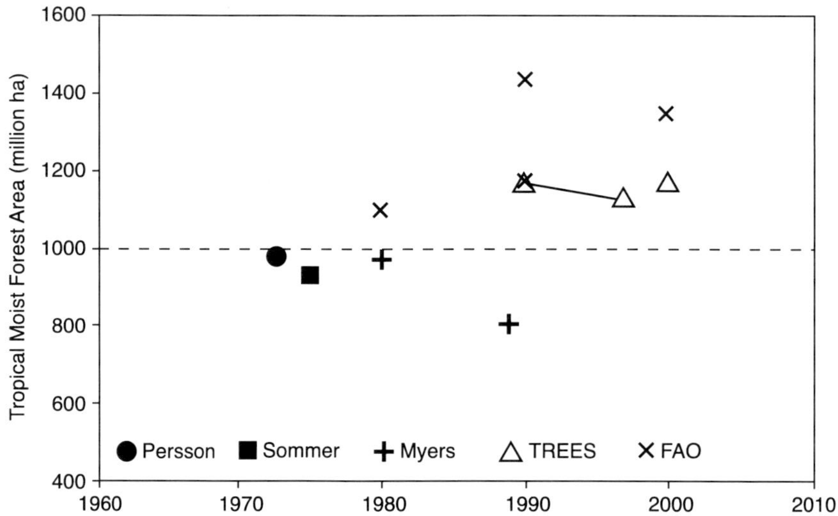


**Fig. 2.2** Estimates of tropical moist forest area 1970–2000 (million hectares) (Grainger 2008), based on data in Lanly (1981); FAO (1993, 2001); Persson (1974); Myers (1980, 1989); with TREES data from Achard et al. (2002); Eva et al. (2002a, b), Mayaux et al. (2003); Stibig et al. 2003, 2004) Sommer (1976)

Uncertainty about the pan-tropical trend undoubtedly reflects the large errors involved in estimating national forest areas in numerous countries (House et al. 2003; Houghton 2005). But it is also possible that deforestation is being offset by various types of forest expansion, or forestation (Wiersum 1984), and the case studies in other chapters in this volume suggest that in parts of some countries the downward trajectory has indeed been reversed. The two explanations could even be linked, since uncertainty could also result from difficulties in making accurate estimates when both deforestation and forestation are occurring simultaneously at significant rates. Aggregating unsynchronized forest trends in many countries, each of which is in a different phase of its land use evolution, is not a trivial matter.

If these assertions are correct then a similar uncertainty might be expected for national trends in countries where there is net forestation in some regions while deforestation continues in others. We test this hypothesis in a non-random sample of eight countries discussed in other chapters: Madagascar in East Africa; India and Nepal in South Asia; Vietnam in Southeast Asia; and Costa Rica, Guatemala, Honduras and Mexico in Central America. Our findings do indeed show uncertainty in some national trends.

Uncertainty is said to be inherent when so-called "post-normal" phenomena are analysed using normal scientific methods (Funtowicz and Ravetz 1993). Global climate change appears to be one of these phenomena (Saloranta 2001), and it seems that global forest change is another. In addition to technical difficulties in making accurate large-area measurements, this chapter provides evidence that social factors contribute to uncertainty too. It identifies concept, context and practice as key elements in this, and shows that how forest change is conceptualized and measured depends on the world views and practices of the groups involved, and the spatial scale at which each group operates.

This chapter has five main sections. Section 2.2 outlines a methodology to frame analyses of disciplinary and scalar differences in knowledge construction. Section 2.3 compares the concepts and practices of two scientific disciplines: forest science and land change science. Section 2.4 examines how conceptualization varies with spatial context. Section 2.5 discusses the role of institutions in forest monitoring. Section 2.6 builds on this foundation to compare different narratives for pan-tropical forest plantation area and natural forest area in the eight case study countries.

## 2.2 Methodology

### *2.2.1 Discursive Space*

To frame our analysis of the construction of forest trends, we assume that the totality of knowledge of any phenomenon is represented by a two-dimensional *discursive space* (Fig. 2.3).

From a poststructuralist perspective, conceptualization is framed by the world views, or discourses, of the actors involved. A discourse is "a specific ensemble of ideas, concepts, and categorizations that are produced, reproduced and transformed in a particular set of practices and through which meaning is given to physical and social realities" (Hajer 1995). Discourses are an aid to exercising power, as they determine through the use of language what is considered 'normal' in a society.

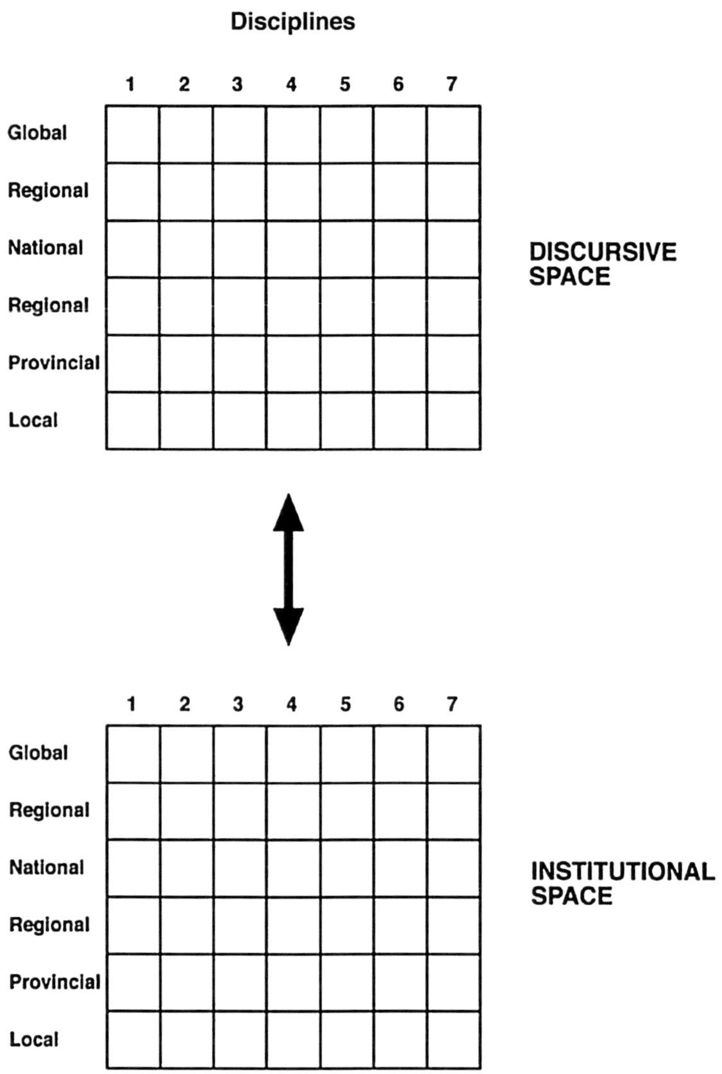


**Fig. 2.3** The role of discursive and institutional spaces in knowledge construction by scientists from different disciplines and stakeholders at different spatial scales

One dimension of discursive space, along the horizontal axis, comprises knowledges constructed by different scientific disciplines. Each column represents the terms and relationships used to construct knowledge within the discourse of one discipline. The second dimension, along the vertical axis, comprises contextual knowledges constructed at different spatial scales. Each row includes the terms and relationships in the discourses of actors with expertise derived from living and working at that spatial scale.

The two dimensions of knowledge must be integrated to describe a phenomenon completely, and any partiality in constructing knowledge within discursive space will lead to uncertainty. So however rigorously scientific knowledge is constructed, each discipline only partially describes a phenomenon and may be more reliable at some scales than others. This explains why disciplinary boundaries must be transcended to understand 'post-normal' phenomena.

### *2.2.2 Institutional Space*

The reproduction of actors' discourses is inseparable from that of their institutions (Hajer 1995). The term institution is often used as a synonym for an organization, but it really refers to a common practice of all members of an organization or any other group. Rationalist theories treat institutions as mainly constraints on human behaviour (e.g. North 1990), but others see them as more multifaceted, in being not so much external impositions on human activity as created and sustained by it. More formally, Crawford and Ostrom (1995) defined institutions generically as "enduring regularities of human action in situations structured by rules, norms and shared strategies, as well as by the physical world", thereby recognizing the importance of context.

So *institutional space* displays the institutions of each of the groups of actors participating in discursive space. Each column represents the institutions of a particular discipline, while the rows comprise the institutions of stakeholders at different scales. At each scale multiple institutions may be present.

## 2.3 Concepts and Practices of Two Scientific Disciplines

Land change scientists are skilled in analysing changes in trajectory in national forest trends, but have until now often depended on forest scientists for data. So our analysis begins by focusing on the horizontal dimension of discursive space and the concepts and practices of these two disciplines.

### *2.3.1 Forest Science Concepts and Practices*

Cultivating and harvesting trees is for foresters a normative way of using forested land. In their traditional 'productivist' discourse (Dolman 2000), trees are grown to produce timber for conversion into products. This discourse is exclusive, demanding that a certain 'forest space' be totally devoted to material uses, to the exclusion of other uses, such as environmental protection and farming, and actors who might wish to engage in these uses (Mather 2001). Forest scientists are concerned about the erosion of forest space by deforestation, but they do not engage in detailed studies of why it occurs.

The origins of forest science are contextual, and can be traced to Germany in the middle of the sixteenth century. Over the next 300 years a series of abstract geometrical, calculus and economic methods were developed there that created a 'scientific forestry' that could be applied to forest space (Westoby 1987, 1989). The use of a discounting formula to calculate the 'optimal economic rotation' of a plantation, including the costs involved in tending a forest crop over many decades, arguably represented the culmination of this conceptual development (Faustman 1849).

The application of forest science to tropical forests began in the 1850s, soon after its basic set of practices had been assembled. Widespread and prolonged experiments by colonial foresters over the next 100 years were a valuable learning experience (Troup 1952). These showed that various silvicultural systems, in which forestation in natural forests is intentionally manipulated by foresters, could be effective, but were too labour intensive to be economic (Whitmore 1975). This led from 1960 onwards to an increasing professional focus on artificial forest plantations, and to the accumulation of knowledge about this, initially synthesized by Evans (1982).

Knowledge continues to be accumulated within forest science discourse, but is effectively limited to systematizing practices. Wiersum (1984) discussed the general principles of a potential "conceptual framework for a general forestation theory", but this merely comprised sets of strategies, tactics and operations to ensure successful forestation. He edited a collection of studies of forestation experiences, presented to a symposium convened to mark the centenary of forestry education at the University of Wageningen. Another valuable collection was compiled by Mather (1993).

Any group which commissions or engages in a scheme to monitor a phenomenon will design it to maximize the collection of data to provide the information it needs. Forest scientists mainly work at local scale, studying such key issues as the growth of trees and how this affected by different management methods. They use data on trends in national and pan-tropical forest areas to place their studies against a wider background, or measure 'progress' in achieving key policy goals, for example meeting forestation targets or improving the sustainability of forest management by controlling deforestation. So while papers in forestry journals may include tables showing such trends, these data are mainly for background purposes, as the bulk of the analysis depends on empirical data collected in field studies. Forest scientists working for FAO's Forestry Department share the same discourse as their colleagues in government forestry departments, so it is within this framework that they decide how to produce international compilations of national forest statistics supplied by governments.

### *2.3.2 Land Change Science Concepts and Practices*

Land change science recently emerged from efforts by geographers, economists, sociologists and others to explain the social, economic and political reasons for land use and land cover change through empirical study and theoretical development (Rindfuss et al. 2004). Its data needs differ from those of forest science, and ideally comprise annual values of forest area and other attributes for every country.

#### 2.3.2.1 The Forest Transition Model

Particular attention has been paid to studying the reversal in trajectory of the trend in national forest area, from decline to rise. One conceptualization is the *forest transition* model (Fig. 2.4) (Mather 1992). Its popularity stems from the apparent simplicity of the U-shaped curve near the point of transition. Initial studies focused on temperate countries (e.g. Mather et al. 1998; 1999; Mather and Fairbairn 2000; Mather 2004), but evidence for transitions in the tropics is now being reported too (Rudel et al. 2005; Mather 2007). While theorization of the forest transition has not been neglected (Mather and Needle 1998), top priority has been given to empirical studies (Mather 2007)

#### 2.3.2.2 An Alternative Forest Transition Model

The currently dominant land change science discourse assumes that forestation follows shortly after deforestation. If a distinct reversal in trajectory cannot be identified then land change scientists working within this discourse might infer that either data are uncertain, or that the land change pattern is aberrant.

An alternative model, which would not automatically lead to such an inference, was suggested by this author some time ago (Grainger 1995). It divides the forest transition curve into two parts: the decline in forest area, termed the *national land use transition*, and the rise after the forest transition, termed the *forest replenishment period* (Fig. 2.5). The two are separated by an interregnum of variable extent. In certain conditions they combine to give a U-shaped curve, but in the general case they need not do so.

The justification for separating the two parts of the curve is that it combines two functions relating to different market demand curves. The national land use transition curve is merely the inverse of a curve showing how agricultural area expands in response to rising demand for food as a country develops. It tapers off as the limits of land suitability are reached and farming intensifies following investment in

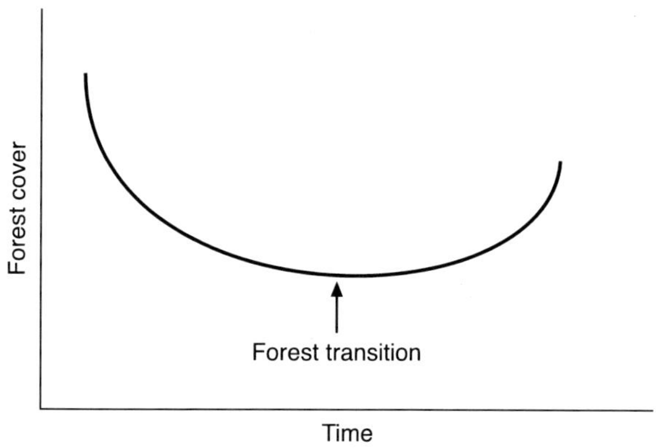


**Fig. 2.4** The forest transition model (Mather 1992)

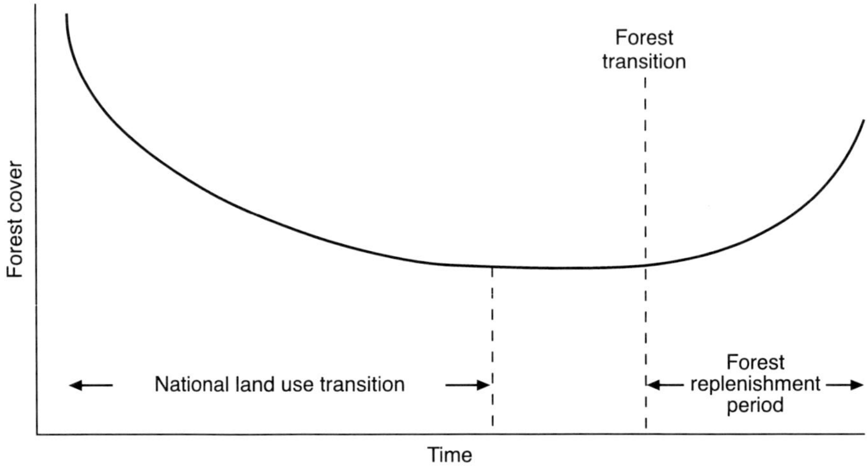


**Fig. 2.5** An alternative conceptualization of the forest transition model (Grainger 1995)

improved technologies (Drake 1993). Forest replenishment, on the other hand, is a response to changing demand for wood, and non-market environmental services supplied by forests. Natural reforestation may occur if farmland is abandoned as it becomes unproductive or uneconomic. The forest transition marks a switch from the dominance of agricultural institutions to that of forestry institutions.

A U-shaped curve around the forest transition could be seen if the national land use transition follows what this author has called a 'normative scenario'. Deforestation will end once market mechanisms have allocated all land to its optimum use. A large amount of forest would be left (Fig. 2.6), but as the spatio-temporal land allocation trend is invariably piecemeal, with some marginal farmland cleared before all of the most productive land is identified, the end of deforestation could be followed by natural reforestation on abandoned marginal land. The conditions required for this scenario include the presence of institutions within which: (a) market forces can operate; and (b) the non-market values of environmental services provided by forests affect land allocation decisions.

The U-shaped curve is deformed if these conditions do not apply, and national forest cover falls much lower than in the normative scenario. In what was termed the 'critical scenario' (Fig. 2.6), deforestation does not stop until either (a) environmental services collapse, and floods and other hazards put pressure on governments to intervene: or (b) wood supplies become restricted and a rise in price allows forest protection and forestation to become more profitable. A possible lower limit of 0.1 ha per capita was suggested by Grainger (1993) for this scenario, based on the forest area needed to supply mean internal domestic wood demand. Of the eight case study countries discussed in other chapters, Nepal is close to this limit (Table 2.1). Only India and Vietnam have fallen below it, but seem to have passed through their forest transitions. They also have the highest proportions of plantations in Total Forest area of all eight countries (Table 2.2), and so wood market forces seem to be operating there.

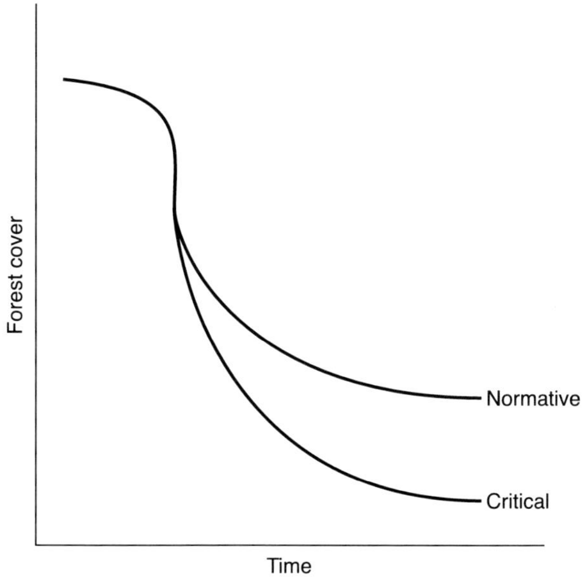


**Fig. 2.6** Normative and critical national land use transition scenarios (Grainger 1995)

**Table 2.1** Forest area (hectares) per capita and percent forest cover in the eight case study countries, calculated using lowest national forest area in FRAs 2000 and 2005 (million hectares) (FAO 2001, 2006a)

| | Lowest forest area | Percent forest cover | Forest area per capita |
|---|---|---|---|
| Madagascar | 11.4 | 19.6 | 0.74 |
| India | 31.5 | 10.6 | 0.03 |
| Nepal | 3.6 | 26.3 | 0.15 |
| Vietnam | 6.8 | 20.9 | 0.09 |
| Costa Rica | 1.8 | 35.3 | 0.46 |
| Guatemala | 2.7 | 25.0 | 0.24 |
| Honduras | 4.7 | 42.0 | 0.75 |
| Mexico | 54.9 | 28.8 | 0.56 |

**Table 2.2** The share of forest plantations in total forest area in the eight case study countries 1990 and 2000 (percent) (FAO 1993, 2001)

| | 1990 | 2000 |
|---|---|---|
| Madagascar | 2.3 | 2.6 |
| India | 30.1 | 50.9 |
| Nepal | 2.1 | 2.6 |
| Vietnam | 26.9 | 17.3 |
| Costa Rica | 0.0 | 10.0 |
| Guatemala | 2.9 | 6.9 |
| Honduras | 0.0 | 1.9 |
| Mexico | 0.2 | 0.5 |

NB. 'Total forest' comprises both Natural Forest and Forest Plantations.

Mather (1992) recognized that different types of national land use transitions occur, showing that while the USA experienced a transition similar to the normative scenario, that of France was closer to the critical scenario. The two scenarios correspond to the 'economic development' and 'forest scarcity' pathways proposed for forest transitions by Rudel et al. (2005). In the second pathway they used the 0.1 ha per capita limit as a benchmark.

#### 2.3.2.3 Incorporating Uncertainty into the Forest Transition Model

Uncertainty about the timing of forest transitions can result from discontinuities in the transition trend because:

1. National forest area declines in distinct phases of decline and rise, owing to changes in economic or political conditions.
2. The end of the national land use transition and the start of forest replenishment are separated by an extended interregnum, because people continue farming on marginal land or commercial forestation is delayed by a lack of wood price signals.
3. Forest area rises in distinct phases of decline and rise after the forest transition, owing to changes in economic or political conditions.
4. Deforestation continues in forest-rich regions of a country, but forestation occurs elsewhere in response to either local or global wood market forces.

#### 2.3.2.4 Data Needs in Land Change Science

Land change scientists therefore need different kinds of data from forest scientists. They require data on trends in national and pan-tropical forest areas that are accurate and preferably of high frequency. Some land change scientists have even used an alternative set of annual statistics in FAO's Production Yearbook, despite their inferior quality (Barbier and Burgess 2001). This approach began with Allen and Barnes (1986) and continued for a long time (e.g. Dietz and Adger 2003).

## 2.4 Contextual Conceptualizations of Forest Change

Conceptualizations of forest change also vary with spatial scale, along the vertical dimension of discursive space. These differences are complicated by interactions between scales.

### *2.4.1 Differences Between Scales*

At national scale, the conceptualization of forests by governments and the foresters who work for them varies between countries according to culture, prevailing state discourse and other factors. This leads to hundreds of different national classification systems (FAO 2001).

At international scale, to combine forest area estimates from many countries to give a uniform data set in its Forest Resource Assessments (FRAs), FAO must adopt a coherent global classification system which, while reflecting its own discourse, is comprehensive enough to all these different national systems.

Some differences between systems are just *semantic*, that is the meaning of terms is the same even though different names are used. More fundamental structural differences are termed *ontological*. Here an ontology refers to "an explicit, partial account of a conceptualization" (Guarino and Giaretta, 1995), not a "theory of what can be known", as in philosophy (Johnston 1986). Global ontologies are vital to produce uniform global data sets, but dispensing with national ontologies, which reflect contextual knowledge, can reduce resolution (Cruz et al. 2004). Forest data can be interpreted within different ontologies to form different pictures, or narratives, of forest area and how this changes over time. A *narrative* is a meaningful totality of past, present and future events (Barton 2000).

FAO has actually changed its global forest ontology in successive FRAs (Fig. 2.7). FRA 1980 distinguished Natural Woody Vegetation from Forest Plantations, divided the former into All Forest and Shrubs, split All Forest into Broadleaved and Coniferous, and classified Broadleaved Forest as Closed, Bamboo or Open (Lanly 1981). FRA 1990 had a simpler ontology, which combined Closed Forest and Open Forest as Natural Forest, though national areas of Closed Broadleaved Forest were listed for many countries (FAO 1993). FRA 2000 combined Natural Forest and Forest Plantations to give a new statistic of Total Forest (FAO 2001). This reflects the productivism of FAO's discourse in treating the two land cover types as equivalent sources of timber, despite fundamental differences in biodiversity and other features. Total Forest was retained in FRA 2005, but divided into Primary, Modified Natural and Semi-Natural Forest, and Productive and Protective Plantations. FRA 2005 retained another change in FRA 2000 too: widening the definition of 'Forest Plantations' to include rubber plantations (FAO 2006a).

### *2.4.2 Interactions Between Scales*

Political discourses at global or national scale have for a long time been combined with forest science discourses to influence how actors behave at national or local scales. For example, forest science was introduced to tropical contexts when Western countries colonized Africa and Asia, and became known there as *scientific forestry* (Lanz 2000). Necessarily, it was imposed together with imperial discourses that subjugated local people to the desires and perceptions of the colonial power. In Africa, forestation was a crucial tool within colonial environmental discourse, which assumed that the unregulated spread of agriculture was leading to the formation of desert (Fairhead and Leach 1998).

After failing to exclude people from forest spaces, in the 1970s many tropical foresters tried to remove the physical and conceptual 'walls' that enclosed their activities,

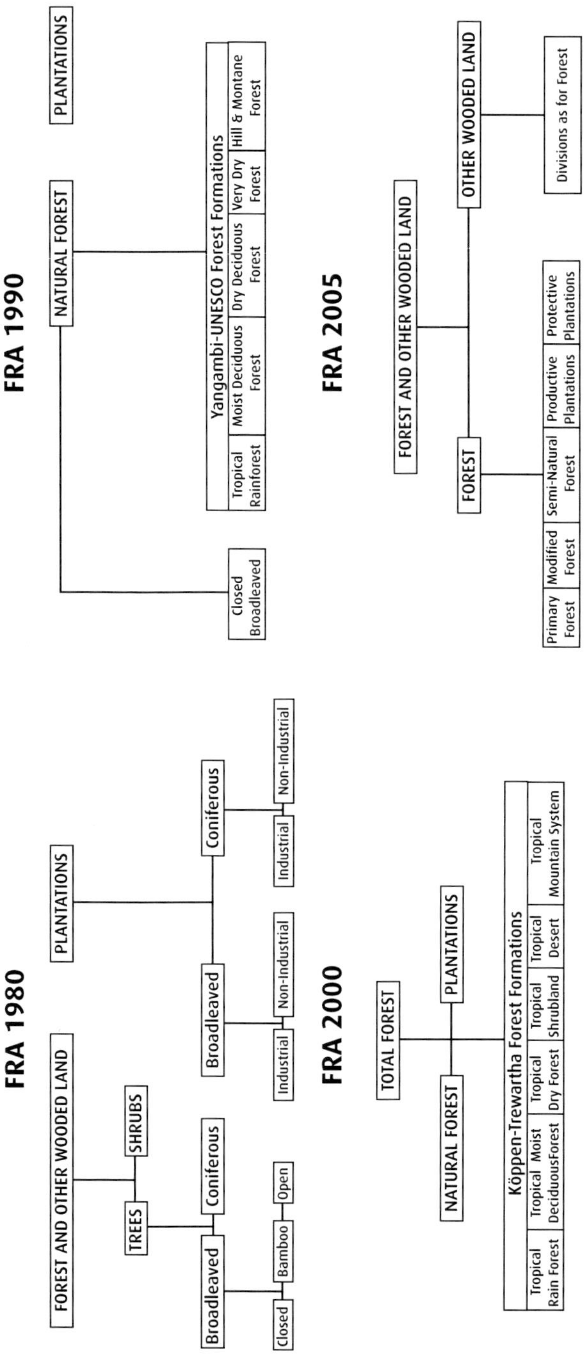


**Fig. 2.7** Global forest ontologies in FAO Forest Resources Assessments (FRAs) 1980, 1990, 2000, and 2005 (Grainger 2007)

seeking instead to involve 'outsiders' in forest protection and restoration. Under the *community forestry* discourse that emerged from this rethink, the role of forest plantations was extended to supplying rural peoples' needs for fuelwood, food and fodder (FAO 1978). Again, the new discourse was combined with a political discourse, in this case the community-based development discourse that also became popular at the time (Korten 1980).

The 1970s also saw the imposition on developing countries of a globalist environmental discourse. Emanating from the UN Conference on the Human Environment in 1972, this put the needs of the planet above the needs of people. Following the UN Conference on Desertification (UNCOD) in 1977, degraded tropical lands were recognized as an aberration from the norm that needed economic and environmental rehabilitation through forestation (OTA 1983). Up to 758 million hectares of land was thought to be in need of restoration (Grainger 1988).

When the significance of human-induced global climate change was recognized in the 1990s, the idealism of environmental globalism was combined with a political discourse that sought to maintain global economic activity, by creating new terrestrial carbon sinks to offset the carbon dioxide emissions on which this relied. Assessing the potential for forestation to sequester carbon was an early task of the new science of global change. According to Sedjo and Solomon (1989), 465 million hectares of new tropical forest plantations with an annual growth rate of 15 $m^3$/ha were required to absorb the net increase in atmospheric carbon dioxide content from all sources. But since plantation growth rates varied with environmental contexts, almost 600 million hectares might be needed in practice (Grainger 1991).

## 2.5 Forest Monitoring Practices

Achieving the measurements with comparable accuracy that are needed for regular forest monitoring requires the consistent and repeated practices that constitute institutions.

### *2.5.1 State Forestry Institutions*

Collecting statistics is a routine function of state forestry departments to meet criteria for information quality and frequency. The institutions used reflect those of forest science but vary according to each government's discourse. For example, the Indian government is technocentric, and uses satellite images to monitor forests. The government of Vietnam, however, has a socialist ideology, and relies on the members of each local commune to collect forest statistics.

Traditionally, the gap between surveys has been at least 10 years. Such a frequency meets the needs of government forestry departments but is too low for land change scientists, as it may underestimate both deforestation and forestation. Clearances occurring after one survey has taken place will not be visible on satellite images by the time of the next survey if forest has regenerated naturally on cleared land.

### 2.5.2 International Forestry Institutions

The institutions which FAO uses for forest monitoring consist of two main types. First, formal institutions. FAO follows a rule-based process of decision making typical of all UN organizations, and as one of the most important rules is respect for state sovereignty FAO depends on national statistics supplied by its member states, rather than monitoring states itself.

Second, informal institutions, such as the practices of the forestry profession. It is interesting how its methods reflect the geometrical underpinnings of forest science. Thus, FAO has routinely projected *forwards* the result of the last national forest survey for each country from the year in which the survey was carried out to a common reporting year for all countries, for example 1990 for FRA 1990. This usually involves linear projection, by extrapolating a line joining the areas found in the last two surveys. But the more time has elapsed since the last survey, the higher the errors associated with the projected area.

FAO also revises earlier estimates to be compatible with the latest ones. To produce the three trends in Fig. 2.1, FAO revised estimates for 10 years before (and in FRA 2005 for 15 years before as well) by projecting *backwards*. It usually interpolated between the last two surveys. But if reliable earlier data were not available the last survey might have been the sole starting point. This has three drawbacks:

First, FAO generally assumes that the latest survey is the most accurate. This is understandable, given the wider use of remote sensing measurements. But an increase in forest area resulting from greater accuracy is difficult to distinguish from one resulting from forestation since the last survey, shown by the dotted line in Fig. 2.8.

Second, before projection a subjective decision must be taken about whether forest area is falling or rising. The usual assumption is deforestation, but as forestation

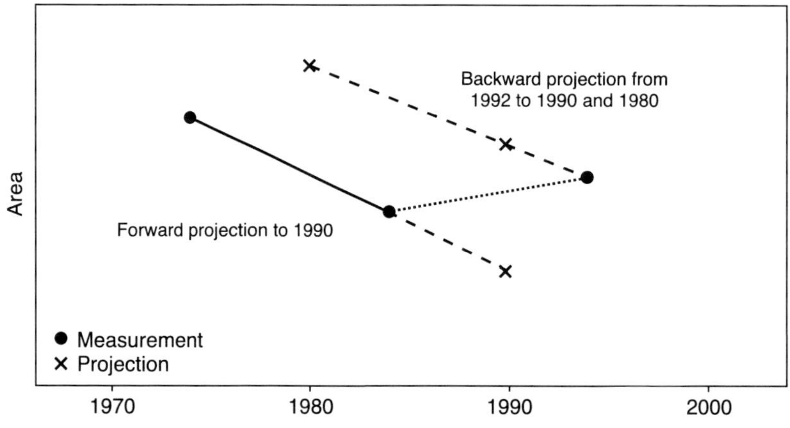


**Fig. 2.8** Projection methods in FAO Forest Resources Assessments, showing: a forward projection to 1990 from two earlier surveys (*lower line*); a backward projection to 1980 and 1990 from a later survey (*upper line*); and how assumptions of increasing survey accuracy can obscure a reversal in trajectory (*dotted line*)

becomes more widespread this is less justifiable. For Madagascar, Nepal, Costa Rica, Guatemala, Honduras and Mexico the downward trend for 1990–2005 in FRA 2005 was at a *higher* mean area than that for 1990–2000 in FRA 2000 (see Section 2.6.2.2). Yet the inference was that earlier estimates were too low, not that forestation had occurred.

Third, projection methods may change and lead to inconsistencies with previous estimates. In FRA 2005, for the first time, the governments of countries themselves chose which national forest survey to use to estimate forest areas in 1990, 2000 and 2005. So they could interpret differently the survey used in FRA 2000 or choose another survey entirely. They also now made the forward and backward projections themselves, and this could lead to differences from estimates in FRA 2000 too.

### *2.5.3 Land Change Science and Global Change Science Institutions*

FAO's institutions constrain the suitability of FRA statistics for land change and global change scientists. Ideally, they would employ their own institutions to collect data and analyse them to construct knowledge. Yet while they can construct knowledge using data obtained in local and even national studies in individual countries, they lack an equivalent set of institutions covering every country in the world. So, despite having access to global data in the form of satellite imagery, they do not have the institutions to process them on a global scale and interpret them with reference to contextual knowledge. If they did they could construct forest narratives themselves just as FAO does, but in conformity with their own quality criteria, not those of FAO.

## 2.6 Analysing Evidence for Forestation in Forest Change Narratives

The expansion of forest plantations and regeneration of natural forest are two of the key mechanisms involved in national forest transitions, and the latter also plays an important role in influencing the trajectory of the trend in pan-tropical natural forest area. This section builds on the preceding analysis of how forest area estimates are constructed and interpreted to assess the evidence for forestation provided by alternative forest change narratives and the extent to which this explains current uncertainty.

### *2.6.1 Trends in Pan-Tropical Forest Plantations Area*

Forest plantations in the tropics are only equivalent in area to some 2% of the area of natural tropical forest. Given their small extent and well known locations, area estimates should be more accurate than for natural forest, and show a continued increase over time. In fact, they exhibit a similar inconsistency to that for natural forest in Fig. 2.1. Taking the latest FAO estimate in each FRA, the Forest Plantation area in 90 tropical countries apparently increased sixfold from 11.5 million hectares in 1990 to 67.5 million hectares in 2000, but then halved to 34.7 million hectares in 2005 (Table 2.3).

Errors do play a role in explaining such inconsistencies. Governments usually report to FAO the area planted with trees, not the area that actually survives. Consequently, FAO routinely deducts 30% from all reported areas to correct for high mortality, but this does not wholly remove uncertainty (FAO 2001).

However, the social aspects of monitoring have an equally important influence. Thus, the sharp rise from 1990 to 2000 was driven by plantations in India almost doubling in area from 18.9 to 32.6 million hectares. The subsequent decline between 2000 and 2005 is explained by FRA 2005 reporting only 3.2 million hectares of plantations for the same country in 2005 (FAO 2006a). Yet the "missing plantations" did not simply vanish: the Indian Government responded to the ontological refinement imposed on it in FRA 2005 by reallocating them to the new category of 'semi-natural forest' (FAO 2006b). The 3.2 million hectares remaining as forest plantations consisted of non-indigenous species.

The doubling of Africa's plantation area between 2000 and 2005 was due solely to expansion in one country – Sudan. According to FRA 2000, it had only 0.6 million hectares of Forest Plantations in 2000 (FAO 2001), but FRA 2005 reported 6.1 million hectares in 1990, declining to 5.4 million hectares in 2005. There is no easy way to compare the accuracy of the two estimates, since the FRA 2005 estimate of forest area was produced by combining a national Landsat survey from 1972 with a remote-sensing survey made in 2000 for just three of the country's 12 provinces (FAO 2006c).

FAO's decision in FRA 2000 to change the definition of 'Forest Plantations' to include rubber plantations as well as timber plantations has also contributed to uncertainty. Statistics about rubber plantations are collected by government agriculture departments, not forestry departments, and country reports for FRA 2005 suggest that the latter find it difficult to obtain these statistics and combine them with forestry statistics. Moreover, in recent decades large areas of rubber plantations, particularly in Malaysia, have been replaced by oil palm plantations. So end users who are unaware of this could mistakenly interpret the 383,000 ha fall in Forest Plantations area in

**Table 2.3** Trend in the area of forest plantations in 90 tropical countries 1980-2005, according to successive FAO Forest Resource Assessments (FRAs) (1,000 ha) (Lanly 1981;,1993, 2001, 2006a)

| | FRA 1980 | FRA 1990 | FRA 2000 | FRA 2005 |
|---|---|---|---|---|
| | 1980 | 1990 | 2000 | 2005 |
| Africa | 1,780 | 3,000 | 4,573 | 9,499 |
| Asia-Pacific | 5,111 | 32,153 | 54,716 | 16,726 |
| Latin America | 4,620 | 8,636 | 8,188 | 8,482 |
| Total | 11,511 | 43,789 | 67,478 | 34,707 |

NB. The total estimates for 1980 and 1990 are directly from the publications identified. Those for FRAs 2000 and 2005 were calculated by this author from national statistics in those publications.

Malaysia between 1990 and 2005 (FAO 2006a) to mean that the area of timber plantations had declined sharply owing to deforestation.

As in the case of natural forest area, in each FRA FAO has revised its earlier estimates of forest plantation area to be consistent with its latest ontological and definitional changes. So in the revised narrative published in FRA 2005, Forest Plantations area in our 90 countries rose from 29.0 million hectares in 1990 to 32.7 million hectares in 2000 and 34.7 million hectares in 2005.

A different narrative emerges, however, if estimates in Table 2.3 for 2000 and 2005 are corrected for these ontological and definitional changes to retain consistency with estimates in FRAs 1980 and 1990 (Table 2.4). In our new narrative, Forest Plantations in 90 tropical countries rose steadily from 11.5 million hectares in 1980 to 62.5 million hectares in 2005. This involved deducting 9.3 million hectares of rubber plantations reported by FAO in FRA 2000; and 3.7 million hectares reported in seven country reports for FRA 2005 (FAO did not provide a separate estimate, though a special review of forest plantations claimed that rubber plantations still covered 9 million hectares (FAO 2006d)). The reclassification of 31.5 million hectares of Indian forest plantations as 'Semi-Natural Forest' was reversed and the higher Sudan estimate retained.

## *2.6.2 Trends in Forest Area in Eight Tropical Countries*

We now analyse three forest change narratives constructed using FRA data for eight tropical countries studied in other chapters, to see if they replicate the reversals in subnational trajectories described there. The first set of narratives comes directly from statistics included in the FRAs. The second, still partly reflecting global perceptions, draws on survey reports cited in FRAs. The other, more rooted in national contexts, is based on surveys cited in reports submitted to FAO by governments for FRA 2005.

### 2.6.2.1 A Summary of Case Study Findings

Of all the case studies presented in this volume, those covering significant areas in eight tropical countries have been selected for discussion in this chapter.

**Table 2.4** Trend in the area of forest plantations in 90 tropical countries 1980–2005, revised to exclude rubber plantations in FRAs 2000 and 2005 and include Indian timber plantations reclassified as 'Semi-Natural Forest' in FRA 2005 (1,000 ha)

| | FRA 1980 | FRA 1990 | FRA 2000 | FRA 2005 |
|---|---|---|---|---|
| | 1980 | 1990 | 2000 | 2005 |
| Africa | 1,780 | 3,000 | 4,000 | 9,159 |
| Asia-Pacific | 5,111 | 32,153 | 47,869 | 44,884 |
| Latin America | 4,620 | 8,636 | 7,953 | 8,482 |
| Total | 11,511 | 43,789 | 59,823 | 62,525 |

By analysing satellite images for 1984, 1993 and 2000, Elmquvist and Tengö, in Chapter 13, show that for a 550,000 ha area of southern Madagascar, the rate of deforestation of dry tropical forests declined between 1984–1993 and 1993–2000, while the rate of natural forestation rose.

In Chapter 7, Nagendra finds evidence from satellite images for natural and artificial forestation between 1989 and 2000 in the Terai plains of Nepal and in its middle hills between 1990 and 2000. She identifies a similar trend in the east of the Indian state of Maharashtra between 1989 and 2001. Through a discriminating analysis of a range of forest maps based on satellite imagery, Meyfroidt and Lambin, in Chapter 14, show that national forest cover in Vietnam rose from 25–31% in 1991–1993 to 32–37% in 1999–2001, though deforestation was still continuing in some areas.

In Chapter 10, Daniels analyses satellite images for 1975, 1987 and 2000 of a 550,000 ha area of northwest Costa Rica which includes both protected areas and private land. She reports substantial natural and artificial forestation, especially on former grassland. Bray, in Chapter 5, reviews a large number of reports of forest change in Central America. He finds clear evidence for both forestation and deforestation in Costa Rica, Guatemala, Honduras and Mexico.

#### 2.6.2.2 Narratives Constructed in FAO Forest Resource Assessments

Only for one of our eight countries, Vietnam, discussed in Chapter 14, do FRA statistics appear to provide firm evidence that the decline in Natural Forest area has been reversed. It rises consistently in both FRA 2000 and FRA 2005 (Table 2.5). Similar trends are evident in Total Forest area, which also includes Forest Plantations. These FAO reports were the primary source of data used by Mather (2007) to test for a reversal in forest area trajectory in Vietnam.

FRA statistics suggest that trajectories may have been reversed in two more countries. FRA 2000 reports a decline in Natural Forest area in India (Chapter 7) between 1990 and 2000. However, FRA 2005 reports a rise, and then a fall from 2000 to 2005, while Total Forest area rises in both FRAs 2000 and 2005. Natural Forest in Costa Rica (Chapter 10) declined between 1990 and 2000 in both FRAs 2000 and 2005, but in FRA 2005 it rose slightly from 2.36 to 2.39 million hectares between 2000 and 2005 (not evident in Table 2.5), a pattern repeated by the Total Forest statistic. Trajectory reversal was inferred from FRA statistics for both India and Vietnam by Mather (2007) and for India alone by Rudel et al. (2005).

For the other five countries continuing declines in Natural Forest and Total Forest areas are reported in FRAs 2000 and 2005. On this evidence, sub-national forestation narratives presented in other chapters for seven of our case study countries may not be replicated at national scale.

This conclusion comes into question, though, when we examine how these statistics were assembled. For Nepal, discussed in Chapter 7, differences between the trends in FRAs 2000 and 2005 were slight. Yet while information in the two FRAs implies that both trends are based on the same 1994 survey, the Nepalese Government actually adopted a more complex projection approach for FRA 2005 (FAO 2006e).

**Table 2.5** Trends in Natural and Total Forest area in eight case study countries, based on estimates in FAO Forest Resource Assessments (FRAs) (million hectares) (FAO 1993, 2001, 2006a)

| | Natural forest | | | | | | | Total forest | | | | |
|---|---|---|---|---|---|---|---|---|---|---|---|---|
| FRA | 1980 | 1990 | 2000 | 2000 | 2005 | 2005 | 2005 | 2000 | 2000 | 2005 | 2005 | 2005 |
| Date | 1980 | 1990 | 1990 | 2000 | 1990 | 2000 | 2005 | 1990 | 2000 | 1990 | 2000 | 2005 |
| Madagascar | 13.2 | 15.8 | 12.6 | 11.4 | 13.4 | 12.7 | 12.5 | 12.9 | 11.7 | 13.7 | 13.0 | 12.8 |
| India | 57.2 | 51.7 | 44.5 | 31.5 | 62.0 | 64.7 | 64.5 | 63.7 | 64.1 | 63.9 | 67.6 | 67.7 |
| Nepal | 2.1 | 5.0 | 4.6 | 3.8 | 4.8 | 3.8 | 3.6 | 4.7 | 3.9 | 4.8 | 3.9 | 3.6 |
| Vietnam | 10.1 | 8.3 | 6.8 | 8.1 | 8.4 | 9.7 | 10.2 | 9.3 | 9.8 | 9.4 | 11.7 | 12.9 |
| Costa Rica | 1.5 | 1.4 | 2.1 | 1.8 | 2.6 | 2.4 | 2.4 | 2.1 | 2.0 | 2.6 | 2.4 | 2.4 |
| Guatemala | 4.1 | 4.2 | 3.3 | 2.7 | 4.7 | 4.1 | 3.8 | 3.4 | 2.9 | 4.7 | 4.2 | 3.9 |
| Honduras[a] | 3.5 | 4.6 | 6.0 | 5.3 | 7.4 | 5.4 | 4.7 | 6.0 | 5.4 | 7.4 | 5.4 | 4.7 |
| Mexico | 45.3 | 48.6 | 61.4 | 54.9 | 68.0 | 64.5 | 63.2 | 61.5 | 55.2 | 69.0 | 65.5 | 64.2 |

[a]Since forest plantation area is very small, Natural Forest and Total Forest are essentially the same.

The 2000 estimate was made by projecting forward a line joining the results of a 1978/79 survey (5.6 million hectares) and a 1999 expert consultation (area not stated). The 2005 estimate was made by projecting forward a line joining the 1978/1979 and 1999 areas plus the finding of the 1994 survey (5.5 million hectares). The 1990 estimate was produced by interpolating between the 1978 and 1999 areas.

For Honduras (Chapter 5), notes in FRA 2005 imply that government estimates were based on an earlier survey, from 1990, than that used by FAO for FRA 2000 (1995), so that differences between the two trends presumably derived from subjective evaluations of survey reliability. Yet the corresponding country report reveals a different picture, showing that estimates simply drew on national forest statistics for those years, and there was no information about which (if any) surveys these were based on (FAO 2006f). Indeed, a 1995 forest map mentioned in FRA 2000 is the only clear reference in an FRA to any national forest survey ever being carried out in the country. So neither FRA statistics nor survey data cited in country reports provide evidence that forest decline has ended in Honduras.

For India (Chapter 7), the rise in Natural Forest area from 1990 to 2005 reported in FRA 2005 starts from a point 17 million hectares higher than the area reported for 1990 in FRA 2000 (Table 2.5). This is linked to the Indian Government's decision to reclassify some Forest Plantations as Semi-Natural Forest (see above). When Mather (2007) discussed the evidence for a reversal in India's forest trajectory he referred to national as well as FAO statistics. He too spotted the plantation anomaly. So the FRA Natural Forest statistic is not a reliable indicator for testing for reversal. As all reclassifications took place within the Total Forest statistic it is not affected by them and is a more reliable indicator. Its trend suggests there has been a switch from forest decline to forest rise, though the 2000 and 2005 estimates are identical. The Indian country report suggests a more cogent reason for using the Total Forest statistic as an indicator: it corresponds to the main statistical output of the approximately biennial satellite surveys of India's forests undertaken since 1982 (FAO 2006b).

#### 2.6.2.3 Constructing Alternative Narratives Using Survey Reports in FRAs

An alternative method for constructing national forest change narratives is not to use FRA statistics themselves but data from actual surveys in each country reported in FRAs. Circumventing problems with FRA statistics in this way is possible since our focus is on tracing national trends. So all the encumbrances required to produce estimates for all countries in the same year can be removed.

Our dataset, summarized in Table 2.6, lists the date of each survey used to produce statistics in FRAs 1980, 1990, 2000 and 2005, and any other information included in the FRAs. Only FRAs 1980 and 2000 published the actual survey findings (in FRA 2000 these are listed in a table of "latest national statistics"). So in the absence of other evidence, we assume that (a) the FRA 1990 survey found more forest than the estimate given for 1990 (indicated by a ">" symbol in Table 2.6); and (b) the finding of the survey used for FRA 2005 is identical to the estimate for the year closest to the survey year, that is 2000 or 2005. When specific survey data are available from

**Table 2.6** Reconstructing trends in Natural Forest area for selected countries based on original survey findings quoted in FAO Forest Resource Assessments (FRAs) (million hectares) (Lanly 1981; FAO, 1993, 2001, 2006a)

| Source | FRA 1980 | | FRA 1990 | | FRA 2000 | | FRA 2005 | |
|---|---|---|---|---|---|---|---|---|
| | | | Survey date | Area | Survey date | Area | Survey date | Area |
| Madagascar | 1955 | 16.8 | 1975 | >15.8 | 1996 | 11.6 | 2004 | 12.6 |
| India | – | – | 1988 | >51.7 | 1997 | 63.7 | 2001 | 64.8 |
| Nepal | 1964 | 6.4 | 1979 | >5.0 | 1994 | 4.3 | 1994 | 4.8 |
| Vietnam | 1974 | 9.2 | 1987 | >8.3 | 1995 | 8.3 | 2003 | 10.2 |
| Costa Rica | – | – | 1987 | >1.4 | 1997 | 2.1 | 2000 | 2.4 |
| Guatemala | – | – | 1988 | >4.2 | 1999 | 2.8 | 2003 | 3.8 |
| Honduras | – | – | 1986 | >4.6 | 1995 | 6.0 | – | – |
| Mexico | 1973 | 52.5 | 1978 | >48.5 | 1993 | 58.5 | 2002 | 64.5 |

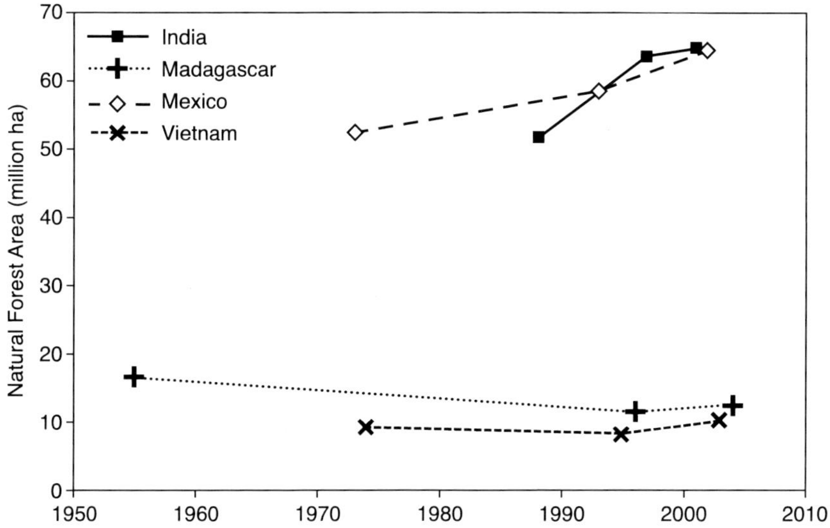


**Fig. 2.9** Forest area narratives for four tropical countries based on survey reports listed in FAO Forest Resources Assessments

FRA 1980, and they are compatible with the assumptions made about the surveys used for FRA 1990, the latter have been omitted from Fig. 2.9. Otherwise, for convenience, these surveys are given the same values as those listed, on the understanding that they are interpreted as in Table 2.6.

The results of our analysis again show a reversal in trajectory in Vietnam, but that forest area in India and Mexico has been rising since records began (Fig. 2.9). The same applies to Costa Rica. A continuing decline is indicated for Nepal, though since the latest survey was in 1994 the trajectory may have changed since then. Similarly, although Honduras appears to have an upward trajectory, this conclusion is only tentative, being based on just two estimates, the last of which was in 1995.

A reversal in trajectory is also inferred for the other two countries. In Madagascar, discussed in Chapter 13, Natural Forest area in the 1980s presumably exceeded the 15.8 ha listed in FRA 1990, which was apparently projected from a 1975 survey. This is supported by a reference in FRA 1980 to a survey that found 16.8 million hectares of forest in 1955. "The latest national statistic" listed in FRA 2000 showed 11.6 million hectares remaining in 1996, while the estimate of 12.6 million hectares in 2005 was apparently based on a 2004 survey. This implies that forest area fell from 1990 to a minimum of at least 11.6 million hectares by 1996, before rising again to 12.6 million hectares by 2005 (Fig. 2.9). For Guatemala (Chapter 5), according to FRA 2000 the "latest national statistic" was 2.8 million hectares for 1999. As a national forest inventory was the basis for an estimate of 3.8 million hectares of forest in 2003 – though its mapping methods were not listed in the FRA 2005 country report – forest area appears to have been rising in the country since 1999.

The three apparent reversals inferred here all crucially depend on single survey results cited in FRA 2000. Moreover, while the trajectories for four countries are supported by pre-1980 surveys cited in FRA 1980, no such surveys were available at the time for other countries. So it was not possible to confirm a reversal in trajectory for India and Costa Rica, and only infer a continuing rise.

#### 2.6.2.4 Constructing Contextual Narratives Using Survey Evidence in FRA Country Reports

A third method involves identifying patterns in the time series formed by combining actual surveys recognized as valid by governments in their country reports for FRA 2005 (Table 2.7). Some trends are shown in Fig. 2.10.

The set of forest area estimates to which the government of Vietnam referred in its country report includes the survey finding of 8.3 million hectares quoted in FRA 2000 for 1995. This appears to have been the minimum in the forest trend, since the estimate for 2000 was 9.7 million hectares. The 2003 figure of 10.0 million hectares confirms the reversal in trajectory. By comparison, FRA 1980 reported that in 1974 a Landsat survey revealed the presence of 9.2 million hectares of forest. However, an important caveat is that subsequent estimates are based not on national remote sensing surveys but on local forest statistics collected by each commune and submitted to the government (FAO 2006g). Meyfroidt and Lambin provide independent, objective survey evidence in Chapter 14 to support them, but suggest that the turning point was a few years earlier. Mather's (2007) assessment of the reversal in forest trajectory relied solely on FRA data, though he did refer to estimates made earlier in the twentieth century.

India's country report summarizes the findings of national forest surveys undertaken using satellite imagery every 2–3 years since 1982, when 64.1 million hectares of Total Forest were identified. It confirms the reversal in trajectory suggested by the Total Forest statistic in FRAs 2000 and 2005. Reversal apparently occurred in 1994 at an area of 63.3 million hectares (FAO 2006b).

**Table 2.7** Reconstructing trends in Natural Forest area for selected countries based on original survey findings quoted in FAO Forest Resource Assessments (FRAs) or obtained from other sources summarized in FRA 2005 country reports (million hectares) (Lanly 1981; FAO, 1993, 2001, 2006a, b, e, f,g, h, i, j, k)

| Source | FRA 1980 | | Surveys referred to in country reports to FRA 2005 | | | | | |
|---|---|---|---|---|---|---|---|---|
| | | | Survey date | Area | Survey date | Area | Survey date | Area |
| Madagascar | 1955 | 16.8 | – | – | 1996 | 13.3 | – | – |
| India[a] | – | – | 1982 | 64.1 | 1994 | 63.3 | 2000 | 67.6 |
| Nepal | 1964 | 6.4 | 1978 | 5.6 | 1994 | 4.3 | – | – |
| Vietnam[b] | 1974 | 9.2 | 1990 | 8.4 | 1995 | 8.3 | 2003 | 10.0 |
| Costa Rica | – | – | 1992 | 2.5 | 2000 | 2.4 | – | – |
| Guatemala[c] | – | – | – | – | – | – | 2003 | 4.1 |
| Honduras[b] | – | – | 1990 | 7.4 | 2000 | 5.4 | 2005 | 4.7 |
| Mexico | 1973 | 52.5 | – | – | 1993 | 68.7 | 2002 | 65.6 |

[a]India's forest survey findings combine Natural Forests and Forest Plantations and are reported here as such for consistency.[b]Figures for Vietnam and Honduras come from national statistics, not specified surveys.

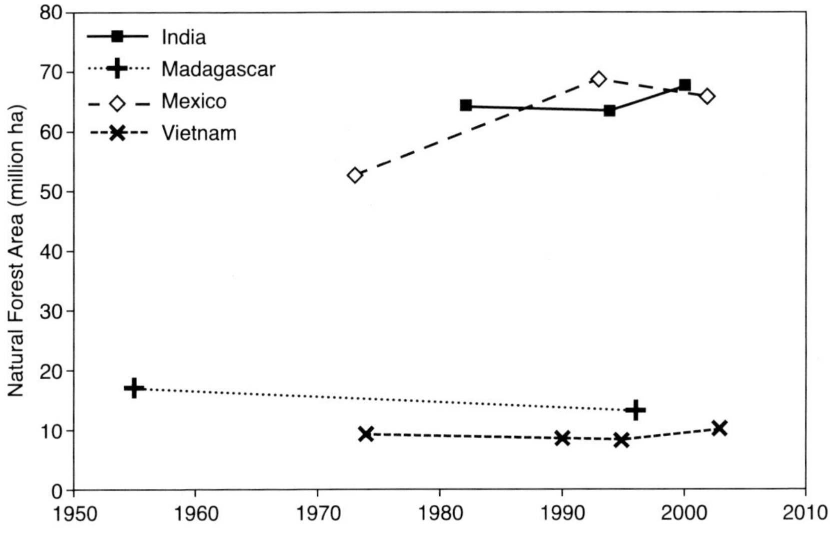


**Fig. 2.10** Forest area narratives for four tropical countries based on survey reports listed in country reports to FAO Global Forest Resources Assessment 2005

For Nepal, the country report mentions 4.3 million hectares of forest found in a survey that according to FRA 2000 took place in '1994'. Yet this was just the mean year, as the survey actually lasted from 1988 to 1998. Another survey reported 5.6 million hectares for 1978–1979 (listed as 1978 in Table 2.7 for convenience). This was less than the 6.4 million hectares found in a 1964 survey cited in FRA 1980.

The combination of these surveys only indicates a continuing decline until the 1990s. A more recent survey is required to provide evidence for a subsequent reversal in trajectory (FAO 2006e).

The contents of the country report for Honduras were summarized above. No reliable data were available to provide a basis for its FRA 1980 estimate. Subsequent national statistics imply continuing deforestation, though more recent and more reliable data are needed (FAO 2006f).

For Madagascar, the country report refers to the 1996 estimate listed in FRA 2000 but not to the 1975 and 2004 surveys cited in FRAs 1990 and 2005, respectively. The 2000 and 2005 estimates were projected forward from this 1996 survey, but the government of Madagascar interpreted this differently from FAO. Instead of using the figure of 11.6 million hectares of Natural Forest given in FRA 2000, which was the basis for identifying a reversal of trajectory in Table 2.6, it used 13.3 million hectares (FAO 2000h). This is consistent with a continuing decline in forest area, not a reversal in trajectory (Fig. 2.10). A more recent survey is therefore needed to determine whether a reversal has subsequently occurred.

For Costa Rica the 1990 figure of 1.4 million hectares listed in FRA 1990 appears to have been an anomaly, because the country report refers to an estimate of 2.5 million hectares from 1992, based on an aerial photographic survey. No earlier survey was mentioned in FRA 1980 to provide a point of reference. The most recent survey finding is 2.36 million hectares for 2000, based on a Landsat study. Crucially, the estimate of 2.39 million hectares for 2005 was made by projecting forward from the 2000 survey, assuming there would be reforestation in mangrove forests and no further deforestation elsewhere. So while deforestation seems to be proceeding slowly in Costa Rica, no evidence for a change in forest trajectory is available from the country report (FAO 2006i), and the best inference is that the trajectory is uncertain.

For Mexico (Chapter 5), the "latest national statistic" of 58.5 million hectares in 1993, listed in FRA 2000, seems to have been an anomaly, because according to the country report a Landsat survey found 68.7 million hectares in 1993. The next (and latest) survey, also made using Landsat, was undertaken in 2002 and found 65.6 million hectares. The country report does not refer to the 1978 survey used to construct the FRA 1990 estimate for 1990, but FRA 1980 mentions a 1973 Landsat survey which found 52.5 million hectares. So from these findings it is not possible to reach a firm conclusion about the direction of Mexico's forest trajectory and whether it has changed (Fig. 2.10) (FAO 2006j).

The country report for Guatemala states that the estimate of 3.8 million hectares for 2005 in FRA 2005 was made by projecting forward from the 4.05 million hectares found in a 2003 national forest inventory made with FAO assistance. No more information is available on how this was measured. The 1999 statistic of 2.8 million hectares, cited in FRA 2000, was crucial in suggesting a reversal in trajectory. Yet it appears from the country report to have been derived from an assessment by a group of experts, not a remote sensing measurement. In preparing its submission to FRA 2005, the Guatemalan government clearly assumed continuing deforestation. It used the deforestation rate for 1992 to 1999 estimated by the expert group to project backwards from the 2003 survey to 2000 and 1990, and forwards to 2005. As no earlier surveys were referred to in the country report, or in FRA 1980, only a single data point is listed in Table 2.7. The trajectory of the forest area trend in Guatemala is therefore uncertain (FAO 2006k).

The use of this third method therefore supports claims for a reversal in forest trajectory in India and Vietnam (the latter with qualifications about the apparent lack of recent surveys). It suggests that deforestation is still continuing in Madagascar, Nepal and Honduras. It sheds no new light on the trajectories for Costa Rica, where FRA statistics implied reversal, or for Guatemala and Mexico, where they did not. So these trajectories must be regarded as uncertain.

## 2.7 Conclusions

This chapter has analysed available data, obtained from the Forest Resources Assessments (FRAs) of the UN Food and Agriculture Organization (FAO), on national trends in the eight tropical countries for which regional evidence for forestation has been provided in other chapters. Only for two countries, India and Vietnam, has it found reliable evidence that forestation is replicated at national scale.

At the start of the chapter it was suggested that current uncertainty about the trajectory of the long-term trend in pan-tropical forest area may result not only from inaccuracies in forest monitoring, but also from the difficulty of making estimates when the rates of deforestation and forestation are both significant. Our analysis supports this, revealing similar uncertainty about the trajectories of forest area trends in five of our sample of eight countries: Costa Rica, Guatemala, Honduras, Madagascar and Mexico (Table 2.8).

Context plays its part, as uncertainty appears greatest in national forest narratives constructed at national scale than at global scale. For five of the eight countries, the narrative communicated by FRA statistics is one of continuing decline. Only when all manipulations made at global scale to national survey data are removed, and the narratives of national governments are also considered, does uncertainty appear.

Uncertainty, of course, is a matter of perception. This chapter has argued that judgements about the reliability of forest area narratives differ according to the discourse of the group evaluating the data on which these are based. In contrast to the forest scientists who construct FRA statistics, our analysis was made from within the descriptive discourse of land change science, which has different data needs and quality criteria. A 'forest transition' model in which the reversal in the trajectory of national forest area follows a U-shaped curve is widely accepted in this discipline. Although uncertainty about the trajectory is not expected by those who have adopted the standard model, we have proposed an alternative model which is more compatible with uncertainty as net deforestation comes to an end and net reforestation begins. In this model the two parts of the transition curve may be separated by a 'turning zone', not a turning point. So while the interpretation of data varies between disciplines, it can also vary within a discipline.

**Table 2.8** A comparison of forest area narratives in selected countries based on three alternative data constructions: FRA Trends, FRA Survey Data and National Constructions of Survey Data

| | Method 1: | Method 2: | Method 3: | Overall |
|---|---|---|---|---|
| | FRA statistics | FRA survey reports | Country survey reports | Overall assessment |
| Madagascar | Decline | Reversal | Decline | Uncertain |
| India | Possible reversal | Rise | Reversal | Possible reversal |
| Nepal | Decline | Reversal | Decline | Decline |
| Vietnam | Reversal | Decline | Reversal | Reversal |
| Costa Rica | Possible reversal | Rise | Uncertain | Uncertain |
| Guatemala | Decline | Reversal | Uncertain | Uncertain |
| Honduras | Decline | Decline | Decline | Uncertain |
| Mexico | Decline | Uncertain | Uncertain | Uncertain |
| Summary: | | | | |
| Decline | 5 | 2 | 3 | 1 |
| Reversal | 1 | 3 | 2 | 1 |
| Rise | 0 | 2 | 0 | 0 |
| Possible reversal | 2 | 0 | 0 | 1 |
| Uncertain | 0 | 1 | 3 | 5 |

Although many land change scientists admit that FRA statistics have their limitations, they feel they must use them since no other data are available. Our analysis demonstrates the fallacy of this argument. It shows how different national forest narratives can be constructed if the empirical data on which FRA statistics are based are analysed in different ways. So applying institutions linked to a land change science discourse to knowledge constructed within another scientific discourse with different institutions will not necessarily produce reliable knowledge. Land change (and global change) scientists may only get forest data that satisfy their quality criteria when they have the global institutions to match these.

Another challenge concerns theory. The continuous curve model of the forest transition needs far stronger theoretical justification. Our analysis suggests that a U-shaped curve may occur as a special case under certain conditions, but that a clear reversal of trajectory may not be seen in the general case, where the trajectory in the 'turning zone' may exhibit similar uncertainty to the pan-tropical trend. Our argument has been based on an analysis of processes. A better theory is needed to confirm this. It will also help land change and global change scientists to collaborate to model the likely global and pan-tropical outcomes of aggregating unsynchronized forest transition curves in multiple forest types in many countries, each of which is in a different phase of its land use evolution.

Uncertainty about trajectories of long-term trends in global and pan-tropical forest areas provides an incentive for a new initiative to remove the data limitations under which land change scientists work, *and* tackle the theoretical void at the heart of this science. For if the latter is not addressed quickly we shall not understand patterns emerging when higher quality data sets become available.

## References

Allen JC, Barnes DF (1986) The causes of deforestation in developing countries. Ann Assoc Am Geogr 75:163–84

Barbier EB, Burgess JC (2001) The economics of tropical deforestation. J Econ Surv 15:413–433

Barton E (2000) Sanctioned and non-sanctioned narratives in institutional discourse. Narrative Inq 10:341–375

Crawford SE, Ostrom E (1995) A grammar of institutions. Am Pol Sci Rev 89:582–600

Cruz IF, Sunna W, Chaudhry A (2004) Semi-automatic ontology alignment for geospatial data integration. In: Egenhofer, MJ, Freksa, C, Miller, HJ (eds) Proceedings of geographic information science, third international conference, GIScience, Adelphi, MD. Lecture Notes in Computer Science, Springer, Berlin, Vol 3234, pp 51–66

Dietz S, Adger WN (2003) Economic growth, biodiversity loss and conservation effort. J Environ Manag 68:23–35

Dolman P (2000) Biodiversity and ethics. In: O'Riordan T (ed) Environmental science for environmental management. Prentice Hall, Harlow, England, pp 119–148

Drake WD (1993) Towards building a theory of population-environment dynamics: a family of transitions. In: Ness G, Drake WD, Brechin SR (eds) Population-environment dynamics. University of Michigan Press, Ann Arbor, MI, pp 305–356

Achard F, Eva HD, Stibig H-J, Mayaux P, Gallego J, Richards T, Malingreau J-P (2002) Determination of deforestation rates of the world's humid tropical forests. Science 297: 999–1002

Eva HD, Miranda EE de, Bella CM di, Gond V, Huber O, Sgrenzaroli M, Jones S, Coutinho A, Dorado A, Guimerarães M et al (2002) A vegetation map of South America. Joint Research Centre Publication EUR 2015, European Commission, Luxembourg

Evans J (1982) Plantation forestry in the tropics. Oxford University Press, Oxford

Fairhead J, Leach M (1998) Reframing deforestation. Routledge, London

FAO (1978) Forestry for local community development. FAO Forestry Paper No 7. UN Food and Agriculture Organization, Rome

FAO (1982) Tropical forest resources. FAO Forestry Paper No 30. UN Food and Agriculture Organization, Rome

FAO (1993) Forest resources assessment 1990: Tropical countries. FAO Forestry Paper No 112. UN Food and Agriculture Organization, Rome

FAO (2001) Global forest resources assessment 2000. FAO Forestry Paper No 140. UN Food and Agriculture Organization, Rome

FAO (2006a) Global forest resources assessment 2005. FAO Forestry Paper No 147. UN Food and Agriculture Organization, Rome

FAO (2006b) Global forest resources assessment 2005. National report No 001: India. UN Food and Agriculture Organization, Rome

FAO (2006c) Global forest resources assessment 2005. National report No 212: Sudan. UN Food and Agriculture Organization, Rome

FAO (2006d) Global forest resources assessment 2005. Special study on forest plantations. UN Food and Agriculture Organization, Rome

FAO (2006e) Global forest resources assessment 2005. National report No 192: Nepal. UN Food and Agriculture Organization, Rome

FAO (2006f) Global forest resources assessment 2005. National report No 174: Honduras. UN Food and Agriculture Organization, Rome

FAO (2006g) Global forest resources assessment 2005. National report No 124: Vietnam. UN Food and Agriculture Organization, Rome

FAO (2006h) Global forest resources assessment 2005. National report No 021: Madagascar. UN Food and Agriculture Organization, Rome

FAO (2006i) Global forest resources assessment 2005. National report No 106: Costa Rica. UN Food and Agriculture Organization, Rome

FAO (2006j) Global forest resources assessment 2005. National report No 189: Mexico. UN Food and Agriculture Organization, Rome

FAO (2006k) Global forest resources assessment 2005. National report No 003: Guatemala. UN Food and Agriculture Organization, Rome
Faustman M (1849) Calculation of the value which forest land and immature stands possess for forestry. In: Gane M (ed) and Linnard W (trans) (1968) Institute Paper No 42. Commonwealth Forestry Institute, University of Oxford, Oxford
Funtowicz SO, Ravetz JR (1993) Science for the post-normal age. Futures 25:735–755
Grainger A (1988) Estimating areas of degraded tropical lands requiring replenishment of forest cover. Intl Tree Crops J 5(1,2):31–61
Grainger A (1991) Overcoming constraints on assessing feasible afforestation and revegetation rates to combat global climate change. In: Sargent C (ed) Proceedings of technical workshop to explore options for global forest management, Bangkok, April 1990, pp 196–208. Office of the National Environment Board, Government of Thailand/US Environmental Protection Agency
Grainger A (1993) Population as concept and parameter in the modelling of tropical land use change. In: Ness G, Drake WD, Brechin SR (eds) Population-environment dynamics. University of Michigan Press, Ann Arbor, MI, pp 71–101
Grainger A (1995) The forest transition: an alternative approach. Area 27:242–251
Grainger A (2007) The influence of end-users on the temporal consistency of an international statistical process: the case of tropical forest statistics. J Off Stat 23:553–592
Grainger A (2008) Difficulties in determining the long-term trend in tropical forest area. Proc Natl Acad Sci USA 105:818–823
Guarino N, Giaretta P (1995) Ontologies and knowledge bases: towards a terminological clarification. In: Mars N (ed) Towards very large knowledge bases: knowledge building and knowledge sharing. IOS Press, Amsterdam, pp 25–32
Hajer MA (1995) The politics of environmental discourse: ecological modernization and the policy process. Clarendon Press, Oxford
Houghton RA (2005) Aboveground forest biomass and the global carbon balance. Global Change Biol 11:945–958
House JI, Prentice IC, Ramnkutty N, Houghton RA, Heimann M (2003) Reconciling apparent inconsistencies in estimates of terrestrial CO2 sources and sinks. Tellus 55B:345–363
Johnston RJ (1986) Philosophy and human geography, 2nd edn. Edward Arnold, London
Korten DC (1980) Community organization and rural development: a learning process approach. Public Adm Rev 40:480–511
Lanly JP (ed) (1981) Tropical forest resources assessment project (GEMS): Tropical Africa, Tropical Asia, Tropical America, 4 Vols. UN Food and Agriculture Organization/UN Environment Programme, Rome
Lanz TJ (2000) The origins, development and legacy of scientific forestry in Cameroon. Environ Hist 6:99–120
Mather AS (1992) The forest transition. Area 30:117–124
Mather AS (1993) Afforestation: policies, planning and progress. Belhaven, London
Mather AS (2001) Forests of consumption: postproductivism, postmaterialism, and the postindustrial forest. Environ Plan C 19:249–268
Mather AS (2004) Forest transition and the reforesting of Scotland. Scott Geogr J 120:83–98
Mather A (2005) Assessing the world's forests. Global Environ Change 15:267–280
Mather AS (2007) Recent Asian forest transitions in relation to forest-transition theory. Int For Rev 9:491–501
Mather AS, Fairbairn J (2000) From floods to reforestation: the forest transition in Switzerland. Environ Hist 6:399–421
Mather AS, Needle CL (1998) The forest transition: a theoretical basis. Area 30:117–124
Mather AS, Needle CL, Coull JR (1998) From resource crisis to sustainability: the forest transition in Denmark. Int J Sust Dev World Ecol 5:182–193
Mather AS, Fairbairn J, Needle CL (1999) The course and drivers of the forest transition: the case of France. J Rural Stud 15:65–93

Mayaux P, Bartholomé E, Massart M, Van Cutsem C, Cabral A, Nonguierma A, Diallo O, Pretorius C, Thompson M, Cherlet M et al (2003) A land cover map of Africa. Joint Research Centre Publication EUR 20665, European Commission, Luxembourg

Myers N (1980) Conversion of tropical moist forests. US National Research Council, Washington DC

Myers N (1989) Deforestation rates in tropical forests and their climatic implications. Friends of the Earth (UK), London

North DN (1990) Institutions, institutional change and economic performance. Cambridge University Press, Cambridge

OTA (1983) Sustaining tropical forest resources: reforestation of degraded lands. Office of Technology Assessment, Congress of the United States, Washington DC

Persson R (1974) World forest resources. Royal College of Forestry, Stockholm

Rindfuss RR, Walsh SJ, Turner BL II, Fox J, Mishra V (2004) Developing a science of land change: challenges and methodological issues. Proc Natl Acad Sci U S A 101:13976–13981

Rudel T, Coomes OT, Moran E, Achard F, Angelsen A, Xu J, Lambin E (2005) Forest transitions: towards a global understanding of land use change. Global Environ Change 15:23–31

Saloranta TM (2001) Post-normal science and the global climate issue. Climatic Change 50:395–404

Sedjo RA, Solomon AM (1989) Climate and forests. In: Rosenberg NJ, Easterling WE, Crosson PR, Darmstadter J (eds) Greenhouse warming: abatement and adaptation. Resources for the Future, Washington DC, pp 105–119

Sommer A (1976) Attempt at an assessment of the world's tropical forests. Unasylva 28(112–113):5–25

Stibig HJ, Malingreau JP (2003) Forest cover of insular southeast Asia mapped from recent satellite images of coarse spatial resolution. Ambio 32:469–475

Stibig HJ, Achard F, Fritz S (2004) A new forest cover map of continental Southeast Asia derived from satellite imagery of coarse spatial resolution. Appl Veg Sci 7:153–162

Troup RS (1952) Silvicultural systems, 2nd edn. Clarendon Press, Oxford

Westoby JC (1987) The purpose of forests. Basil Blackwell, Oxford

Westoby JC (1989) Introduction to world forestry. Basil Blackwell, Oxford

Whitmore TC (1975) Tropical rain forests of the far east. Clarendon Press, Oxford

Wiersum KF (1984) Introduction: towards a global forestation strategy. In: Wiersum KF (ed) Proceedings of an international symposium on strategies and designs for afforestation, reforestation and tree planting. University of Wageningen, Netherlands, September 1983, pp 7–25. Pudoc, Wageningen

# Chapter 3
# Three Paths to Forest Expansion: A Comparative Historical Analysis

**Thomas K. Rudel**

## 3.1 Introduction

During the past 15 years land cover change has become a more pressing issue as environmental crises like global warming and species extinctions, with clear links to landscape changes, have grown more severe. Calculations of carbon emissions make it clear that tropical deforestation makes a substantial contribution to the emissions of greenhouse gases that drive global warming, accounting for anywhere from 15% to 25% of annual greenhouse gas emissions by humans (IPCC 2007). Efforts to slow deforestation over the past 30 years have been

T.K. Rudel (✉)
Departments of Human Ecology and Sociology, Rutgers University, NJ, USA
e-mail: rudel@aesop.rutgers.edu

H. Nagendra and J. Southworth (eds.), *Reforesting Landscapes: Linking Pattern and Process,* Landscape Series 10,
DOI 10.1007/978-1-4020-9656-3_3, 

largely ineffectual (Chomitz 2007; Rudel 2005). In this context, apprehensive observers have begun to promote the expansion of other forests as a partial solution to environmental problems because it seems like a more achievable goal than stopping tropical deforestation. Under these circumstances it becomes useful to outline the human conditions that foster forest expansion.

Forest expansion occurs in three diverse sets of social circumstances. In one setting forests regenerate spontaneously, usually after people have left the land for one reason or another. In a second setting forests expand because people plant large blocks of land with single species of trees, creating forest plantations. In a third situation people engage in agro-forestry, increasing the density of fruit bearing and other types of trees in landscapes near their homes. The following pages describe each set of circumstances and the associated sets of causal conditions in some detail. A final section describes policy initiatives that would encourage or expedite the expansion of forests.

The methodological approach for this analysis is broadly comparative historical, distinguishing between three types of forest expansion prevalent in different places at different times. These ideal types, spontaneous forests, planted forests, and household agro-forests, are to some degree analytic conveniences. They represent extreme versions of frequently occurring landscapes. Particular forests will often represent mixed cases. For example, when a commercially valuable, planted species is an understory plant, like coffee, cacao, or rubber, a form of 'rubber jungle', 'coffee jungle', or 'cacao jungle' may emerge in which naturally regenerating shade trees mix with planted understory trees. Similarly, household agro-forests may begin to resemble small forest plantations when one species of trees predominates among the plantings near the house.

To distinguish between the first two types of forest expansion involving spontaneously regenerating forests and large planted stands of trees, I draw upon cross-national data on forests collected by the Food and Agricultural Organization of the United Nations (FAO) as well as case studies of changing forest cover in disparate locales throughout the world. The FAO data, collected under the auspices of their Forest Resources Assessment (FRA) program, comes from individual countries who submit it in response to requests from FAO. The quality of this data has improved over the last two decades, with more of it now based on remote sensing analyses (Downton 1995), but there is still some unevenness in the quality of the data from country to country. FAO analysts have also periodically revised their definitions and methods for estimating forest cover, and, to provide for consistency across countries and through time, they have re-estimated earlier estimates of forest cover using the new methods (Grainger 2008). This analysis uses FAO data to document the very large differentials by continent in rates of change in forest cover. These differences are unlikely to be affected substantially by the small changes in the data introduced through periodic redefinitions and recalculations by FAO staff. With this cautionary note in mind, the FAO data, assembled in Table 3.1, charts basic changes in forest cover by continent for the 1990–2005 period.

Table 3.1 displays two distinguishable patterns of forest expansion. Europe embodies one of these patterns. It features forests (in column six) that are spreading into old, now abandoned, agricultural fields through spontaneous regeneration of trees. This is the typical pattern of 'old field succession' that North American ecologists analyzed in

**Table 3.1** Regional patterns of change in forest cover and forest plantations

| | (1) | (2) | (3) | (4) | (5) | (6) |
|---|---|---|---|---|---|---|
| | Forest Area, 1990 | Forest Area, 2005 | Change in Forest Area, 1990–2005[a] | Area in Plantations, 1990 | Change in Plantations, 1990–2005 | Change in Natural Forest Area, 1990–2005[b, c] |
| Africa | 1,517,626 | 1,378,519 | −1391.06 | 323.95 | +27.57 | −1349.22 |
| Asia | 1,660,412 | 1,652,323 | −80.88 | 1600.52 | +623.65 | −593.89 |
| Europe | 2,753,117 | 2,786,594 | +334.77 | 644.14 | +192.96 | +174.96 |
| North America | 4,179,329 | 4,150,317 | −290.11 | 898.08 | +583.75 | −596.92 |
| Oceania | 3,020,914 | 2,931,528 | −893.86 | 606.75 | +346 | −893.86 |
| South America | 6,852,446 | 6,396,461 | −4559.84 | 823.30 | +296 | −5496.10 |

Source: FAO (2005).

[a] All of the figures in the table are in thousands of hectares.

[b] Column six represents the change in forest area without the growth in the extent of forest plantations.

[c] The changes in the extent of plantations, in Column 5, and natural forest, in column 6, should add up to the overall change in forest cover, in column 3. This calculus does not work because the population of countries with data on forest cover, columns 1–3, is somewhat larger than the population of countries, columns 4-6, with data on forest plantations.

great detail in the twentieth century. A second pattern of forest expansion runs across all of the regions (see column five). In every region forest plantations have increased in extent between 1990 and 2005. This pattern of forest expansion is relatively absent in Africa and particularly prevalent in East and South Asia. A third pattern of forest expansion, artisanal in its *modus operandi*, prevails around the houses of smallholders in densely populated and peri-urban settings in the developing world. It cannot be tracked with national accounts data. The following pages review the character, origins, and driving forces of all three patterns of forest expansion.

## 3.2 Expansion Through Spontaneous Regeneration

When people abandon agricultural lands in places with moist climates, woody vegetation first encroaches and later envelopes the old fields. The new growth takes a wide variety of forms. In some well watered places secondary forests, featuring pioneer species native to the region, quickly take hold, creating in less than 15 years a recognizable secondary forest. In places with dry seasons forest recovery proceeds more slowly, if it occurs at all. Recurrent fires sweep through these lands, destroying seed sources. In some cases the fires destroy so many seeds and seed sources that forests never really regenerate. Instead, a low, fire resistant vegetation emerges, referred to variously as 'bush fallow', 'farm bush', or 'scrub growth'. Fire resistant, invasive species like the shrubs *Chromolaena odorata* (Jack in the bush) and *Pteridium aquilinum* (Bracken fern) often come to dominate these landscapes, creating a low and dense woody formation that does not fit the classic definition of a forest, but clearly represents something other than agricultural land. The prevalence of invasive species in these disturbed landscapes reflects the large exchanges of genetic material that have occurred with the globalization of human societies during the past 500 years (Crosby 2003).

Spontaneous reforestation has occurred at least as long as humans have practiced agriculture in wooded regions, and it usually occurs after upheavals and transformations in the surrounding human societies. While spontaneous reforestation occurs most frequently in settings where people practice shifting cultivation, it sometimes occurs in circumstances where people do not return, as they do in routines of shifting cultivation, to cultivate the land after a fallow period. In these instances the spontaneous regeneration of forests represents a long-term expansion in forests. This type of spontaneous reforestation occurred in depopulated rural districts following the Black Death in medieval Europe (Herlihy 1997; Poos 1991). In the aftermath of the 30 Years War during the seventeenth century forests encroached on fields when diminished populations of German peasants began fallowing more of their lands more frequently (Slicher van Bath 1966).

When do people abandon these lands? Who abandons these lands, and where does abandonment occur? The historical record provides partial answers to these questions. It is clear that spontaneous reforestation is more likely in some settings than in others. It occurs more frequently on less fertile lands than on more fertile lands. The Piedmont region of the American South saw extensive spontaneous reforestation from 1935 to 1975 when cotton farmers decided to abandon the relatively infertile red clay soils in the region because they could no longer afford the heavy expenditures for fertilizers necessary to produce cotton on these lands (Rudel and Fu 1996). Fields in topographically rugged settings are much more likely to revert to forest than are fields in flat, lowland settings (Mather and Needle 1998). Working topographically rugged lands imposes a variety of extra costs on farmers. The sloped land makes it difficult to cultivate these lands with machinery. Because these fields are typically located farther from roads than lowlands, transporting agricultural produce from these lands to markets costs farmers more. Finally, the location of these lands on hillsides makes it less likely that they will contain alluvial soils, so they may have less fertile soils than lowland fields. For all of these reasons land abandonment and the spontaneous spread of forests occurs most frequently in mountainous places. This generalization applies across a range of different geographical scales. Farmers in the northeastern United States allowed individual fields with the steeper slopes to revert to forest (Goldthwait 1927). Communities in Southeast Asia abandoned upland agriculture before they abandoned lowland agriculture within their borders (Tachibana et al. 2001).

Physical factors interact with social factors in sometimes complicated ways to influence the extent and spatial patterning of spontaneous reforestation. Latin America has long had a latifundia – minifundia complex which has featured smallholders living on hillsides while large landowners farmed the more fertile valley bottoms (Barraclough and Collarte 1973). Recently, mountainous regions and hillside districts in Latin America have begun to lose people at a greater rate and experience more spontaneous reforestation than do the lowland regions of Latin America (Aide and Grau 2004). Coffee cultivation in interior Puerto Rico illustrates how these interactions take place. Coffee grows best at mid-elevations in the tropics, so coffee cultivation in Puerto Rico during the first half of the twentieth century concentrated in the mountainous interior of the island. Coffee was then as now a labor intensive crop, so over time a considerable population of poor cultivators congregated in these upland communities where they eked out an existence cultivating coffee. After World War II the industrialization of coastal Puerto Rico under the auspices of Operation Bootstrap and the advent of rapid air travel to the continental United States pulled poor cultivators off of these lands. Factory owners provided better wages and urban communities offered more amenities, so people left the land, migrating either to cities on the island like San Juan or cities on the continent like New York. They left behind in the highlands a failing agricultural economy in which labor scarcities made it impossible to cultivate coffee. Widespread land abandonment began occurring, and old coffee groves began reverting to secondary forests (Rudel et al. 2000).

A variant on this pattern of spontaneous reforestation on the lands of poor uplanders unfolded on the island of Palawan in the Philippines during the 1990s. Palawan, like Puerto Rico, features a narrow coastal plain surrounding a mountainous interior. Poor farmers earned their livelihoods by alternating periods of work as wage laborers on coastal rice farms with work growing corn and other foodstuffs on their own small farms that they had recently carved out of forests in the uplands. During the 1990s the owners of many of the lowland rice farms installed irrigation systems. The installation of the irrigation systems saddled these farmers with debt and lengthened their growing seasons by making water to irrigate the rice paddies available for longer periods of time. Under these circumstances lowland farmers could cultivate three crops of rice in a year instead of just two crops. To pay off their debts and maximize the return on their investment most rice farmers moved to triple cropping. This change increased their need for labor, so they began to employ the upland smallholders for longer periods of time each year. The uplanders in turn spent less time on their upland farms and reduced the amount of acreage that they farmed in the uplands. These lands promptly reverted to forests (Shively and Martinez 2001).

In both Puerto Rico and Palawan labor scarcities generated by vigorous economic growth elsewhere pulled farm workers off of the land and, in so doing, set the stage for the spontaneous regeneration of forests in old fields. This pattern appears to apply more generally in regions where wealthier societies with abundant alternative economic opportunities make farm labor expensive which in turn makes agriculture on marginal lands unprofitable. In a global analysis of forest transitions during the 1990s this labor scarcity path to forest expansion applied to Europe and the Americas where relatively abundant alternative economic opportunities reduced the supply of farm labor. It did not apply in either Africa or Asia where large rural populations and a relative lack of urban economic opportunities made farm labor more available (Rudel et al. 2005).

In the latter settings people may migrate elsewhere to work, but they make so little as migrant laborers that they must somehow find a way to continue farming the lands that they just left. For example, migrant laborers on the eastern slope of the Andes in Peru continued during the 1980s to work coffee groves on their small farms during periodic visits home because they made so little as migrant laborers elsewhere in Peru. Inevitably, the coffee groves suffered from neglect, experiencing erosion problems in particular (Collins 1988). These people are caught in resource degrading poverty traps (McPeak and Barrett 2001), and forests do not recover or expand under these conditions even though the cultivators leave the land for considerable periods of time.

Other factors sometimes contribute to the abandonment of marginal agricultural lands and their subsequent reoccupation by forests. Technological advances in agriculture make prime agricultural lands even more productive. These advances increase the size of harvests, depress agriculture prices, and induce the abandonment of marginal agricultural lands. Christened ‘the Borlaug hypothesis’ after Norman Borlaug, an influential Green Revolution scientist and early proponent of this thesis, it implies that technological advances can reduce rates of tropical deforestation (Angelsen and Kaimowitz 2001). The reforestation of the American South between 1935 and 1975 illustrates this dynamic. Counties that relied on crops whose productivity per acre increased disproportionately (corn, cotton, peanuts) during this period saw a turnaround in forest cover trends in which deforestation in earlier periods gave way to spontaneous reforestation. Counties that relied on crops whose productivity did not increase as much (rice, oranges) during this period experienced continued deforestation (Rudel 2001).

In more general theoretical terms the spontaneous reforestation occurring in the now abandoned fields of the wealthier nations conforms to several theoretical formulations about the environment and society. It supports arguments by economists about the existence of environmental Kuznets curves in which environmental abuses initially increase with economic development, only to level off and actually decline at higher levels of economic development (Stern et al. 1996). It conforms to closely related historical generalizations about the forest transition in which forests decline in extent during an initial stage of economic development and then increase in extent during a second stage when societies become economically affluent and people migrate to urban areas (Mather and Needle 1998).

## 3.3 Expansion Through Tree Plantations

The hand of man (or, more frequently, woman) sometimes plays a more visible role in the expansion of forests. People decide to grow trees, much as they would grow a crop of corn, planting row after row of the same species of tree in a confined area. People refer to these tree plantings as forest plantations. They have increased by more than seven times in developing countries between 1980 and 2000 (Del Lungo et al. 2001). Communities, corporations, and government agencies have all created forest plantations. Villagers organized into forest protection committees have created many new plantations in South and East Asia (Singh 2002; Muller and Zeller 2002). Government agencies, usually the forestry department, but sometimes other agencies, like the military in Ecuador, create forest plantations. Corporations also create plantations, sometimes infamously, as with the ill-fated Jari project in the Brazilian Amazon in which the American billionaire, Daniel Ludwig, replaced natural forests with fast growing stands of eucalyptus and tropical pine (Smith et al. 1995). Other corporations have established plantations on more disturbed and degraded sites like the pulpwood producing Aracruz plantation in southern Brazil (Marchak 1995). Smallholders are less likely to create forest plantations because they do not want to devote all of their land to a single crop (trees) that may not yield an income for 15 years.

As might be expected, tree plantations exhibit low levels of biodiversity, especially when the planted trees are a hardy tree like *Eucalyptus diversicolor* from the Eucalyptus genus. These trees contain toxins that kill undergrowth that might compete with the tree and appropriate large amounts of the ground water that otherwise might nourish plants in the understory. Forest plantations with other dominant trees sometimes exhibit surprisingly large amounts of biodiversity in their understories. Plant communities in the younger plantations do not equal the biodiversity in a naturally regenerating forest, but in older plantations biodiversity does approach that of a naturally regenerating secondary forest (Lugo 1992).

Increases in the planting rates sometimes have important off-farm effects. In places like large parts of the Indian subcontinent where natural and planted forests co-exist, expansion in the extent of planted forests has reduced the rates of harvest in nearby natural forests (Kohlin and Parks 2001). In other settings the exact opposite effect can occur through process referred to as 'leakage' (Lee et al. 2004). Forest plantation owners evict smallholders from the site of the plantation, and then the displaced smallholders carve a new farm out of nearby natural forests, thereby nullifying any environmental services that the forest plantation might have provided.

A glance at column five in Table 3.1 indicates that forest plantations increased in extent on all of the world's land masses between 1990 and 2005. The increases were not, however, uniform across the continents. People in Africa planted only small amounts of land in forest plantations while people in Asia expanded the extent of their forest plantations in major ways between 1990 and 2005. With the recent surge in planting, plantation forests have come to comprise a significant fraction of all forests in places humid enough to support natural forests. By 2000 plantation forests constituted nearly half of all of the forests on the Indian subcontinent. The plantation owners vary considerably by continent. While large landowners own two-thirds of the land in forest plantations in the Americas, they only own 5% of the land in forest plantations in Asia (FAO 2005).

The continental differences in rates of growth in forest plantations reflect in part market dynamics involving the increasing scarcity of forest products in societies that have already experienced extensive deforestation. Satellite images that show concentric zones of denuded and deforested lands around major cities in Africa and Asia portray in visual terms the scarcities of wood poles for house construction and firewood or charcoal for cooking that have emerged during the last two decades of the twentieth century in these places (Foster and Rosenzweig 2003). These images suggest that people establish forest plantations in places with few natural forests, around large cities or in arid environments like those that characterize North Africa.

A market response to these scarcities in the form of forest plantations only occurs under conditions where consumers generate sufficient demand for forest products. Consumers must be numerous and wealthy enough to pay for forest products. First, large concentrations of people in cities, a condition more easily met in South Asia than in sub-Saharan Africa, would create a potential population of consumers. Second, people must have the income to pay for forest products before the increased scarcity of forest products will translate into more demand for forest products and the higher prices that might prompt landowners to plant trees on their lands. This condition may not be met in many places in sub-Saharan Africa where poor households go to great lengths to meet their subsistence needs outside of the markets through time consuming routines of collecting a wide variety of forest products from the lands around their homes (Cavendish 2000). The smaller populations and greater poverty in sub-Saharan African societies relative to South and East Asian societies may explain why landholders have converted more extensive areas into forest plantations in South and East Asian nations.

As with instances of spontaneous reforestation, labor supplies play an important role in the expansion of forest plantations. Because the maintenance of a forest plantation requires less labor that the cultivation of a similar sized agricultural field (Bentley 1989), landholders save on labor when they convert lands from agriculture into a forest plantation. While the conversion of agricultural lands into forest plantations saves on labor, it does not eliminate the need for labor on these lands because forest plantations, unlike lands undergoing spontaneous reforestation, require some inputs of labor. Cultivators must plant the trees, spray them on occasion with herbicides, thin them, and eventually harvest them. The need for timely labor makes it more likely that people will convert lands to forest plantations in places that continue to have a ready supply of farm laborers. The continuing presence of rural laborers has probably contributed to the ability of Indian villages to expand the extent of the forest plantations that they co-manage with the state (Poffenberger et al. 1996). At the same time the departure of some younger workers for urban areas may relieve human pressure on the land and free up marginal agricultural lands that can then be devoted to forest plantations. This type of demographic dynamic may explain in part why Vietnamese and Chinese villages have been able to increase the extent of their forest plantations so rapidly since 1990 (Muller and Zeller 2002).

Governments have begun to play an important role in the expansion of forest plantations through political economic reforms that give communities and individuals incentives to establish forest plantations. The most common reform has involved the devolution of control over the management and proceeds from forest plantations from the state to the village. Indian politicians devolved partial control over state forests to communities during the late 1980s (Hussain et al. 1999). Beginning in 1990, reformers in eight West African countries succeeded in getting similar reforms enacted in their countries (Wily 2002). In East and South Asia these forest reforms have accelerated the creation of forest plantations (Mather 2007).

International political economic forces, in the form of the Clean Development Mechanism (CDM) of the Kyoto Protocol, should also begin to accelerate the expansion of forest plantations. Because the newly created forests sequester carbon, they contribute in a marginal way to climate stabilization. Through an emissions trading scheme, wealthy countries can receive credits for emissions reduction by financing the creation of carbon sequestering forest plantations in developing countries. In effect the landowners in developing countries would receive payments for the environmental services delivered by the recently planted forests. In practice only seven of the first 500 projects authorized under the CDM through mid-2007 have involved forest plantations (CIFOR 2007). While this type of international political economic rationale for increases in forest plantations is conceivable, the recent increases in the extent of forest plantations have been driven entirely by local and national political economic considerations.

## 3.4 Expansion Through Household Agroforests

While large scale forest plantations remain rare in sub-Saharan Africa, another kind of forest expansion has begun to occur in densely settled, peri-urban districts outside of some large African cities. Large numbers of smallholders have begun planting individual, often fruit bearing, trees in strategic spots on their smallholdings. They might, for example, plant a line of trees along an irrigation ditch in a semi-arid district like Machakos, just outside of Nairobi, Kenya (Tiffen et al. 1994; Holmgren et al. 1994). The products from the trees, fruit or wood, get sold in the nearby city. Household agroforests also exist in Asia and the Americas. Smallholders in the Dominican Republic, north of Santo Domingo, have planted a fast growing exotic,

*Acacia Mangium*, sometimes mixed in with fruit trees, in their backyards and established small scale sawmills to provide lumber for the Dominican construction industry (Rocheleau et al. 2001).

Over time these landscapes take on an almost forested appearance. The small farms outside of Nairobi, for example, contained an average of 59 fruit trees per hectare (Tiffen et al. 1994). The small scale of the tree planting and the mixture of land uses on the lands planted with trees makes it inappropriate to refer to the planted trees as forest plantations, but aggregated across the plots of neighboring farmers, these plantings can create an agro-forest where no forest existed earlier. Because smallholders obtain their seed stock from a wide variety of sources and plant a wide variety of trees on their lands, these lands exhibit high levels of agrobiodiversity. Given their proximity to cities, these agro-forests have begun to create incipient green belts around some African cities.

Given the economic and environmental promise of tree planting, community activists decided to promote it, beginning in the 1970s. Wangari Maathai, a Kenyan scientist and community activist founded the Green Belt Movement (GBM) in 1977. At its inception the movement consisted of groups of women who established tree nurseries, using funds provided by the GBM. They planted seedlings from the nurseries on their own farms and gave away other seedlings to nearby farmers. ICRAF (International Center for Research on Agro-Forestry), with headquarters in Nairobi, currently sponsors projects designed to enhance smallholder agricultural productivity throughout sub-Saharan Africa. A significant proportion of these projects promote agro-forestry, most commonly in the form of trees planted as living fences or on bunds.

The anticipated acceleration of efforts to sequester carbon through CDM projects raises questions about how one might extend the payments to the low income smallholders who create agro-forests on small plots of land. The payments for the environmental services provided by smallholders will not be large, so they will have to be integrated into the ongoing exploitation of the land in ways that allow smallholder to earn some income from the land in other ways. Given the long chain of hands through which the payments might pass before they reach small scale tree planters, governance issues in these programs will be challenging (Boyd et al. 2007). These challenges notwithstanding, the prospect of providing cash payments to poor people for climate stabilization activities holds out the hope of benefiting both the very poor and the global environment.

## 3.5 Conclusion: Historical Conditions and Policies for Encouraging Forest Expansion

Of the three historical paths to forest expansion outlined above, the spontaneous reforestation path remains perhaps the most precarious because it assumes that the returns to agricultural labor will remain less than the returns to non-farm labor, as was the case throughout most of the twentieth century. If this condition should shift, with agricultural commodity prices rising faster than prices for labor, then farmers in the affluent societies could expand farmland at the expense of the spontaneously regenerated secondary forests adjacent to their farmlands. This surmise underscores how the historical trajectories of forest expansion described here depend upon underlying, sometimes ephemeral, historical conditions.

Another underlying condition involves the salience of market dynamics in the forces driving forest expansion. All three paths of forest expansions have their origins in market dynamics. Labor scarcities in rural areas figure centrally in the spontaneous reforestation of old fields in the northern hemisphere. Forest product scarcities in urban areas have played a central role in the creation of forest plantations and household agro-forests. With the emergence of a global, post Kyoto effort by governments and NGOs to reduce carbon emissions through sequestration strategies, one might speculate that the period in which market dynamics alone have driven forest expansion may be coming to a close.

Governments and NGOs have already played a facilitating role in expediting the growth of forest plantations and household agro-forests by giving rural residents access to free seedlings and allowing them to keep some of the revenues from forest products grown on lands that they manage or own. In contrast governments have done little to facilitate spontaneous reforestation. Policy changes tied to a post-Kyoto climate change agreement could change this pattern. REDD (Reducing emissions from deforestation and degradation) payments for averted deforestation could include payments for spontaneous reforestation. While conservation set aside programs, designed to reduce soil erosion on agricultural lands may have encouraged some spontaneous regrowth of forests, the much more extensive price support and agricultural subsidy programs in North America and Western Europe have allowed farmers on marginal lands to continue to cultivate these lands. In so doing, these programs have probably prevented the reversion of much larger areas to forest than actually occurred during the twentieth century. The elimination of these subsidies would change the mix of crops cultivated and encourage, depending on price signals in markets, the reversion of at least some of agricultural lands to forest.

A new generation of international CDMs could include incentives and institutional mechanisms that allow the poor from the rural South, with control over small parcels of land, to participate in programs that pay for forest expansion. These programs would presumably target the creation of agro-forests and forest plantations on cleared and degraded lands. Unlike so many reforms promoted by environmentalists, a significant number of these reforms do not attempt to reverse existing trends in society, rather they try to accelerate trends, like plantation expansion, that are already under way. For this reason forest reforms stand a good chance of succeeding in increasing forest cover, and, in so doing, contributing in a small but measurable way to climate stabilization.

## References

Aide M, Grau H (2004) Globalization, migration, and Latin American ecosystems. Science 305(24 September):1915–1916

Angelsen A, Kaimowitz D (2001) Agricultural technologies and tropical deforestation. CAB International, Wallingford

Barraclough S, Collarte J (1973) Agrarian structure in Latin America: a resume of the CIDA land tenure studies of Argentina, Brazil, Chile, Colombia, Ecuador, Guatemala, and Peru. Lexington Books, Lexington, MA
Bentley J (1989) Bread forests and new fields: the ecology of reforestation and forest clearing among small woodland owners in Portugal. J For Hist 17:188–195
Boyd E, Gutierrez M, Chang M (2007) Small-scale forest carbon projects: adapting CDM to low-income communities. Glob Environ Chang 17:250–259
Cavendish W (2000) Empirical regularities in the poverty–environment relationship of rural households: evidence from Zimbabwe. World Dev 28:1979–2003
Chomitz K (2007) At loggerheads? Agricultural expansion, poverty reduction, and environment in tropical forests. World Bank, Washington, DC
CIFOR (Center for International Forestry Research) (2007) Carbofor: CIFOR's Carbon Forestry Research. Accessed at http://www.cifor.cgiar.org/carbofor, May, 2007
Collins J (1988) Unseasonal migrations: the social causes of environmental destruction. Princeton University Press, Princeton, NJ
Crosby A (2003) The Columbian exchange: biological and cultural consequences of 1492, 2nd edn. Greenwood, Westport, CT
Del Lungo A, Vuorinen P, Carle J (2001) Preliminary analysis of global trends in forest plantation development, 1980–2000. FAO, Rome
Downton M (1995) Measuring tropical deforestation: development of the methods. Environ Conserv 22:229–240
Food and Agricultural Organization of the United Nations (2005) Forest Resources Assessment, 2005. Forestry Division, Rome. Accessed at www.fao.org/forestry/fra2005. May, 2007
Foster A, Rosenzweig M (2003) Economic growth and the rise of forests. Q J Econ 118:601–637
Goldthwait J (1927) A town that has gone downhill. Geogr Rev 17:527–552
Grainger A (2008) Difficulties in tracking the long-term global trend in tropical forest area. Proc Natl Acad Sci 105:818–823
Herlihy D (1997) The black death and the transformation of the West. Harvard, Cambridge, MA
Holmgren P, Masakha E, Sjoholm H (1994) Not all African land is being degraded: a recent survey of trees on farms in Kenya reveals rapidly increasing forest resources. Ambio 23:390–395
Hussain T, Bibi P, Kaushal P (1999) We are all part of the same 'Kudrat': community forest management in Rajaji National Park. For Trees People 38:35–38
Intergovernmental Panel on Climate Change Fourth Assessment Report (2007) Working group 1 report: the physical science basis. Accessed at http://www.ipcc.ch/WG1_SPM_17Apr07.pdf. May 2007
Kohlin G, Parks P (2001) Spatial variability and disincentives to harvest: deforestation and fuelwood collection in South Asia. Land Econ 77:206–218
Lee H, McCarl B, Schneider U, Chen C (2004) Leakage and comparative advantage implications of agricultural participation in greenhouse gas emission mitigation. UWO Department of Economics Working Papers. University of Western Ontario, London, Ontario
Lugo A (1992) Comparison of tropical tree plantations with secondary forests of similar age. Ecol Monogr 62:1–41
Marchak P (1995) Logging the globe. McGill-Queens University Press, Montreal
Mather A (2007) Recent Asian forest transitions in relation to forest transition theory. Int For Rev 9:491–502
Mather A, Needle C (1998) The forest transition: a theoretical basis. Area 30:117–124
McPeak J, Barrett C (2001) Differential risk exposure and stochastic poverty traps among East African pastoralists. Am J Ag Econ 83:674–679
Muller D, Zeller M (2002) Land use dynamics in the central highlands of Vietnam: a spatial model combining village survey data with satellite imagery interpretation. Ag Econ 27:333–354

Poffenberger M, McGean B, Khare A (1996) Communities sustaining India's forests in the twenty-first century. In: Poffenberger M, McGean B (eds), Village voices, forest choices: Joint Forest Management in India. Oxford, Delhi
Poos L (1991) A rural society after the black death: Essex 1350–1525. Cambridge University Press, Cambridge, UK
Rocheleau D, Ross L, Morrobel J, Malaret L, Hernandez R, Kominiak T (2001) Complex communities and emergent ecologies in the regional agroforest of Zambrana-Chacuey, Dominican Republic. Ecumene 8:465–492
Rudel T (2001) Did a green revolution restore the forests of the American South? In: Angelsen A, Kaimowitz D (eds) Agricultural technologies and tropical deforestation. CAB International, Wallingford
Rudel T (2005) Tropical forests: regional paths of destruction and regeneration in the late twentieth century. Columbia University Press, New York
Rudel T, Coomes O, Moran E, Achard F, Angelsen A, Xu J, Lambin E (2005) Forest transitions: towards a global understanding of land use change. Glob Environ Chang 15:23–31
Rudel T, Fu C (1996) A requiem for the southern regionalists: reforestation in the South and the uses of regional social science. Soc Scienc Q 77:804–820
Rudel T, Perez-Lugo M, Zichal H (2000) When fields revert to forests: development and spontaneous reforestation in post-war Puerto Rico. Prof Geogr 52:386–397
Shively G, Martinez E (2001) Deforestation, irrigation, employment, and cautious optimism in southern Palawan, the Philippines. In: Angelsen A, Kaimowitz D (eds) Agricultural technologies and tropical deforestation. CAB International, Wallingford, UK
Singh N (2002) Federations of community forest management groups in Orissa: crafting new institutions to assert local rights. For Trees People. #46.
Slicher Van Bath B (1966) The agrarian history of western Europe, a.d. Edward Arnold, London, pp 500–1850
Smith N, Alvim P, Serrão E, Falesi I (1995) Amazonia. In: Kasperson J, Kasperson R, Turner B (eds) Regions at risk: comparisons of threatened environments. United Nations University Press, Tokyo
Stern D, Common M, Barbier E (1996) Economic growth and environmental degradation: the environmental Kuznets curve and sustainable development. World Dev 24:1151–60
Tachibana T, Nguyen T, Otsuka K (2001) Agricultural intensification versus extensification: a case study of deforestation in the northern hill region of Vietnam. J Environ Econ Manage 41:44–69
Tiffen M, Mortimore M, Gichuki F (1994) More people, less erosion: Environmental environmental recovery in Kenya. Wiley, New York
Wily L (2002) The political economy of community forestry in Africa – getting the power relations right. For Trees People. #46

# Chapter 4
# A Tri-Partite Framework of Forest Dynamics: Hierarchy, Panarchy, and Heterarchy in the Study of Secondary Growth

**Stephen G. Perz and Angelica M. Almeyda**

S.G. Perz (✉)
Department of Sociology, University of Florida, FL, USA
e-mail: sperz@ufl.edu

A.M. Almeyda
Department of Anthropological Sciences, Stanford University, CA, USA

H. Nagendra and J. Southworth (eds.), *Reforesting Landscapes: Linking Pattern and Process*, Landscape Series 10,
DOI 10.1007/978-1-4020-9656-3_4, 

## 4.1 Introduction

As a complement to the enormous literature on deforestation, there is now an emerging body of work on the interrelated topics of secondary vegetation (herbaceous or woody vegetation on previously cleared land), reforestation (growth of woody vegetation on previously cleared land), secondary forests (woody vegetation on previously cleared land, still distinguishable from mature forest), and forest recovery (woody secondary vegetation that increasingly approximates historical and contemporary mature forest in the same region), which we summarize with the term secondary growth (the emergence of vegetation on previously cleared land, whether via plantations or natural plant succession). There is growing documentation that whereas forests continue to fall, especially in tropical regions, secondary growth is expanding in temperate regions, and furthermore even in areas with deforestation, there is often considerable land under secondary growth (other chapters, this volume, especially Grainger).

Consequently, there is an expanding literature on the theoretical explanations for forest decline as well as secondary growth. As with the case of deforestation, most initial efforts to understand secondary growth drew on established theoretical frameworks originally developed for another topic, usually some long-studied focus within a discipline. The result in both cases has been a cacophony of explanations, some mutually incommensurate, most framed in distinct language due to the diverse assumptions being made. This state of affairs has resulted in confusion for audiences in other disciplines and has driven efforts to re-examine the assumptions and interrogate the blind spots of established theoretical frameworks. Such efforts constitute a means of forging integrative theoretical approaches that are more open to multiple explanations and more adaptable to specific contexts.

This chapter suggests an integrative approach for theoretical explanations for land use and land cover change (LULCC), featuring the case of secondary growth. The core of this chapter focuses on a tri-partite framework that draws on hierarchy theory from landscape ecology, the notions of "adaptive cycles" and "panarchy" from complex systems theorists, and the concept of "heterarchy," previously applied in archaeology and business management (Perz 2007, 2008). We deploy these components as heuristics by borrowing them from other purposes and applying them to understanding the causation behind secondary growth. Together, a hierarchical framework with panarchic dynamics that results in a heterarchical structure of causation constitutes a "tripartite framework of forest dynamics" (TFFD). These heuristics can be criticized for being too abstract (cf. Walker 2007). We therefore employ examples involving secondary growth in the Brazilian Amazon to illustrate the TFFD, and conclude with a discussion of methodological applications.

We suggest that a TFFD provides a theoretical framework in which we can situate ostensibly competing explanations for secondary growth, and move beyond limitations in theoretical approaches such as grand theorizing, reductionism, and context-specificity (Perz 2007, 2008). We offer the TFFD as a means organizing explanations for secondary growth, rather than as a replacement. One of our goals is to avoid red herring debates about theoretical primacy among explanations, which results from reductionism in the presence of competing explanations. We also suggest the TFFD as a means of avoiding the universalist pretensions of isomorphic grand theorizing as well as the haphazardness of context-specificity. By offering a framework in which various explanations can be arranged and evaluated, we can see to what extent a given explanation applies across cases, while recognizing that the configuration of explanations itself may vary from place to place. This allows for systematic comparisons in a common framework while acknowledging uniqueness.

## 4.2 A Tri-Partite Framework of Forest Dynamics

By now there is no shortage of theoretical perspectives seeking to provide explanations for LULCC (Gutman et al. 2004; Lambin and Geist 2006; Moran and Ostrom 2005). A major challenge is how to evaluate multiple theoretical perspectives across many cases, in order to select some explanations and abandon others (Lambin and Geist 2006). There is increasing attention going to the need to find some means of integrating theoretical arguments in a broader framework that can also specify factors that make certain theoretical perspectives particularly relevant in specific contexts while considering a standard set of explanations across many cases (Lambin and Geist 2006).

There are several requirements of integrative frameworks which pose serious challenges to theoretical integration. First, an integrated framework must account for the different levels of scale on which distinct causal processes operate. In terms of theory, this has as much to do with the operation of actual processes as with the resolution and extent of observation. In terms of socioeconomic explanations, it is crucial to distinguish among various social actors, from individuals to communities to governments to international coalitions, for they have different arenas of influence, diverse interests and operations, and varying LULCC impacts.

Second, an integrated framework must have some means for specifying the direction of causal relationships, of sorting out direct and indirect causation, and thereby specifying chains of causal processes and conditioning factors. Reductionist approaches and assumptions of isomorphic causation and universal applicability help little in explaining forest dynamics across different contexts. An integrative framework therefore must provide some structure in which a given set of explanations can be situated for consideration, and retained if they gain empirical support.

Third, an integrative framework must account for complex dynamics of forest cover change. Forest dynamics themselves are variable, and this requires some explanation. Forest change may be fast or slow, localized or widespread, and thus exhibits complex spatio-temporal patterns. In terms of theoretical explanations, there is a need to evaluate what accounts for rapid or slow forest change. Further, it is important to pay attention to feedback effects, for land use decisions yield ecological as well as socioeconomic outcomes that can change decision contexts.

And fourth, an integrated framework must incorporate the possibility that key causal factors themselves may change over time. Switching of causation has primarily been addressed via empirical observation, though it is seen by some as consistent with sufficiently abstract theories as well. Predicting changes in causation remains a dubious proposition, but any framework that seeks to account for forest change still needs to be able to accommodate changes in causal processes to explain forest dynamics.

Here we suggest a TFFD that addresses these requirements (Perz 2007, 2008). We discuss hierarchy theory as a means of addressing the scale and causal sequence requirements; we draw from the notions of adaptive cycles and panarchy to account for slow-fast dynamics in causal factors and non-linearities in forest change; and we deploy the concept of heterarchy to recognize that causal factors and causal structures may change over time. If the TFFD we suggest is not an elegant formulation, neither are the dynamics nor the causation involved in forest change. We are well aware of criticisms of the three components of the TFFD, and address criticisms in our remaining text. Specifically, we respond to criticisms of the initial components of the TFFD by incorporating the later components, and we deal with broader criticisms of our approach in the conclusion.

### *4.2.1 Hierarchy Theory*

Hierarchy theory emerged out of landscape ecology as a means of organizing numerous biological processes in terms of the spatio-temporal scale ("level") on which they operate so as to better organize and understand their relationships, from the cellular to the ecosystem level (O'Neill, et al. 1986; Allen and Starr 1988; Ahn and Allen 1996). This was based on the observation that many biological processes that operate on a small scale tend to be rapid and frequent, whereas those operating on larger scales tend to be slow and infrequent. Further, the operation of small-scale biological processes is periodically affected by the larger-scale processes, following the rhythm of the larger-scale processes. The key insight of a hierarchical perspective is that entities operating on one scale, especially larger scales, condition the operation of entities on other scales, especially smaller scales. As a result, hierarchy theory provides a structural approach to understanding ecosystems as nested sets of entities, with cells operating within organs, organs within organisms, organisms within ecosystems, and so forth.

A hierarchical perspective can be viewed as a heuristic (cf. Abbott 2004) and applied not only to understand the structure of ecosystems, but also causal pathways. If we apply the hierarchy heuristic to the causation behind forest change, we can deploy hierarchy theory as a framework that can elucidate the organization of processes affecting secondary growth. That is, large-scale determinants of LULCC condition the effects of smaller-scale determinants, which then directly affect forest change as via secondary growth.

A hierarchical approach offers several advantages as an integrative theoretical approach. First, a hierarchical framework explicitly recognizes that there are multiple causal and conditioning processes operating on different spatial scales.

Whereas scale is often featured in theoretical discussions of LULCC (Lambin and Geist 2006), discipline-based theoretical perspectives often focus on one scale or another. Acknowledgment of scale-specificity in causation allows us to situate a variety of explanatory factors, regardless of disciplinary origin, in a hierarchical framework by locating explanations on scales they feature.

Second, hierarchical frameworks organize causal processes in a specific way in order to sort out proximate, intermediate, and distant mechanisms. Causal proximity is also emphasized in the LULCC literature (Geist and Lambin 2002; Lambin and Geist 2006). Hierarchical frameworks view the locus of LULCC as the result of specific land use decisions, making micro-scale factors tied to land users proximate causes, with larger scale factors that influence land users intermediate causes, and macro-scale mechanisms distant causes. By organizing micro-, meso-, and macro-scale processes in terms of causal proximity to an outcome, we can forestall theoretical debates about causal primacy, and avoid captivity to a theoretical framework confined to one discipline or another. We note that hierarchical frameworks need not be limited to three scales, and can accommodate more if necessary.

Third, given the emphasis on spatial scale as a means of organizing causal processes, hierarchical frameworks can explain multi-scale spatial variations in secondary growth. Hierarchical frameworks thus link the scale of variation in secondary growth to the causal factors operating on that scale. Micro-scale variations in land cover are most likely due to proximate variations among land users; meso-scale variations are due to differences in intermediate-scale factors; and macro-scale variations are due to distant, large-scale causes.

The complex causation behind LULCC, and the many socioeconomic explanations offered to account for secondary growth, has made hierarchical frameworks attractive to social scientists (Gibson et al. 2000; Wood and Porro 2002; Perz 2002; Warren 2005). Social science explanations for secondary growth focus on identification of relevant social actors ("agents") who operate on specific spatial scales ("arenas of influence"). Examples of social actors and their arenas of influence include farm families with rural properties or logging firms with timber concessions. Such social actors constitute distinct "decision units" with specific land areas over which their decisions directly influence forest change.

Focusing on proximate social actors allows us to identify specific decisions they may take that directly affect the areal extent and age composition of secondary growth. In the Brazilian Amazon, there are several such land use decisions: (1) let land lie in fallow, (2) abandon a plot within a property, (3) abandon an entire property, (4) allow secondary growth in pastures, (5) modify forest cover (via selective timber extraction) , (6) plant valuable trees, or (7) engage in new or expanded land use (Perz and Walker 2002; see also Crews-Meyer, this volume). Whereas the first five decisions will result in an increase in secondary growth via plant succession, the sixth involves reforestation via more artificial means, and the seventh will either not affect secondary growth (if new land use is on land not previously under secondary vegetation) or will reduce it (if new land use is on land formerly under secondary vegetation). These decisions are illustrated in Fig. 4.1.

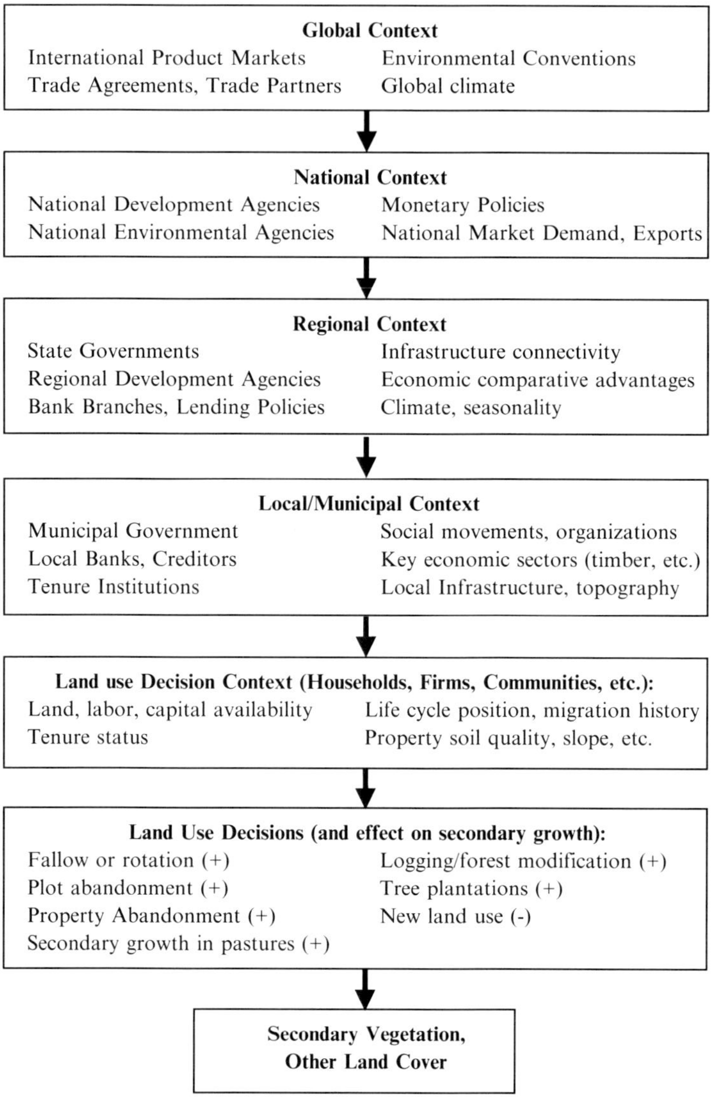


**Fig. 4.1** A hierarchical framework of causal agents influencing secondary vegetation and other land covers (only top-down causation shown; bottom-up feedbacks also possible)

That said, the broader context in which decisions are made is crucially important. It is therefore necessary to account for the "decision context" of proximate social actors. The most immediate decision context for social actors is defined by their own characteristics. In economic terms, the "decision latitude" of social actors reflects their land, labor and various capital assets. Such assets help determine their goals with land use decisions as a means of achieving those goals, which then defines the logic of their land use decisions and the consequent land use patterns. Family land use is greatly influenced by their available labor; logging firms make decisions based on their capitalization. See the land use decision context box in Fig. 4.1.

In turn, the land use decisions of micro-level social actors are partly dependent on the larger scale context in which they operate. At this point the contributions of a hierarchical perspective become evident, for we can view social actors who make decisions that directly affect land use as micro-level agents operating in a broader socioeconomic and biophysical context that affects their decisions. Households and logging firms operating in a given locality face contextual circumstances defined by local politics, infrastructure quality, etc. that reflects the operation of municipal governments and other agents who have larger arenas of influence (see Fig. 4.1).

For purposes of exposition, we follow Vayda (1985) who proposed "progressive contextualization" as a means of identifying causes behind an outcome. In progressive contextualization, one begins by observing the outcome of interest – in our case, secondary growth – and then proceeds to identify proximate causes – in our case, land use decisions. One then "progressively contextualizes" those decisions by looking to their immediate context, such as household or firm characteristics, and then situating those decision contexts in ever larger, more distant contexts, such as at the community or municipality levels to identify key characteristics of those entities that can help account for the behavior of the proximate decision units. Through progressive contextualization, we can identify factors that account for land use decisions by looking to the structural context of decisions in order to identify facilitators and constraints. This in turn allows identification of cross-scale causation that affects land use decisions.

Progressive contextualization can be used heuristically to populate a causal hierarchy by theoretically identifying causal factors operating on ever larger scales. A key contribution of progressive contextualization is that via induction it considers progressively larger-scale contexts at which to identify causal factors which then become situated on distinct scales in a hierarchy of causation. This inductive process allows us to organize a set of causes, from the micro-level to various meso- and macro-levels. Decisions to allow secondary growth by households, firms, and other micro-level social actors can be situated in a local context at the municipal level, a regional context of the Amazon basin itself with specific policy programs and comparative advantages, the national context of Brazil with a particular policy environment, and the global context with commodities markets and trade agreements (Fig. 4.1).

Per the intent of progressive contextualization, there is no requirement that the causal factors be identified on a predetermined spatial scale, and it is not necessary that specific explanations, tied to certain disciplines or theoretical perspectives, be found as crucially important. In this sense, a causal hierarchy does not privilege one sort of explanation over another, but rather provides the structure in which various explanations can be situated and related to each other.

While progressive contextualization works from the "bottom up" so to speak, that is from the micro-level to the meso- and macro-levels, it does so to identify causation running "top-down," with the consequence that the resulting hierarchy of causation highlights mechanisms operating on the larger scales that influence what happens on smaller scales (Fig. 4.1). In a hierarchical perspective, the reforestation in the Amazon is the result of a variety of processes operating on multiple scales (Perz and Skole 2003). Global opportunities and constraints condition the national context in Brazil, which in turn influences the regional context of the Amazon, local municipalities, and eventually proximate social actors making land use decisions. Global demand for tropical timber can influence national policies for the timber sector, regulations on timber extraction in the Amazon, municipal politics concerning the logging sector, and the behavior of logging firms and colonists with timber on their land, resulting in forest modification that fosters the subsequent emergence of secondary growth in canopy gaps opened by logging activity (Fig. 4.1).

The emphasis on top-down processes has generated criticism of hierarchy theory for its perceived structural determinism. It is important to emphasize that distant, large-scale processes are indirect causes, whose effects can be modified by intermediate and proximate determinants that serve as intervening variables. While it may be the case that high prices on tropical timber in global markets would drive more timber extraction and secondary growth in modified forests, the effect of such demand will vary from place to place. Some countries have more tropical timber than others; some regions as well. Spatial variation can emerge at each level in a hierarchy of causation, such that a global timber demand effect would manifest itself more in some countries, regions, localities, and properties than others, due to biophysical variations in the spatial distribution of timber stocks and responses by social actors with timber and the capital to extract it. As causation cascades down a hierarchy, to the extent that factors operating on smaller scales are important, they will differentiate the effects of the higher-scale process, yielding spatial variation in land use decisions and secondary growth. To the extent that factors on multiple scales exert moderating effects, they create complex patterns of variation on those scales, such that secondary growth exhibits national, regional, local, and property-level variation.

We have emphasized socioeconomic explanations, particularly economic factors, but we also recognize that other social factors can also be relevant, as are biophysical factors. This complicates the hierarchy of causation. It is easy enough to progressively situate local, regional, national and global markets via distribution chains, or to do the same for policies across administrative levels. But if we attempt to consider both socioeconomic and biophysical factors at the same time, we begin to encounter scale mismatches. If a specific causal mechanism runs from regional policies to municipal markets, then the picture is reasonably straightforward, for municipalities can be situated neatly within administrative regions, a presumption of a hierarchical framework. But if there are multiple causes tied to spatial units that don't neatly fit one within another, such as cultural areas or watersheds, then the hierarchy of causation less clear because causation does not pass neatly among nested levels of scale.

Another issue with hierarchical frameworks is the tendency to emphasize cross-scale ("vertical") causation at the expense of attention to within-scale ("horizontal") causation. While global conditions may well influence national, regional, local, and micro-level processes, it is just as likely that multiple local or micro-level factors may be operating at the same time and influencing each other. Elections in many frontier areas of the Amazon are often determined on the basis of favors to specific constituencies, especially favors that influence their ability to market produce, as via road maintenance (Perz et al. 2007). Consequently, a local political outcome may result from local road maintenance during the campaign that facilitates new land use and the reduction of secondary growth. Further, land use decisions are made simultaneously with respect to different land covers, such that household decisions about timber, crops, pasture, and off-farm activities may all affect the extent of secondary growth on a property (Perz et al. 2006).

Another issue concerning cross-scale causation is the assumption that factors at a given scale only directly affect those on the scale immediately below, and necessarily have only indirect effects on several scales below. This raises the issue of whether mechanisms at intervening scales always serve to moderate the effects of a large-scale process. This is not necessarily the case. Movement of the timber frontier, driven by global demand for tropical timber, yields local shifts in the emergence of secondary growth as timber firms move, making decisions directly on the basis of ongoing global demand.

Related to questions of top-down causation is the issue of neglecting bottom-up processes. If individual landowners form producer associations to improve market access via economies of scale for their production, they may be prompted to expand their fields at the expense of secondary vegetation, reducing it at the local or perhaps even the regional level. Such mechanisms constitute bottom-up feedback effects that originate with individual property owners but result in larger-scale effects on secondary growth.

These examples however reveal another important critique of hierarchical frameworks, namely that they emphasize causal structures and spatial variations at a given time but handle dynamics awkwardly. There is a tendency to fall back on assumptions about dynamics borrowed from landscape ecology, that is, small-scale processes operate faster, and thus can change faster, than larger-scale processes. This assumption is questionable since large-scale social actors can change behavior quickly in some instances, such as national policy changes after landslide elections. The dynamics of secondary growth require more attention than hierarchy theory by itself can offer. To address this last critique, we turn to the second component of our TFFD.

### *4.2.2 Adaptive Cycles and Panarchy*

A key issue in the study of forest dynamics concerns the temporal scales on which secondary growth can vary over time. Whereas forest transition theory rightly emphasizes that there are long-run trends in primary and secondary forest cover (Walker 1993; Mather and Needle 1998), remote sensing analyses have also called attention to important short-term changes in the Amazon (Lucas et al. 1993; Steininger 1996), as well as medium-term changes over a decade or more (Skole et al. 2004).

It becomes evident that theory about secondary growth, and other types of LULCC, needs to account for short- as well as medium- and long-term dynamics. In this section, we draw on notions of "adaptive cycles" and "panarchy" to account for the dynamics of secondary growth on multiple temporal scales, as well as changes in causal processes behind those dynamics.

The "adaptive cycle" was theorized to account for slow and fast functioning of ecosystems and their components (Gunderson et al. 1995; Gunderson and Holling 2002). An adaptive cycle has four stages. The first is the "r" stage, also called "growth" or "accumulation," which occurs when the variables that define an ecosystem are tightly linked and there is considerable accumulation of biomass and nutrients. As the cycle continues, the ecosystem moves to the "K" stage, also known as "conservation," which involves a slow-down in growth as the system exhausts available nutrients. This slow phase eventually gives way to the "Ω" stage, also called "creative destruction," when connectivity among ecosystem properties drops and the accumulated natural capital is liquidated, freeing it up for conversion into something else. This involves much faster dynamics than r or K. The creative destruction phase soon is followed by the "α" stage, or "reorganization," when available nutrients and biomass again accumulate in a context of an incipient rise in linkages among system variables. Reorganization involves rapid dynamics which eventually slow in a new r stage as available nutrients decline and biomass again accumulates while linkages among system variables again become established.

A classic illustration of the adaptive cycle from ecology is secondary vegetation succession itself on a forest plot that has been cleared (Berkes and Folke 2002). Pioneer species of plants colonize the plot and grow for several years (in the r stage). Later, slower-growing species come to dominate and the vegetation structure changes more slowly (in the K stage). Eventually, a person cuts the plot down or a fire burns the plot in a short period of time (per the Ω stage). After this, the remaining seed bank yields new sprouts and vegetation (in the α stage), which gives way to a new r stage.

As with hierarchy theory, we view the adaptive cycle as a heuristic (cf. Abbott 2004) and apply certain aspects to the causal processes behind secondary growth. Rather than focus on the adaptive cycle of ecosystems and their components, we focus on the adaptive cycles of the components in a causal hierarchy. Each component in a causal hierarchy can exhibit slow or fast operations, which we account for with the heuristic of the adaptive cycle. This provides a means of accounting for switches between slow and fast operations among the social actors and other entities in a causal hierarchy.

The content of the operations of specific agents then becomes important to consider. With respect to proximate causes of secondary growth, those operations involve land use decisions. Proximate agents define land management in terms of production goals, which resemble the α stage, when land use decisions are taken and new land covers appear. Agents then pursue those goals, per the r stage, where a land use system becomes established and may expand. This continues until the system encounters a constraint that hinders further implementation (K stage), when operations slow and change ceases. Further, when a crisis and/or opportunity arises, whether induced externally or via an internally planned change of goals (as in Ω), previous land use may be discontinued or rapidly modified. The agent then envisions new production goals (α again), when new land use decisions are taken. For landholders in the Amazon, this adaptive cycle may involve the seasonal routine of land use decisions, such as the clearing of vegetation, planting, and harvesting of annual crops, or the breeding, calving and culling of cattle, both of which occur in synchrony with an annual cycle.

However, a given causal agent may have more than one adaptive cycle. This reflects the fact that a given social actor does not do just one thing, and may have various operations, each with its own temporal rhythm. A given causal agent can thus exhibit a set of adaptive cycles, some with longer periods than others (Almeyda 2004; Gallopín 2002). While farm households may cultivate following an annual cycle, they also engage in implementing their farming system over a period of many years, following the demographic life cycle of the family over a generation (Perz and Walker 2002; Perz et al. 2006).

To elaborate an Amazon example applied to secondary growth, consider a young colonist family which arrives on the frontier and clears forest to establish a land claim and grow annual crops for food security. During the course of the annual cycle of planting and harvesting, fast changes occur when land use decisions are implemented, particularly preparation of plots (by clearing out mature or secondary vegetation) and during harvest; slower changes occur as crops grow and after harvest when secondary vegetation emerges. In addition, over several years, land productivity declines and weeds invade, which requires clearing of other plots of land and results in advanced secondary growth on the initial cultivated plot. This longer cycle depends more on the length of time until soil fertility declines and on the ability of the household to clear more land, which tends to occur periodically every several years. The longer adaptive cycle still involves land use decisions by the household, but punctuates the annual decision-making by expanding the total area cleared of mature forest as well as the area with secondary growth. The fast dynamics on this cycle occur in those years when it is necessary to abandon a given plot and clear a new one; the slow dynamics proceed in between the abandonment/clearing years. The combined result is seasonal fast–slow changes in the area under secondary growth, punctuated by periodic increases (once every several years) in secondary growth due to fallowing or plot abandonment. In this sense, a given causal agent with multiple adaptive cycles may exhibit slow–fast behavioral changes that result in non-linear changes in secondary growth over time.

All that said, Gunderson and Holling (2002) emphasize that shifts among slow and fast dynamics of a given causal agent also reflect responses to changing contextual circumstances. This is a crucial point: slow–fast changes in one agent's operations can result in additional changes in the operations of other agents entrained in the causal hierarchy. Just as hierarchical frameworks theorize top-down causal cascades, adaptive cycles of agents in causal hierarchies prompt one to theorize that slow–fast changes in large-scale agents can entrain slow–fast changes in lower-scale agents. In this way, we can link hierarchies of causation to slow–fast dynamics in the adaptive cycle of a given causal agent. Shifts among slow and fast operations of various agents in a causal hierarchy thus provide an additional theoretical basis that accounts for slow and fast dynamics in secondary growth.

By explicitly incorporating fast–slow dynamics in causal agents, adaptive cycles set a causal hierarchy in motion, which leads us to our rendition of the concept of a panarchy (Gunderson and Holling 2002; Holling 2004). For our purposes, a panarchy amounts to a causal hierarchy with agents that exhibit slow and fast operations over time.

It is useful here to reflect back on the assumption of hierarchy theory that agents operating on larger scales tend to operate more slowly, with small-scale agents operating more rapidly. This is not to foreclose on the possibility of rapid change at a large scale. The adaptive cycle makes explicit that all agents, regardless of the level they occupy in a hierarchy, have slow and fast dynamics, so while global agents may operate slowly most of the time, abrupt changes can also occur periodically. That said, large-scale agents tend to be very complex, involving large aggregations such as populations or complicated structures such as bureaucracies, which tend to exhibit considerable inertia and are difficult to rapidly redirect. The global climate system does not change as fast as local weather; by the same token, it generally takes more time for the UN general assembly to gather for important decisions than a local government commission. While all agents in a causal hierarchy can exhibit fast operations, on average large-scale causal agents go through their adaptive cycles more slowly, with less frequent periods in r and K than small-scale agents.

Considered as a whole, different agents in a causal hierarchy will be at various stages of their respective adaptive cycles. This results from the differing lengths of adaptive cycles of various agents due to their different operations and logics. Seasonality, household life cycles, market price fluctuations, election cycles, and other factors that can influence land use decisions do not have the same temporal rhythms. But the mere recognition that different agents have adaptive cycles of differing periods bears important implications for the hierarchy of causation and the resulting dynamics in secondary growth.

One key implication is that at a given moment, agents in a causal hierarchy are not necessarily at the same stage in their adaptive cycle. That is, there is not necessarily synchrony among agents. Asynchrony leads to a crucial point: the adaptive cycle stage of a large-scale agent in the causal hierarchy can affect the stage of adaptive cycles of other agents at smaller scales in the panarchy (Holling et al. 2002). The asynchrony of adaptive cycles in a causal hierarchy implies that punctuations wrought by various large-scale agents entering fast stages will occur periodically, at variable intervals. Improved roads, growth in urban markets, new lines of credit, rising demand for timber, declining prices for key cash crops, and other changes in the contexts of land use decisions can occur periodically, and greatly alter the goals and logic of proximate decision units (Perz 2002).

This leads to specific arguments about the effects of slow–fast switches in operations of one agent on the pace of operations of other agents. In brief, if one agent in a causal hierarchy enters the $\Omega$ stage and exhibits rapid operations that change, this slow–fast shift may require other agents, causally entrained, to similarly enter a stage of fast operations, as a means of responding and adapting (see Fig. 4.2, panel a). The dynamics that result among lower-scale agents amount

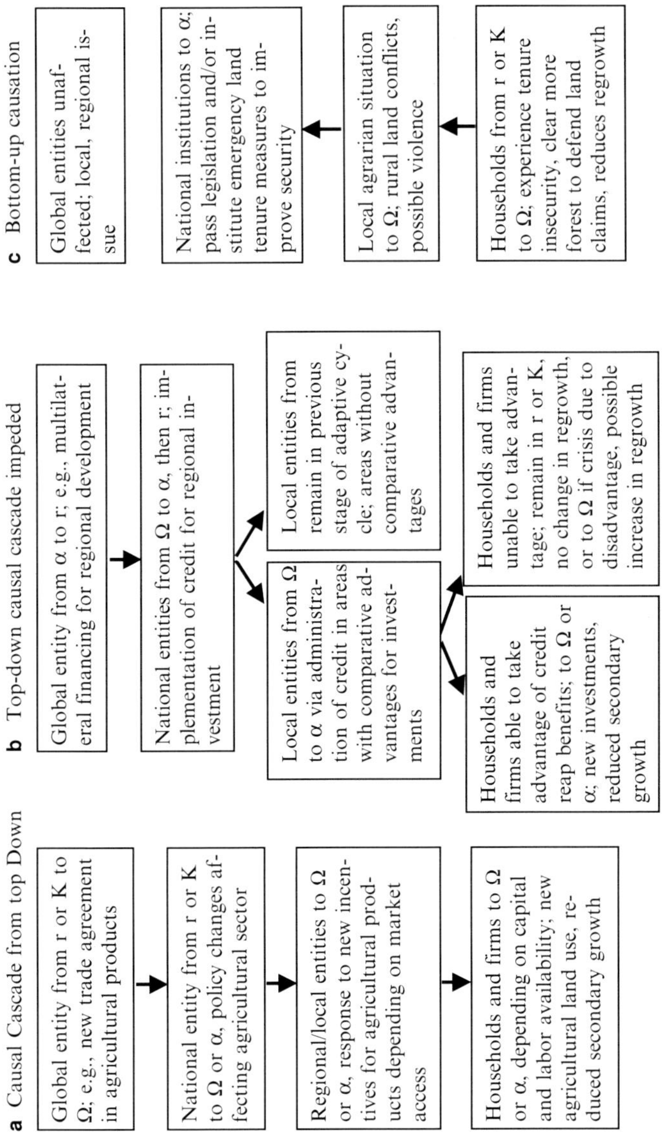


**Fig. 4.2** Adaptive cycles and panarchic dynamics in hierarchical causation behind secondary growth: (**a**)–(**c**)

to "punctuated adaptation" to changes cascading down the causal hierarchy. Lower-scale agents may also enter the Ω stage in the case of a real crisis for which lower-scale agents are unprepared or to which they are incapable of responding, or perhaps the α stage if the rapid large-scale changes present opportunities for which lower-scale agents are prepared, or to which they are capable of responding (Fig. 4.2, panel a).

Cascading changes from global, national, regional and local agents ultimately impact proximate agents, who at a given time will be in diverse stages in their adaptive cycles (cf. Westley 2002). A change in the operation of agents atop a causal hierarchy catalyzes alterations in the operation of agents below, such that their stage in key adaptive cycles is changed to α or Ω, regardless of what stage those agents were in before. Sudden shifts in global product prices and bank lending policies may alter the availability of capital to rural producers in the Amazon, affecting their decision logics and prospects for continuing or expanding production, which then alters prospects for fallowing and other decisions affecting secondary growth. Such shifts can "punctuate" household adaptive cycles, prompting households to "jump" from the r stage immediately into the Ω stage, skipping a prospective K stage (Fig. 4.2, panel a). This implies that agents do not necessarily pass obediently through all stages of their adaptive cycles.

That said, lest we be too carried away with structural determinism, it is important to also recognize that the responses of smaller-scale agents to cascading punctuations in the causal hierarchy may be differentiated. That is, cascading slow–fast shifts may affect some localities more than others, and may impact some landholders more than others (see Fig. 4.2, panel b). If crop price drops are more important for producers in some places than others, only producers in the affected municipalities will experience jumps in their adaptive cycles. Similarly, to the extent that there are inequalities among producers, such that some have more diversified livelihood portfolios, more bank credit, or more capital, the responses of producers in the same locality can also vary (Fig. 4.2, panel b).

Against structural determinism, we can push back farther via panarchy. While the foregoing examples emphasized "top-down" cascades of causation involving shifts in stages of adaptive cycles, there can also be "bottom-up" cascades (Fig. 4.2, panel c). To this point we have made much of asynchronies in adaptive cycles among agents in causal hierarchies. Now we want to emphasize the importance of synchronies in adaptive cycles among many causal agents. Periodic convergences in the stages of adaptive cycles of micro-level agents can generate bottom-up feedbacks and result in changes in the operations of larger-scale agents. Simultaneous tenure insecurity among many landholders can result in rural violence that in the Amazon historically begets state action via land titling (Schmink and Wood 1992). Tenure security may in turn motivate new land use that reduces secondary growth. In this example, convergence of many micro-level actors (land claimants) at Ω (via rural violence in the struggle over land claims) led to higher-level agents (the state) to shift into α (legislation permitting emergency land titling operations). Thus, a bottom-up process involving fast dynamics cascaded up the causal hierarchy, generating faster dynamics at higher levels, which in turn resulted in later shifts in secondary growth (Fig. 4.2, panel c).

Beyond punctuated adaptive cycles due to cross-level interactions, there are more fundamental changes that can occur in a causal hierarchy. One such change involves alterations in the composition of the agents themselves. Turnover in the agents relevant to land use in a given area, as via migration, can alter a causal hierarchy and change land use decision-making and secondary growth. Rapid in-migration during the 1970s and 1980s to the Amazon constituted new populations with resource management practices that differed from indigenous and other traditional groups, which led to non-traditional land use systems that involved extensive land cover conversion (Schmink and Wood 1992). Changes in the profile and composition of relevant land users in turn altered land cover dynamics, such as via the expansion of pasture, followed by widespread land degradation (Serrão and Homma 1993) and emergent secondary growth (Perz and Skole 2003). Similarly, changes in the institutional context can alter land use decisions. In Brazil, the creation of extractive reserves, agro-extractive settlements, and other innovative land tenure categories have either legitimated traditional practices or imposed new rules on pre-existing land tenure types (for other examples, see chapters by Grainger and Crews-Meyer, this volume).

Panarchic dynamics can bring about yet more fundamental changes, up to and including alterations in the locus of key causal processes or even changes in the causal structure itself. Brazil's new constitution in 1988 led to fiscal decentralization, which had the effect of relocating responsibility for many state services from federal agencies to municipal governments (Souza 1997). The result was that municipalities gained more control over state functions tied to local development. That is, institutions tied to distribution of public resources were shifted from a high, distant level in the causal hierarchy to a lower, more proximate level. This shift made it easier for local landholders to demand state services and infrastructure as political favors to facilitate expanded production systems (Toni and Kaimowitz 2003). Decentralization of state functions also spurred the creation of new municipalities in the Brazilian Amazon during the 1990s ("municipalization") as local players sought to carve out new spaces for local control over state resources. Along with decentralization, municipalization made local circumstances more important for land use decisions than before.

In sum, adaptive cycles and panarchy provide several theoretical explanations as to why secondary growth exhibits short-, medium- and long-term dynamics: (1) slow–fast dynamics in adaptive cycles, (2) multiple adaptive cycles with different temporal scales in a given causal agent, (3) varying temporal periods among adaptive cycles of different causal agents, (4) asynchrony in adaptive cycles among different causal agents, (5) cascading causation that involves shifts and jumps ("punctuations") in the stages of adaptive cycles among many causal agents, whether via top-down or bottom-up cascades, (6) differentiated responses of smaller-scale causal agents due to their heterogeneity in preparedness or capacity to deal with punctuations, (7) changes in the composition of causal agents, and (8) changes in the causal structure, via alterations in the proximity of certain causes. Rapid dynamics in secondary growth, especially on smaller spatial scales, occur due to shifts from slow to fast operations by micro-level land users. Rapid dynamics also result when large-scale agents enter fast stages and entrain cascades of rapid operations among other causal agents. Conversely, slow dynamics can result from a pre-eminence of agents at "slow" stages in their adaptive cycles, and when rapid large-scale processes are inhibited by intervening agents in the causal hierarchy.

### 4.2.3 Heterarchy

The concluding portion of the foregoing section raised the issue of structural changes in the causal hierarchy due to panarchic dynamics. If shifts in the causal structure are frequent, or if causation frequently "skips" over levels in a hierarchy, then the assumption of strictly hierarchical causation becomes less accurate. In this section we therefore present the notion of "heterarchy" as an emendation to hierarchical assumptions. We view heterarchical causal structures as the result of panarchic dynamics applied to hierarchical frameworks.

Heterarchy emerged out of general systems thought and chaos theory as a response to reductionism and post-modernism in the analysis of complex dynamic systems (Beekman and Baden 2005). Like hierarchy theory, the notion of heterarchy has been appropriated and adapted by other disciplines. In the social sciences, one application has been in organizational research on the changing relations between multi-national corporations and their subsidiaries (Birkinshaw and Morrison 1995); another has been on globalization and the reorganization of national economies (Amin 2004), particularly in post-socialist economies of the old Soviet bloc (Stark 2001). In each, presumptions of hierarchical, top-down organization have met with empirical difficulties as firms exhibit new forms of flexibility and organization, as economies reorganize horizontally as well as vertically, and as the structures of socialist economic organization give way and become more fluid. The primary application of heterarchy, however, has been in archaeological research on the political and economic organization of pre-historic societies (Crumley 1994; Ehrenreich, et al. 1995). Once understood to have been controlled top-down in hierarchical fashion by elites, these societies are increasingly viewed as having multiple forms of organization, with important lateral and bottom-up organizational mechanisms, and considerable plasticity in their structures when viewed over time (e.g., Rautman 1998; O'Reilly 2003; Scarborough et al. 2003).

We deploy the concept of heterarchy as a heuristic (cf. Abbott 2004) and apply it to causal structures behind LULCC, specifically the case of secondary growth. For our purposes, heterarchy refers to a set of relationships among entities that includes both hierarchical as well as non-hierarchical relationships. Here it is important to highlight the differences between hierarchical relationships – which occur among ranked entities, are highly structured, and largely uni-directional (top-down) – and non-hierarchical relationships – which occur among unranked or variably ranked entities, are flexible, and are multi-directional (including horizontal and bottom-up). Whereas hierarchical frameworks assume that causation proceeds obediently from one level to the next level, heterarchic causation relaxes this assumption (thus the term "unranked"). Further, whereas hierarchical frameworks don't readily handle dynamics in causal structures, heterarchic frameworks embrace such shifts by highlighting reorganization in causal sequences (thus the term "flexible"). And whereas hierarchical frameworks emphasize top-down causal cascades from large- to small-scale agents, heterarchic frameworks also pay attention to horizontal causation and bottom-up cascades (thus the term "multi-directional").

Thus far, our definition of heterarchy has featured relationships between pairs of causal agents, but heterarchy also refers to the overall causal structure that emerges when considering hierarchical and non-hierarchical relationships among many causal agents. When speaking of heterarchic causal structures, it is important to recognize that the concept of heterarchy encompasses the concept of hierarchy but is broader. Heterarchy recognizes hierarchical assumptions as one type of causal structure, but emphasizes the importance of non-hierarchical causation.

We therefore elaborate on our initial definition of heterarchy, to consider not only its relational dimension (between pairs of causal agents) but also its larger structural dimension (the emergent causal structure resulting from relationships among many causal agents). Heterarchy thus refers to the combination of hierarchical and non-hierarchical causal structures that result from panarchic dynamics (i.e., the importance of horizontal causation within a level of scale, shifts in key causal sequences, and top-down as well as bottom-up causation). In other words, there can be specific non-hierarchical relationships among agents in a causal structure, and their presence, even along with hierarchical relationships, constitutes a heterarchic causal structure. A causal heterarchy thus encompasses both hierarchical and non-hierarchical relationships.

Heterarchy provides a means of grappling with critiques of hierarchy theory. One corrective is the equal emphasis heterarchy places on cross-scale and within-scale relationships. Whereas hierarchy tends to privilege top-down processes, heterarchy also recognizes the importance of bottom-up and horizontal as well as vertical relationships. We provided examples of both in previous sections of this chapter for the case of secondary growth. For horizontal (within-scale) processes, we observed that household characteristics affect household land use decisions, so that households with more labor and capital tend to have larger production systems and less secondary growth. For bottom-up processes, we noted that individual decisions to participate in collective mobilization in order to organize producer cooperatives can improve market access, expanding the local land area under production and reducing the area under secondary growth.

However, heterarchy does not mean that all possible relationships, among all scales, running in every direction, are necessarily important. Like hierarchy theory, heterarchy constitutes a framework within which specific theoretical arguments, whatever their disciplinary source, can be situated, if justified. Heterarchy like hierarchy demands a priori theoretical justification or an empirical basis for including a given type of causal linkage, whether it be top-down, horizontal, or bottom-up.

Heterarchy also allows us to deal with the problem of scale mismatch. Hierarchical frameworks tend to incorporate simplifying assumptions about the relevant scales based on disciplinary proclivities (e.g., local, regional, national administrative areas for social scientists), precisely because of complications due to mismatches in the relevant scales when considering causal factors from other disciplines (e.g., watersheds, soil types, vegetation formations, and agroclimatic zones for biophysical scientists). By relaxing the assumption of nested scales in causal hierarchies, heterarchy provides a means of more easily incorporating causation across scales that does not occur among small-scale agents nested within larger-scale agents. Thus a heterarchical causal structure can acknowledge the importance of different soil types, which generally do not follow the boundaries of municipal administrative areas, in order to account for the effects of both soil quality as well as municipal extension or road maintenance resources as they may both affect land productivity and market access, and in turn affect the viability of land use decisions and secondary growth. We can still diagram cross-scale causation and follow causal chains systematically, but without the constraining assumption of purely hierarchical causation.

Heterarchical causation with scale mismatches implies that we must confront multiple sets of scales, that is, socioeconomic and biophysical. Each set of scales can be viewed as a distinct dimension of a multi-dimensional causal structure. Imagine several causal hierarchies as in Fig. 4.1 placed next to each other, with different sets of scales for each, and causal arrows running from one hierarchy to another. That in turn implies that we must then marry those dimensions in order to capture cross-dimensional causation running from socioeconomic to biophysical factors and vice versa. Areas of erosive soils and the presence of seed banks in nearby forest stands at the local level can be particularly problematic for sustained land use, especially among households with limited labor and capital, resulting in soil degradation and secondary growth, particularly among capital/labor-poor households. All this complicates matters, but heterarchy also requires selectivity about which relationships across different dimensions of the causal structure are important; again, theoretical or empirical justifications are necessary.

Perhaps most importantly, heterarchy accommodates panarchic dynamics that modify causal structures. This is a major focus in empirical research informed by the notion of heterarchy, whether via changes in firm-subsidiary relations in multinational corporations, state-firm relations in post-soviet societies, or institutional rearrangements and shifts in elite-populace relations in pre-historic societies. Heterarchies can thus be characterized via different types of changes in causal structures. Here we note three such structural changes and provide examples of each as applied to the causation behind secondary growth. First, the operations of causal agents change via reallocation to different locations in a causal structure, modifying the proximity of key causal factors. Earlier we provided the examples of decentralization and municipalization, both of which shifted responsibility for and control over state services from the national to the local level, with impacts for local land use via differentiating the level of, for example road maintenance services among municipalities. See Fig. 4.3, panel a.

Second, the causal importance or proximity of a given causal agent may change, even if its scale of operation does not. Here a prominent example concerns multilateral banks, who have in recent years undertaken financing of major infrastructure

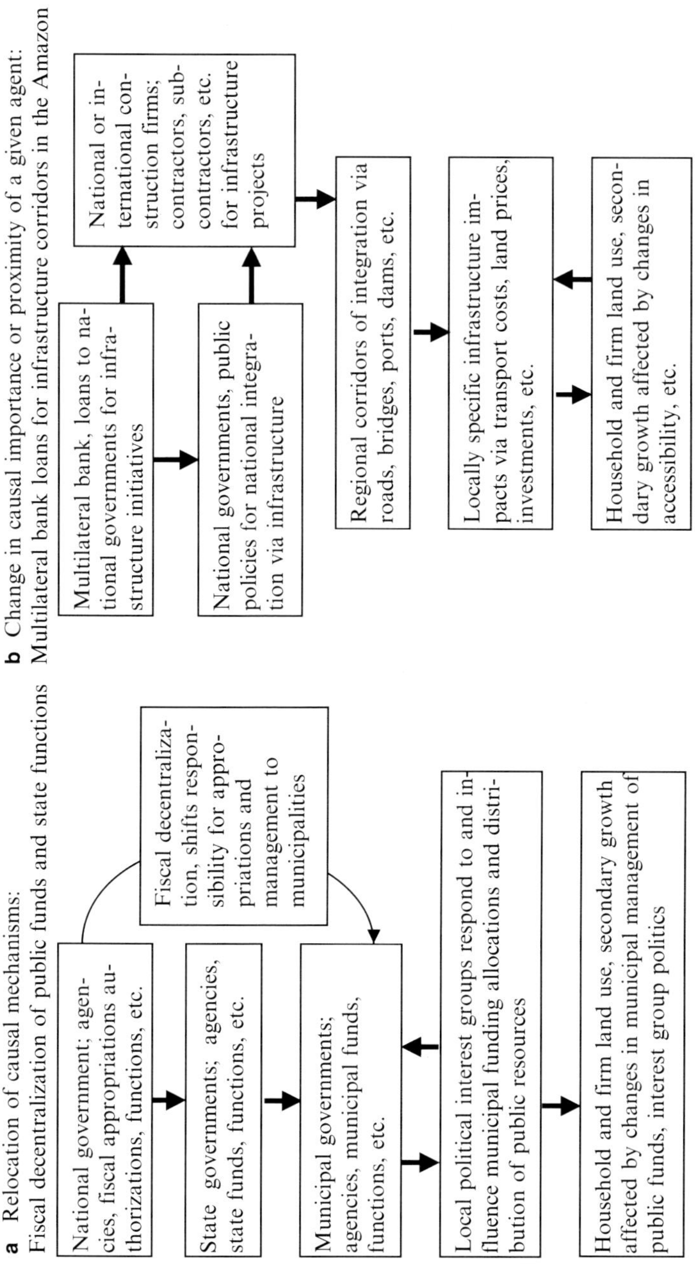


**Fig. 4.3** Heterarchy via structural changes in causal pathways behind secondary growth: (**a**) and (**b**)

upgrades in many parts of South America, including in the Amazon (Mendoza, et al. 2007). Such upgrades have led to a new generation of direct investments of foreign loan funds that have reduced transportation costs and motivated intensified land use along paved road corridors, expanding land use and curtailing secondary growth. In the 1960s and 1970s, infrastructure investments resulted from state initiatives; in the 1980s, major projects were led by state-owned companies; now, the funders and the construction contractors are private entities from outside the Amazon, if not also Brazil. See Fig. 4.3, panel b. Third, the types of operations of a causal agent change, in turn altering the structural relationships to other causal agents. A salient example here concerns the shift toward mechanized agriculture, especially among landholdings previously featuring extensive cattle ranching. Whereas cattle ranching was initially a hedge against inflation and then became oriented to regional markets, soybeans are eminently an export product, which made global markets much more important for land use. The expansion of mechanized agriculture and the heavy capital and chemical inputs to sustain cultivation without multi-year fallow periods curtailed secondary growth in many parts of the Amazon.

In sum, in a heterarchic causal structure, the proximity, operations, structural relationships, and importance of causal agents change over time. The result in terms of the dynamics of the causal structure itself is that a heterarchy rearranges itself as the causal agents in the structure move around, change their operations and relationships, and may become more or less important over time. Viewed in empirical terms, changes in causal structures tend to occur at historically important moments such as elections, policy changes, changes in markets (whether due to trade agreements among countries, major new contracts with buyers, price support modifications, etc.), reorientations in land tenure arrangements, large-scale population movements, or efficacious grassroots movements. Similarly, climatic variability and prospect of climate change also present moments when causation behind land use decisions and secondary growth can change.

## 4.3 Conclusions

Even in places known primarily for deforestation, such as the Amazon, there is considerable secondary growth in previously cleared areas (Perz and Skole 2003), and secondary vegetation is expanding in area over time (Skole et al. 2004). However, cross-site comparisons of the causation behind secondary growth and other types of LULCC reveal complex and variable causation (Lambin and Geist 2006; Moran and Ostrom 2005; see also Rudel, this volume). This constitutes a challenge for theoretical frameworks, which should apply to many cases and also be adaptable to specific contexts (Perz 2007, 2008).

The tripartite framework of forest dynamics (TFFD) responds to the challenges involved in constituting an integrative theoretical approach for LULCC. The first part of the TFFD avoids the extremes of grand theoretic and context-specific avenues to explanation and still offers a systematic and yet flexible approach.

Hierarchy theory provides an encompassing structure in which many theoretical explanations can be situated. It recognizes the scale-specificity of explanations, and in the same stroke organizes them with respect to causal proximity to LULCC. The second part of the TFFD highlights dynamics in causation as a means of accounting for non-linearities in changes in LULCC. Adaptive cycles emphasize the capacity of causal agents to exhibit both slow and fast operations. Panarchy calls attention to different slow-fast periodicities among causal agents, asynchronies in the onset of fast dynamics among different causal agents, cascades of causation generating slow-fast shifts in operations, and other non-linear dynamics affecting causation. The third part of the TFFD stresses the combination of hierarchical and non-hierarchical causation, as well as structural changes in causation behind LULCC. The concept of heterarchy includes both hierarchical (top-down) as well as non-hierarchical (including bottom-up and horizontal) relationships, as well as causation across socioeconomic and biophysical causal structures that are unlikely to be hierarchically nested. The components of the TFFD thus provide heuristics that can be usefully appropriated from other purposes to address challenges facing theoretical integration to better understand LULCC.

There have been several critiques of the type of frameworks on which we draw to constitute the TFFD. One is the "philosopher's stone" critique, namely that the goal of creating an all-encompassing, "unified theory" of something as variable as LULCC is an unrealizable endeavor due to context-specificity. From this perspective, the causation behind LULCC is itself so variable that it cannot be approached systematically. This means that cross-case comparisons are impossible without doing analytical violence to the specifics of the cases being evaluated, because comparisons inevitably result in the loss of understanding of the causation involved in each case.

Conversely, there is the "excessive complexity" critique, which is that a TFFD that is open to any and all explanations becomes unwieldy for not focusing on specific substantive causes a priori, as via reductive or grand theoretic approaches. Here the argument is that it is far more analytically tractable to proceed deductively rather than inductively as via progressive contextualization, and to impose one or a few substantive arguments to be tested. This critique reflects the tensions between logico-deductive approaches to science that emphasize parsimony via the a priori selection of causes to focus analysis, and "holistic" approaches that feature multiple and contingent causation, and therefore require more open inquiry.

Both critiques miss the point of employing a TFFD in which various theoretical explanations may be situated. A key advantage of a hierarchical framework is to be able to incorporate different explanations in different cases, imposing a common assumption about the structure of causation rather than about the substance of the relevant causal processes themselves. The same applies to a heterarchic causal structure, which relaxes strictly hierarchical assumptions about causation. This universalist approach to the structure of causation differs from imposing a standardized expectation of a substantive cause to be important in all empirical cases. The TFFD proposed here thus avoids the proclivities of reductive explanations that propose to fit all cases by ignoring contextual differences (as in e.g. land economics) or by proposing grand theoretical narratives (as in e.g. modernization). Hierarchical and heterarchic causal structures also avoid the haphazardness of context-specific explanations by imposing a standardized structural approach to causation that allows for comparisons. Specific causes may or may not apply across cases, and specific causal structures may or may not be similar, but we can determine that by approaching multiple cases with the same TFFD.

A third critique of the TFFD that comes specifically from social scientists is the "structural functionalist" accusation (cf. Bell 2005). Social scientists often view systems thought as embodying assumptions of structural-functionalism, which has long been criticized for dismissing questions of equity, justice, politics, and conflict. This is also reflected in political economy critiques and related perspectives in debates about modernization and development (Perz 2007, 2008). By contrast, much theorizing of LULCC has emphasized conflict and contestation, as in political ecology, particularly in developing regions like the Amazon (Schmink and Wood 1992). The impression sometimes given is that political ecology is incommensurate with approaches like hierarchy theory (Vayda and Walters 1999). We disagree, and draw on progressive contextualization, suggested by a critic of political ecology (Vayda 1985), as a means of incorporating politics and conflict into explanations for LULCC. We see affinities between the TFFD and political ecology, in that both feature cross-scale processes, which can include cross-scale politics and conflict. Moreover, a heterarchical causal structure also draws attention to bottom-up processes, which can include local mobilization to contest outside agents, as well as reorganization of causal structures, such as via contentious politics.

A fourth critique is the "abstraction problem," which we might also call the "implementation challenge." We have employed hierarchy, adaptive cycles, panarchy, and heterarchy as heuristics and applied them as theoretical tools, which opens questions about how they might be applied for empirical analysis. By extension, the resulting TFFD proposed here faces the same challenge. If we accept the theoretical advantages of the proposed framework, we must turn to the issue of how to apply it via methods and analysis. We offer four suggestions.

Meta-analysis is one way to evaluate theory. By reviewing many publications of empirical work for a range of study cases, we can appraise the scope of applicability of various theoretical explanations. Meta-analysis has become an important tool for evaluation of theory for the land science community (Geist and Lambin 2002; Rudel 2005). An extension beyond previous efforts would be to emphasize the scale dimension, in order to identify the scales on which prominent causes operate and how they interrelate. This would allow identification of different causal structures – hierarchical or heterarchic – that goes beyond the more basic distinction between proximate and underlying causes. However, meta-analysis begs questions about how a TFFD can inform, rather than just organize, empirical research.

One application is via dynamic simulation modeling (see examples in Gutman et al. 2004; Lambin and Geist 2006). The availability of spatially explicit data for socioeconomic and especially biophysical indicators, often derived from satellite images of land cover itself, allows construction of models that incorporate data on various spatial scales. Further, ongoing monitoring efforts, including via satellite remote sensing, afford dynamic modeling with multi-temporal data, increasingly on multiple temporal scales. Such data sources have fed agent-based models which combine spatial data with decision rules to model landscape change (Parker et al. 2003; see examples in Gutman et al., 2004; Lambin and Geist 2006). The limitation of agent-based models however resides with their lack of recognition of causal agents beyond proximate agents who make land use decisions. Multi-agent-based models, which incorporate contagion effects, begin to account for larger-scale processes.

Another approach to modeling that places emphasis on larger-scale contextual effects is multi-level statistical modeling (e.g. Pan and Bilsborrow 2005; Vance and Iovanna 2006; Overmars and Verberg 2006). Most such models assume nested hierarchies in defining actor–context relationships, and most such models employ two or at most three levels. Even then, the data demands can be formidable, especially for social science data that are not pixel-based and not easily aggregated or disaggregated. Further, computational requirements can also be difficult, especially since maximum likelihood estimates may not converge, even for large data sets for the smaller-scale units, if the larger-scale aggregate units are not sufficiently numerous. Nonetheless, multi-level modeling packages have made strides in recent years beyond limiting assumptions of a few levels, and data and computational power allowing, they provide a means of directly testing scale-sensitive theoretical frameworks like the TFFD for empirical cases.

A final avenue is the employment of historical-comparative methods (e.g., Rudel, this volume). Whereas land change science has emphasized deductive methods and quantitative modeling, historical-comparative approaches combine deduction and induction as well as qualitative and quantitative methods. In addition, historical-comparative research in the social sciences seeks to balance generalizability and context-specificity (Tilly 1984; McMichael 1990, 1992), in the same spirit as the TFFD. The approach is to adopt an analytical focus on an outcome, such as LULCC, and then ask questions and collect data in order to derive an interpretation for each of several cases, whether for the same place over time or for multiple sites or both. In practice, this combines theory and method as the methods involve both theory-driven data collection as well as an openness to gathering additional information that is important to understanding specific cases. In this regard, progressive contextualization can be deployed as part of historical-comparative data collection to sort through causes on multiple scales. The important point with historical-comparative methods is that explanations can be fit into a TFFD by recognizing the scale-specificity of causal factors. Deductive and inductive approaches can be employed in series or simultaneously as the analyst identifies explanations for LULCC. Historical-comparative methods extend beyond published academic literature, and thus meta-analysis, by focusing on primary and secondary data sources.

Historical-comparative methods also stress the nature of the comparisons to be made. Analytical decisions must be made a priori to the analysis as to whether the comparisons will highlight uniqueness, commonalities, or situate a given case in a larger structural context (Tilly 1984; McMichael 1990, 1992). The land science community has tended to gloss over some of these distinctions, emphasizing commonalities and uniqueness. Less attention has gone to other forms of comparison, one of which is to situate cases in broader contexts. This can be extended to situating scale-specific causes for a given case in larger scale contexts, as in progressive contextualization and hierarchical (and heterarchical) analytical frameworks. Particular attention to the scale of specific causes in each of several cases constitutes an elaboration of historical-comparative methods via application of the TFFD we have proposed, and would facilitate comparisons of the arrangement of explanations in causal structures for many different cases.

While the focus of a historical-comparative analysis must be decided at the outset, the cases to be selected, key explanatory factors chosen, and the scales of observation are managed provisionally. This then allows specification of the explanations (and the scales on which they operate) in a flexible fashion, not only to allow revisions as additional data are collected but also as causation of LULCC alters over time. Where the structure of LULCC causation is highly plastic, historical-comparative methods have an advantage in themselves being flexible in providing explanations provisionally, allowing for comparisons in causal structures over time.

There is no facile panacea for an ironclad theoretical approach to understand secondary growth and other LULCC, or to empirical applications. We have mentioned critiques of meta-analysis, dynamic simulation models, multi-level statistical models, and historical-comparative methods. We would add that some of the applications we have suggested are either new or have yet to be applied much to the case of LULCC; they remain to be fully tested. For now, we repeat that the TFFD overcomes limitations in other theoretical approaches to explanation of LULCC. While the TFFD harbors liabilities due to its inelegance, that is a reflection of the challenges facing integrative theoretical frameworks seeking more fully understand secondary growth and other LULCC, which alternatives heretofore address but inadequately.

## References


Abbott A (2004) Methods of discovery: heuristics for the social sciences. W.W. Norton, New York

Ahn V, Allen TFH (1996) Hierarchy theory: a vision, vocabulary and epistemology. Columbia University Press, New York

Allen TFH, Starr TB (1988) Hierarchy: perspectives for ecological complexity. University of Chicago Press, Chicago, IL

Almeyda A (2004) Land use and land cover in Inapari, Peru and Assis Brasil, Brazil, Southwest Amazonia. MA Thesis, University of Florida, FL

Amin A (2004) Regulating economic globalization. Trans Inst Br Geogr 29:217–233

Beekman CS, Baden WW (2005) Nonlinear models for archaeology and anthropology. Ashgate, Aldershot, UK

Bell MM (2005) The vitality of difference: systems theory, the environment, and the ghost of parsons. Soc Nat Res 18(5):481–478

Berkes F and Folke C (2002) Back to the future: ecosystem dynamics and local knowledge. In: Gunderson LH, Holling CS (eds) Panarchy: understanding transformations in human and natural systems. Island Press, Washington, DC, pp 121–146

Birkinshaw JM and Morrison AJ (1995) Configurations of strategy and structure in subsidiaries of multinational corporations. J Int Bus Stud 26(4): 729–754

Crumley CL (ed) (1994) Historical ecology: cultural knowledge and changing landscapes. School of American Research Press, Santa Fe, NM

Ehrenreich RM, Crumley CL, Levy JE (eds) (1995) Heterarchy and the analysis of complex societies. American Anthropological Association, Arlington, VA

Gallopín G (2002) Planning for resilience: scenarios, surprises and branch points. In: Gunderson LH, Holling CS (eds) Panarchy: understanding transformations in human and natural systems. Island Press, Washington, DC, pp 361–392.

Geist HJ, Lambin EF (2002) Proximate causes and underlying forces of tropical deforestation. BioScience 52(2):143–150

Gibson C, Ostrom E, Ahn T-K (2000) The concept of scale and the human dimensions of global change: a survey. Ecol Econ 32:217–239

Gunderson LH, Holling CS (eds) (2002) Panarchy: understanding transformations in human and natural systems. Island Press, Washington, DC

Gunderson LH, Holling CS, Light SS (1995) Barriers and bridges to the renewal of ecosystems and institutions. Columbia University Press, New York

Gutman G, Janetos AC, Justice CO, Moran EF, Mustard JF, Rindfuss RR, Skole D, Turner BL II, Cochrane MA (eds) (2004) Land change science: observing, monitoring, and understanding trajectories of change on the earth's surface. Kluwer, Dordrecht

Holling CS (2004) From complex regions to complex worlds. Ecol Soc 9(1): 11. http://www.ecologyandsociety.org/vol9/iss1/art11

Holling CS, Gunderson LH, Peterson GD (2002) Sustainability and panarchies. In: Gunderson LH, Holling CS (eds) Panarchy: understanding transformations in human and natural systems. Island Press, Washington, DC, pp 63–102.

Lambin EF, Geist H (eds) (2006) Land-use and land-cover change: local processes and global impacts. Springer, Berlin

Lucas RM, Honzak M, Foody GM, Curran PJ, Corves C (1993) Characterizing tropical secondary forests using multi-temporal Landsat sensor imagery. Int J Remote Sens 14(16):3061–3067

Mather AS, Needle CL (1998) The forest transition: a theoretical basis. Area 30(2):117–124

McMichael P (1990) Incorporating comparison within a world-historical perspective: an alternative comparative method. Am Sociol Rev 55(3):385–397

McMichael P (1992) Rethinking comparative analysis in a post-developmentalist context. Int Soc Sci J 44:351–365

Mendoza ESG, Perz MS, Nepstad D (2007) Participatory stakeholder workshops to mitigate impacts of road paving in the southwestern Amazon. Conserv Soc 5(3):382–407

Moran EF, Ostrom E (eds) (2005) Seeing the forest and the trees: human–environmental interactions in forest ecosystems. MIT Press, Cambridge, MA

O'Neill RV, DeAngelis DL, Waide JB, Allen TFH (1986) A hierarchical concept of ecosystems. Princeton University Press, Princeton, NJ

O'Reilly DJW (2003) Further evidence of heterarchy in Bronze Age Thailand. Curr Anthropol 44(2):300–306

Overmars KP, Verberg PH (2006) Multilevel modeling of land use from field to village level in the Philippines. Agric Syst 89:435–456

Pan WKY, Bilsborrow R (2005) The use of a multilevel model to analyze factors influencing land use: a study of the Ecuadorian Amazon. Global Planet Change 47(2–4):232–252

Parker DC, Manson SM, Janssen MA, Hoffmann MJ, Deadman P (2003) Multi-agent systems for the simulation of land-use and land-cover change: a review. Ann Assoc Am Geogr 93(2):314–337

Perz SG (2002) The changing social contexts of deforestation in the Brazilian Amazon. Soc Sci Q 83(1):35–53

Perz SG (2007) Grand theory and context-specificity in the study of forest dynamics: forest transition theory and other directions. Prof Geogr 59(1):105–114

Perz SG (2008) forest transitions, environmental social theory, and land science research: reply to Walker. Prof Geogr 60(1):141–145

Perz SG, Walker RT (2002) Household life cycles and secondary forest cover among small farm colonists in the Amazon. World Dev 30(6):1009–1027
Perz SG, Skole DL (2003) Secondary forest expansion in the Brazilian Amazon and the refinement of forest transition theory. Soc Nat Res 16:277–294
Perz SG, Walker RT, Caldas MM (2006) Beyond population and environment: household demographic life cycles and land use allocation among small farms in the Amazon. Hum Ecol 34(6):829–849
Perz SG, Caldas M, Arima E, Walker RT (2007) Socio-spatial processes of unofficial road-building in the Amazon: socioeconomic and biophysical explanations. Dev Change 38(3):529–551
Rautman A (1998) Hierarchy and heterarchy in the American Southwest: a comment on McGuire and Saitta. Am Antiquity 63(2):325–333
Rudel TK (2005) Tropical forests: regional paths of destruction and regeneration in the late twentieth century. Columbia University Press, New York
Scarborough VL, Valdez F Jr, Dunning N (2003) Heterarchy, political economy, and the ancient Maya. University of Arizona Press, Tucson, AZ
Schmink M, Wood CH (1992) Contested frontiers in Amazonia. Columbia University Press, New York
Serrão EAS, Homma AKO (1993) Brazil. In: Sustainable agriculture and the environment in the humid tropics. National Research Council. National Academy Press, Washington, DC, pp 265–351.
Skole DL, Cochrane MA, Matricardi ET, Chomentowski W, Pedlowski M, Kimble D (2004) Pattern to Process in the Amazon Region. In: Gutman G, Janetos AC, Justice CO, Moran EF, Mustard JF, Rindfuss RR, Skole D, Turner BL II, Cochrane MA (eds) Land change science: observing, monitoring, and understanding trajectories of change on the earth's surface. Kluwer, Dordrecht, pp 77–95.
Souza C (1997) Constitutional engineering in Brazil: the politics of federalism and decentralization. St. Martin's Press, New York
Stark D (2001) Ambiguous assets for uncertain environments: heterarchy and postsocialist firms. In: DiMaggio P (ed) The twenty-first-century firm: changing economic organization in international perspective. Princeton University Press, Princeton, NJ, pp 69–104
Steininger MK (1996) Tropical secondary forest regrowth in the Amazon: age, area and change estimation with thematic mapper data. Int J Remote Sens 17:9–27
Tilly C (1984) Big structures, large processes, huge comparisons. Sage Foundation, New York
Toni F, Kaimowitz D (eds) (2003) Municípios e Gestão Florestal na Amazônia. A.S. Editores, Natal, Brazil
Vance C, Iovanna R (2006) Analyzing spatial hierarchies in remotely sensed data: insights from a multi-level model of tropical deforestation. Land Use Policy 23:26–236
Vayda A (1985) Progressive contextualization: methods for research in human ecology. Hum Ecol 11:265–281
Vayda AP, Walters BB (1999) Against political ecology. Hum Ecol 27(1):167–179
Walker RT (1993) Deforestation and economic development. Can J Reg Sci 16:481–497
Walker RT (2007) Forest transition: without complexity, without scale. Prof Geogr 60(1):136–140
Warren WA (2005) Hierarchy theory in sociology, ecology, and resource management: a conceptual model for natural resource or environmental sociology and socioecological systems. Soci Nat Resour 18(5):447–466
Westley F (2002) The devil in the dynamics: adaptive management on the front lines. In: Gunderson LH, Holling CS (eds) Panarchy: understanding transformations in human and natural systems. Island Press, Washington, DC, pp 333–360
Wood CH, Porro R (eds) (2002) Deforestation and land use in the Amazon. University of Florida Press, Gainesville, FL

# Chapter 5
# Forest Cover Dynamics and Forest Transitions in Mexico and Central America: Towards a "Great Restoration"?

**David Barton Bray**

## 5.1 Introduction

The term "forest transition" refers to observed historical processes of forest cover change as societies become more developed and industrialized. A general trend observed in many developed economies is that forest cover declines at a rapid rate

D.B. Bray (✉)
Department of Environmental Studies, Florida International University, FL, USA
e-mail: brayd@fiu.edu

H. Nagendra and J. Southworth (eds.), *Reforesting Landscapes: Linking Pattern and Process*, Landscape Series 10,
DOI 10.1007/978-1-4020-9656-3_5, © Springer Science+Business Media B.V. 2010

during a first phase of economic development, followed by a gradual stabilization of forest cover and subsequent tendencies toward forest recovery (Mather 1992, Rudel et al. 2005). A forest transition is said to have taken place when the rate of forest recovery exceeds the rate of ongoing forest loss. However, historical studies have indicated that these forest recoveries to date have only recovered about half of the forest historically lost, and that a forest transition does not usually begin until historically present forests had been reduced to small areas (Rudel et al. 2005). While forest transitions were originally proposed to have taken place only in industrialized countries, recent studies have proposed different forms of forest transition as having occurring at the level of the entire Brazilian Amazon (Perz and Skole 2003) and at the regional level in Mexico (Klooster 2003). Victor and Ausubel (2000) have gone so far as to argue that, despite the dominance of deforestation narratives, that forest recovery is a global trend and that a "Great Restoration" of the world's forests is underway.

While deforestation is still a major problem in many regions of the world, the growing magnitude of recovering forests and the ecological and livelihood significance of secondary forests in general has been recognized by the FAO, which notes that in many tropical countries the amount of secondary forests is larger than primary forests (FAO 2005:5). However, many national governments are not yet formally collecting statistics on forest recovery and do not take them into account in policy decisions. But the realities of forest cover dynamics are more complex than either studies or narratives of linear deforestation or linear forest recovery suggest. National level statistics can obscure a more complicated dynamic at lower levels, suggesting the need to examine forest cover processes at multiple scales, at local, regional, national, and contiguous international levels. Some regions of individual countries are showing ongoing deforestation, others show forest recovery and others exhibit dynamic landscapes that are nonetheless relatively stable in terms of forest cover area. Further, while public protected areas are usually well-documented, other forest uses that can contribute to the conservation of forest cover, such as shade tree coffee, sustainable forest management for timber, or community conservation areas, are seldom taken into account.

A more complex picture of deforestation and forest recovery, maintenance, and protection is now being filled in by more detailed local and regional studies of forest cover dynamics that emphasize the "context-specificity" or "event ecology" of historical-comparative studies in uncovering subnational trends in forest cover and its drivers (Rudel 2005; Perz 2007). A more sophisticated theoretical framework has been proposed by Perz (2007) who suggests fusing hierarchy theory from ecology and concepts of panarchy from adaptive management theory. He argues that this would allow the incorporation of multiple theoretical perspectives and "afford a means of moving across spatial and temporal scales of observation, as well as recognizing biophysical aspects of forests and the biophysical and socioeconomic forces driving forest dynamics". Among other issues, this introduces the concept of scale, which has been little used in forest transition theory due to the exclusive focus on national scale processes. In the conclusions I will briefly examine the state of forest cover studies in this region from the "hierarchical-panarchic" framework proposed by Perz and references to the scale at which processes take place will be made throughout.

Reviews of forest cover dynamics in the region is made difficult by the conflicting numbers on basic forest cover processes from apparently reliable sources, as we shall see in some of the country case studies. The lack of agreement on basic facts, from deforestation rates to existing forest cover to number of hectares in protected areas, shows the great need for more rigorous and definitive studies and ongoing monitoring. This is made more urgent in a carbon-constrained world where deforestation contributes some 20% to greenhouse gas emissions (Stern 2007). FAO figures are frequently taken as authoritative on deforestation but liberal definitions of forest cover can make them misleading (Rudel 2005). Thus, in most countries in the region it is difficult to determine if a forest transition is taking place.

## 5.2 Forest Change Pathways

With the caveats stated above, I will now examine a selection of the literature on forest cover dynamics and the potential or reality of forest transitions in Mexico and Central America. Mexico and Central America comprise a vast area, and there is little or no published literature available on forest cover dynamics in most subregions. Covering such a large area allows us to see both the national differences and similarities across this broad region that constitutes a truly large landscape. It also allows for a focus on forest masses that cross national boundaries but which may have very different forest cover dynamics in the different national portions of the forest. Most academic and popular narratives on forest cover in the region have focused on forest loss or deforestation as the primary tendency in forest cover (Bray and Klepeis 2005). Indeed, the evidence suggests that in particular periods and places linear tendencies towards forest loss have been or are still predominant. In tropical areas of Mexico and Central America, a major pulse of deforestation occurred from the 1960s to 1980s. However, by the early 1990s this linear wave of deforestation began to subside across the region. It has been replaced by more complex national and regional nonlinear patterns of deforestation, forest recovery, and forest maintenance, in some cases in spite of proximate deforestation pressures (Rudel 2005). For example, national figures show high rates of ongoing deforestation in Mexico, but the pine and oak forests of Mexico's sierras tend to show lower deforestation rates than tropical regions, and rural–urban-international migration appears to be leading to substantial agricultural abandonment as a significant land use component of a forest transition. However, highland Chiapas is a regional case where highland forest deforestation continues to occur at a rapid rate despite substantial outmigration.

Two relatively recent studies have carried out sophisticated metanalyses of the drivers of deforestation in at least parts of this region. Geist and Lambin (2001) established proximate and underlying causes in a meta-analysis of 152 case studies that included Latin America and Rudel (2005) looked at 62 studies in Central

America and the Caribbean, including parts of Colombia and Ecuador, examining processes of both deforestation and forest recovery. For this region plus the Caribbean and parts of Colombia and Ecuador, Rudel proposed four paths that channel forest cover change in the region: (1) small farmers producing cash crops for national urban consumers, (2) new agroexporters responding to world markets, (3) labor out-migration from poor rural areas and (4) a pathway from tourism to ecotourism, land abandonment, and forest regeneration. He found that colonization was a factor in the majority of deforestation trends in the 1970s and 1980s but entirely disappeared by the 1990s in favor a much more diverse set of drivers. In this review, I find no reason to question this finding. However, this conceptualization focuses on drivers and not forest outcomes. As well, drivers which are likely to be the same in practice (e.g. labor out-migration and land abandonment) are placed into two separate pathways. Thus, for this review, I will use a simpler conceptualization which focuses on two broad forest outcomes or pathways, with each pathway having diverse drivers that include drivers of forest maintenance, which have been little considered as such in the literature.

The pathways are (1) "continued deforestation" which frequently has different dynamics in tropical and temperate forest areas, and (2) a "forest recovery, maintenance, and protection pathway" where the differences between tropical and temperate forests are less marked. The *continued deforestation pathway* in the region is proximately driven by (a) population movements and agricultural expansion into tropical frontier areas and (b) local population growth and agricultural expansion in highland temperate forests, including the urban-oriented agricultural expansion described by Rudel (2005). A little documented example of continued deforestation due to export-oriented agricultural expansion is widespread forest loss in the Mexican state of Michoacan for avocado production when US markets were opened in recent years. The results of these processes are continued deforestation, and these are what are commonly reported in the FAO deforestation figures for the countries of the region that can be found in Table 5.1.

**Table 5.1** The continued deforestation pathway: forest area and cover change (1990–2000) (FAO 2005, 136–137)

| | | Forest area 2000 | | Forest cover change 1990–2000 | |
|---|---|---|---|---|---|
| Country | Total land area ('000 ha) | Total forest ('000 ha) | % of land area | Annual change ('000 ha) (–) | Annual rate of change (–) |
| Mexico | 190,869 | 55,205 | 28.9 | 631 | 1.1 |
| Belize | 2,280 | 1,348 | 59.1 | 36 | 2.3 |
| Guatemala | 10,843 | 2,850 | 26.3 | 54 | 1.7 |
| Honduras | 11,189 | 5,383 | 48.1 | 59 | 1.0 |
| El Salvador | 2,072 | 121 | 5.8 | 7 | 4.6 |
| Nicaragua | 12,140 | 3,278 | 27.0 | 117 | 3.0 |
| Costa Rica | 5,106 | 1,968 | 38.5 | 16 | 0.8 |
| Panama | 7,443 | 2,876 | 38.6 | 52 | 1.6 |

The *forest recovery, maintenance and protection pathway* has been less conceptualized, and represents a continuum of processes which lead to the emergence or maintenance of forest cover. These include both forest recovery as usually understood as well as three *forest uses* which tend to maintain or protect existing forests. These include shade tree coffee, sustainable forest management for timber, and protected areas. These outcomes results from a range of drivers beginning with original forest destruction through those that represent decreasing degrees of disturbance in theory, from coffee to selective logging to attempts at full protection in parks. I further elaborate on the drivers of the forest recovery, maintenance, and protection pathway below and a summary can be found in Table 5.2.

1. The forest recovery component of this pathway is most commonly driven by out-migration, whether urban or international, and shifting markets for beef, both of which lead to the abandonment of fields and pastures (Rudel 2005). Agricultural abandonment appears to be led by people with young families in their 20s and 30s who historically would have continued clearing forests to feed their new families, and is a part of world-historical processes of rural–urban migration happening in many regions of northern Latin America. Emigration is an example of a cross-scale phenomenon, which responds to shifts in international economic relations (e.g. wage differentials between rural Mexico and the urban U.S.), but which also originates in individual decisions at the household level. The aggregate of the consequences of these decisions at the village, regional and national level, in response to large-scale macroeconomic forces, can be forest recovery at multiple scales. Despite their importance, only in Guatemala are recent national figures available on forest recovery, although they are available for particular regions in most other countries. A smaller component of forest recovery in most countries is found in plantations. Forest recovery may be occurring more notably in tropical forests, where deforestation was most intense in earlier periods, but is also occurring in temperate forests.
2. The coffee agroforestry pathway component, at one extreme of rusticity, can be more like a forest than a farm (Perfecto et al. 1996). It can thus be an important driver in maintaining forest cover, even if the forest condition has changed. Coffee agroforestry was produced by decades-long responses to world markets which have reshaped the forest use landscape in the region. So-called "sun coffee" with no forest cover has become important in some countries of the region, notably Costa Rica. However, other regions are characterized by what is variously called traditional or rustic coffee, commonly marketed as shade tree or organic. Rustic coffee farms can have biodiversity similar to that of natural forests and can be particularly useful as buffer zones to protected areas. In Central America, about 44% of 812,000 ha in coffee are considered to be "traditional", while in Mexico, there are around 800,000 ha, mostly in traditional shade tree coffee farms (Varangis et al. 2003; Calo and Wise 2005). Remote images have difficulty in differentiating mature shade tree coffee plantations from forest, making it difficult to establish exact extents of traditional coffee agroforests (Southworth et al. 2004a). Coffee will have a tendency to maintain forests at elevations

**Table 5.2** The forest recovery, maintenance, and protection pathway: natural forest recovery, forest plantations, coffee agroforests, certified forests and protected areas in Mexico and Central America (FAO 2005,136–137; World Database on Protected Areas. http://www.unep-wcmc.org/wdpa/, Cited 12 Sept 2007; Forest Stewardship Council. www.fsc.org, Cited 11 Dec 2007)

| | Forest recovery | Forest plantations ('000 ha) | Coffee agro-forests | Total area of FSC certified forests in ha (total forests) | Protected areas in sq km (total PAs) | Percent of National Territory Protected |
|---|---|---|---|---|---|---|
| Mexico | Forest recovery in some regions, no national data | 267 | 800,000 | 787,995 (44) | 196,185 (193) | 8.72 |
| Belize | No regional or national data | 3 | NA | 104,888 (1) | 12,604 (107) | 30.39 |
| Guatemala | Three percent nationally | 133 | 260,000 | 512,321 (13) | 35,859 (161) | 30.76 |
| Honduras | Forest recovery in some regions, no national data | 48 | 237,000 | 49,151 (3) | 29,761 (99) | 20.03 |
| El Salvador | Some national forest recovery documented | 14 | 161,000 | NA | 258 (76) | .93 |
| Nicaragua | No recent national data | 46 | 108,000 | 20,766 (4) | 29,405 (93) | 18.19 |
| Costa Rica | Forest recovery in some regions, no national data | 178 | NA | 76,547 (19) | 17,508 (183) | 23.25 |
| Panama | Forest recovery in some regions, no national data | 40 | NA | 10,878 (8) | 32,795 (62) | 24.59 |

above 700 m, which is considered the lower range for higher quality coffee, but in Mexico and elsewhere significant amounts of coffee are planted below this optimum range, so forests with a montane tropical composition may also be maintained by coffee.

3. The sustainable forest management component, in all countries under study, occurs primarily through community forest management for timber, and when well managed is also an important driver of forest cover stability. Its presence is highly variable between countries. Mexico has by far the largest extent, with an estimated 2,300 community forests, although it is not known how many of these are managed sustainably. Sustainable community forest management in Mexico is a result of historic processes which have led to the conservation of forest cover at multiple scales. Guatemala is the other country where sustainable community forest management has emerged, primarily in the Peten, but also in common property forests in parts of the highlands. Figures on amount of forest under sustainable management is usually not available, so I will take as an indicator the number which have been certified as well-managed by the Forest Stewardship Council (FSC) (Table 5.2). The majority of well-managed forests are in temperate areas, but there is also a notable representation in lowland tropical forests in Quintana Roo and the Peten of Guatemala.
4. Finally, public protected areas emerged as a major component of forest maintenance and protection in the region in recent decades, going from 9% in 1980 to 13% in 1990 and over 22% as of 2006 (Kaimowitz 1996; Table 5.2). According to the World Database on Protected Areas, as of March 2006 Mexico had 193 protected areas with a total of 196,185 sq km representing 8.72% of the national territory, while all of the Central American countries combined had 781 protected areas with 158,193 sq km for a 22.45% coverage for the entire region. Belize and Guatemala have the highest percentages under protection with over 30% each while at the other end El Salvador has only 0.93% protected (see Table 5.2). Much of the land in Central America is in a few large protected areas such as the Maya Biosphere in northern Petén in Guatemala, the Rio Platano Biosphere and Bosawas Biosphere Reserves in the Mosquitia of Honduras and Nicaragua, the Indio-Maiz Reserves in Nicaragua, the transnational Amistad Biosphere Reserve in Costa Rica and the Darien National Park in Panama, in addition to indigenous reserves primarily in Panama and Costa Rica. Private protected areas are also important in Belize and Costa Rica. Many of the figures on protected area also include marine reserves, but these are usually not well-differentiated in the available figures, so figures reported here should not be taken as entirely composed of protected forests. The largest protected areas tend to be in lowland tropical areas.

These pathways are conceptual abstractions that may overgeneralize the actual diversity of forest cover dynamics on the ground. As used here, forest recovery can refer to various stages of growth of secondary succession, and the point on which it becomes detectable in remote images can vary depending on forest type. Forest cover stability may refer to either no significant changes in forest cover in a particular forest mass, or to no changes in total forest cover in a given region, despite ongoing dynamics of forest loss and recovery. Apparent forest cover stability may also occur when there is a degree of ongoing forest degradation or a forest which has changed from earlier conditions (see Evan et al., this volume, on forest condition vs. forest cover in southern Indiana), but unless otherwise specified this broad review does not evaluate forest condition.

There is a considerable degree of uncertainty in the evaluation of landscape dynamics at the multiple scales to be reviewed here. One can find widely varying estimates from apparently reliable sources, as will be reviewed below. Differences in estimates of past and current forest covers can arise because of the definition of forest and forest classes included, whether the studies are based on mapping or sampling, the precision of the estimates, and the sources used (Kleinn et al. 2004). Forest recovery is even more uncertainly documented and may also have varying definitions of what constitutes forest recovery. As well, studies of deforestation frequently include figures on forest recovery in tables which are not discussed or evaluated in the text. This is a symptom of the dominance of a "deforestation narrative", when processes are actually much more complex (Klooster 2003; Bray and Klepeis 2005). In the remainder of this paper I will briefly describe the kinds of forests to be found in the region, the history of forest cover dynamics until the last few decades, review what is known about the status of forest cover dynamics in each country, and the extent to which a "forest transition" may be said to be taking place, and then attempt a synthesis of the dynamics in all the countries under study.

## 5.3 The Forests of Mexico and Central America and the History of Deforestation

The forest cover to be examined here covers a vast and complex area, both in terms of geomorphhololgy and vegetation (see Fig. 5.1). The enormous sierra region, with its pine, pine-oak, montane tropical and cloud forests, begins in the US and occupies most of central and southern Mexico continues on through Guatemala, Honduras, and Nicaragua and parts of El Salvador, with a second unconnected range running from Costa Rica into western Panama (Vreugdenhil et al. 2002). Tropical forest areas, both dry and humid, exist along the coasts of both regions and in much larger areas of the Yucatan Peninsula, southern Mexico, and in the lowland regions of Central America's Atlantic Coast. Two maps produced by National Geographic (National Geographic Society/Cultural Survival 2002; National Geographic Society/Center for the Support of Native Lands 2002) show the location of remaining forests in southern Mexico and Central America, and also provide documentation of a strong relationship between existing forest cover and the presence of indigenous peoples.

The map included here represents 2000 forest cover and shows many scattered masses of temperate and tropical forest in central and south-central Mexico, and a large intact lowland tropical forest mass centered on the southern Yucatan Peninsula in Mexico, the northern Peten in Guatemala, and northern Belize (the "Maya Forest";

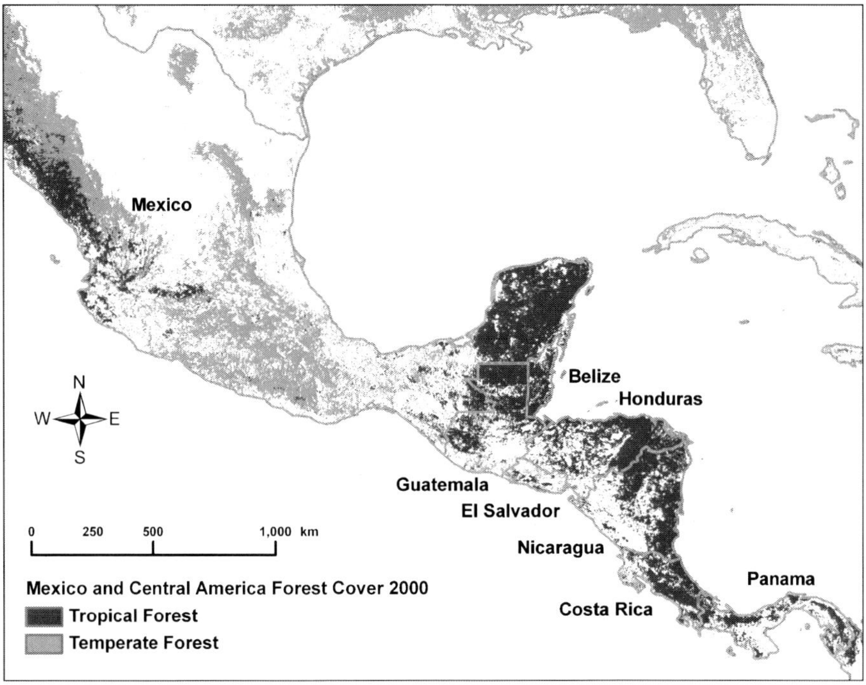


**Fig. 5.1** Tropical and temperate forest cover in Mexico and Central America, 2000 (Latifovic et al. 2002)

Primack et al. 1998). This forest mass has been almost entirely cut off by deforestation from once connected forests in the Lacandon Forest of Chiapas Mexico, to the south, and from forests further south in Honduras. Significant areas of western highland Guatemala still show large forest masses and in the east the Sierra de las Minas range has much intact forest. Further south, the intact highland forests have been reduced to one large area in Honduras and many other isolated islands in Honduras, Nicaragua, and El Salvador. The second large intact tropical forest mass in the region is the Mosquitia between Honduras and Nicaragua, with areas of the southern Atlantic Coast of Nicaragua also having contiguous forests. Tropical forests run along the Atlantic Coast of Nicaragua and Costa Rica, running into the contiguous tropical forest mass in the transnational La Amistad Biosphere Reserve on the Atlantic side of Costa Rica and Panama; this third major transnational forest in the region, unlike the Maya Forest and the Mosquitia, does not appear to have its own name. The "La Amistad" forest then loses connectivity with the large forest mass in the Darien Peninsula on the border with Colombia, which extends into Colombia and thus constitutes the fourth large transnational forest mass in the region.

The maps of remaining forests in the region show the current outcomes of historical oscillations in forest cover at multiple temporal and geographic scales (see also Table 5.1). The forests of southern Mexico and northern Central America have already been through at least one millennia-long civilizational pulse of forest loss and recovery. The region was heavily deforested by the ancient Mayan city-states during the first millennia AD, and then underwent a long recovery process during most of the second millennia AD. A new oscillation of forest cover reduction and partial, complex recovery has been compressed into less than 100 years, beginning in the first decades of the twentieth century but accelerating in the period from the end of World War II to the present, with varying dynamics in different regions (Bray and Klepeis 2005). Less is known about the deep history of highland forests, but on a regional scale a century-long pulse of deforestation and forest recovery as been identified for the Sierra Norte of Oaxaca, which was much more deforested in the second half of the nineteenth century than it is today (Mathews 2003). It may be assumed that other regions of both temperate and tropical forest have undergone similar regional fluxes of forest cover over varying periods, and that current forest cover does not suggest that forest has always been there.

The twentieth century pulse of deforestation in the tropics had its origins in the late nineteenth and early twentieth century with the expansion of export agriculture into lowland areas of both Mexico and Central America, with a moving banana frontier leading to the destruction of broad swaths of tropical forests (Tucker 2000: Rudel 2005). Also beginning in the late nineteenth century, and continuing in successive pulses in various mountain regions of Mexico and Central America, natural forests were modified for the understory crop of coffee (Tucker 2000; Bartra 1996). In Mexico in particular, some of these conversion of natural forests to coffee agroforestry occurred in lower altitude tropical areas not optimal for coffee (Bray et al. 2004). In the post-World War II period, new agricultural export crops like cotton and beef occasioned new waves of deforestation in the tropical areas of Central America. Beginning in the 1960s, growing rural populations in areas of skewed land distributions combined with fears of political unrest to drive large-scale colonization programs in Mexico and several Central American countries (Rudel 2005). In Mexico, both government-directed and spontaneous colonization movements rapidly demolished tropical forests in parts of southern Veracruz, southern Oaxaca, the Lacandon Forest in Chiapas, and southern Campeche and Quintana Roo (Bray and Klepeis 2005).

Deforestation in Central America was rapid and linear during the 1950–1986 period but slowed from an estimated 400,000 ha annually in the 1970s to 300,000 ha annually by 1990, with evidence of further declines and the growth of secondary succession during the 1990s (Kaimowitz 1996). As discussed earlier, one of the few large-scale cross-national efforts to understand the drivers behind these processes of forest change has been that of Rudel (2005), who carried out a Boolean analysis of 62 case studies of deforestation in Mexico, Central America, and the Caribbean. He found significant shifts in the drivers of forest cover change processes since the 1970s. In the 1970s and 1980s colonization was the most important causal element in deforestation, but became totally absent as a causal factor by the 1990s. For the earlier period, this is in agreement with Geist and Lambin (2001) who found the land-migration combination to be the most important factor driving historical deforestation in the region. However, neither study takes into account continued large scale spontaneous colonization pressures in the Petén of Guatemala, possibly the only place in the entire region where colonization pressures remain very strong (Grandia 2006). More recent drivers of deforestation are more mixed and include subsistence and commercial agriculture for local urban areas, cattle ranching at various scales, small farmer colonization (Rudel 2005), as well as demographic growth of existing forest frontier communities. Rudel also confirms the new significance of secondary forests in Central America, where "growth in the extent of secondary forests probably explains why net deforestation began to decline in the region" by the 1990s. Forest plantations, important elsewhere in Latin America, have had little significance in northern Latin America (see Grainger, this volume, and Rudel, this volume). The discussion of forest cover dynamics in each country is partially drawn from Table 5.1, presenting figures on the continued deforestation pathway, and Table 5.2, presenting figures on the forest recovery, maintenance, and protection pathway.

## 5.4 Country-Level Dynamics

### *5.4.1 Mexico*

#### 5.4.1.1 The Continued Deforestation Pathway

FAO figures show that Mexico had 28.9% forest cover in 2000 with a deforestation rate of 1.1% implying an annual loss of 631,000 ha (Table 5.1). The 2000 National Forest Inventory argued for a lower average rate, suggesting that from 1976 to 2000 the average annual deforestation for all forests has been 0.43%. However, this source suggests that the rate of loss in tropical forests has been 0.75%, three times as high as the rate of loss in temperate forests at 0.25%. The hotspots of tropical deforestation were located in the Yucatan Peninsula, Chiapas, along the Pacific Coast, and in northeastern Mexico (Velázquez et al. 2005). This study averages the rate for the entire period, but regional studies suggest that the rhythms of deforestation have varied over this period, likely being higher in the 1970s and dropping by the early 1990s. Examples of very high regional rates for earlier periods include a World Bank report showing an annual tropical deforestation rate of 2% in the mid-1980s; a study of the eight states of southeastern Mexico reporting a 1.9% loss per year from 1977 to 1992, while the northern Lacandon forest reported a rate of loss of 12.4% in the 1960s (Bray and Klepeis 2005). Similarly, a study of seasonally dry tropical forests found only 27% of the original cover intact by 1990, estimating an annual deforestation rate of 1.4% with heavy fragmentation and disturbances (Trejo and Dirzo 2000). Higher deforestation rates in tropical areas (around 2% annually) than in temperate forests (around 1% annually) have been confirmed for the 1980–2001 period for the state of Oaxaca (Velázquez et al. 2003). However, more recent regional studies show countervailing trends, that in some areas of temperate forests there are increasingly high deforestation rates, while many of the hotspot tropical areas seem to show declines in deforestation and increases in secondary forests that could suggest movement toward a forest transition at the regional level.

In the first of several examples of continuing highland deforestation, the region of the Monarch Butterfly Biosphere Reserve in highland Michoacán and Mexico state has shown rapid deforestation over the last several decades due to heavy illegal logging and subsistence agricultural expansion, a countervailing tendency towards reported lower national rates in temperate forests. Over the period 1971–1984 the deforestation rate for the Reserve region was 1.7% annually but accelerated to 2.41% from 1984 to 1999. Inside three sampled areas of the reserve, rates of degradation tripled from 1% to more than 3% between the same two periods (Brower et al. 2002). The montane tropical slopes and pine forests of the Sierra Norte of Oaxaca have also been reported to have deforestation rates of over 3% (Gómez-Mendoza et al. 2006). The temperate forests of highland Chiapas, like Michoacán, show intensifying deforestation in recent years. In 1975 the region was principally characterized by continuous forest cover. But from 1975 to 1990 the deforestation rate was 1.3%, increasing dramatically to 4.8% for the 1990–2000 periods. Other studies also show accelerating deforestation during the 1990s (Ochoa-Gaona and González-Espinosa 2000; Cayuela et al. 2006). While many other temperate forest regions appear to be stable or even expanding, the figures mentioned above suggest the existence of temperate forest hotspots in Mexico.

In contrast, some of the tropical forest hotspots are showing drops in deforestation rates in recent periods. In the Lacandon rainforest, historically the demographic escape valve for highland Chiapas and other areas of Mexico, the rates of forest loss declined from 2.13% annually from 1974–1984 to 1.6% for 1984–1991 (Mendoza and Dirzo 1999). Likewise, the region of the Calakmul Biosphere Reserve in Campeche and Quintana Roo lost 6.2% of the forest from 1969 to 1987, but declined to 2.8% from 1987 to 1997 (Chowdhury and Schneider 2004). The decline in deforestation rates would appear to be due to the end of large scale directed large-scale directed tropical colonization and the discouragement of spontaneous tropical colonization through the declaration of protected areas. For example, southern Campeche and Quintana Roo were the site of large-scale directed colonization projects in the 1970s and 1980s, but no new ejidos were established in the region of Calakmul after 1991.

#### 5.4.1.2 The Forest Recovery, Maintenance, and Protection Pathway

Processes of forest recovery, maintenance and protection are also evident in Mexico. Klooster (2003) analyzes an area of highland Michoacan which, in contrast to the Monarch Reserve Biosphere Reserve, is displaying agricultural abandonment and forest recovery. However, contradictory processes were found, with forest recovery occurring at the same time as forest degradation due to cutting for firewood. Nor is the forest recovery occurring because of "development" as in the classic model of forest transitions, but rather in a process where remittances maintain people in rural areas who are no longer dependent on agriculture. Continued industrial use of firewood for wood-fired artisan production leads to the situation where "Deforestation decline does not guarantee eventual recovery, reforestation can accompany declining forest quality" (Klooster 2003).

The tropical hotspot areas mentioned above showing declines in deforestation rates are also showing some indications of a forest transition, with agricultural abandonment being the apparent principal driver. For example, the Lacandon forest declined from being 91% forested in an earlier period to 63% forested by the late 1990s, but in the same period secondary forest grew from 0.2% to 18.2% of the region. This means that the combination of mature and secondary forests still covered 81.2% of the total area (de Jong et al. 2000). A subregion of the northern Lacandon rainforest, which had been devastated in the 1960s (the 12.4% rate reported above) showed extensive forest recovery. As much as 10% of the subregion of Calakmul has also reverted to secondary forest (Bray and Klepeis 2005). In the montane tropical forests of the Sierra Norte of Oaxaca, particularly the lower elevation tropical forests, there have been losses of 245,809 ha, but there has been forest regrowth on 159,396 ha (Gómez-Mendoza et al. 2006). For the entire state of Oaxaca, forest regrowth occurred in 2.6% of the territory, which only partially counterbalanced deforestation of 8% of the state in the 1980–2001 period (Velázquez et al. 2003).

Of the forest use drivers, Mexico has a reported 800,000 ha in coffee, almost all of it shade tree. A considerable amount of this is in lowland tropical areas suggesting that coffee has been an important factor in maintaining forest cover in both montane tropical regions and at lower elevations. As well, the 14 main coffee-growing regions in Mexico have all been termed biodiversity "hotspots" (Moguel and Toledo 1999). In highland Chiapas it is reported that shade coffee plantations are becoming islands of forest cover isolated from nearby cloud forests. (Cayuela et al. 2006). However, because of low coffee prices and other factors in recent years there have been frequent reports of abandonment or conversion of coffee plots to other crops, producing two different pathways of forest recovery or complete forest loss as a result of the crisis of coffee prices (Blackman et al. 2007).

Community forest management for timber has also played an important role in forest maintenance or even expansion in Mexico. In central Quintana Roo, populations that had been resident since the mid-nineteenth century undertook sustainable logging beginning in the mid-1980s. In this region, the annual deforestation rate for 1976–85 was 0.4%, and fell further from 1984 to 2000 to a negligible 0.1%, the lowest recorded rate of any region of tropical Mexico (Bray and Klepeis 2005). Also from 1984 to 2000, the 12 forest ejidos with the largest logging volume in the region had new deforestation take place on 5,364 ha, but 20,763 ha of previously deforested areas were in various stages of regrowth. During this period, 10% of the total area reverted back to forest (Bray et al. 2004). In the Sierra Norte of Oaxaca, another region dominated by community forest management for timber, pine-oak forests, which represent 25% of the forest cover, have expanded by 3.3% in the 1980–2000 (Gómez-Mendoza et al. 2006). As well, Duran et al. (2005) have shown that community forest regions in Quintana Roo and Guerrero have very low rates of land use change, lower than many protected areas. Mexico has the largest number and hectares in certified forests on community lands in the region, with 44 forests totaling 787,995 ha.

Protected areas have also played an important role in lowering deforestation and maintaining forest cover. Mexico has 193 protected areas including 196,185 sq km for a total of 8.72% of the national territory under protection. Given that as much of 60% of Mexican forests are in the hands of communities, a growing number of formally recognized community protected areas is also a notable new trend (Bray et al. 2008). Protected areas in Mexico seem to be reducing deforestation in many cases, as in the Calakmul region after the establishment of the Calakmul Biosphere Reserve in 1979, noted above. The Montes Azules Biosphere Reserve in the Lacandon, established in 1978, has kept deforestation low within its borders (Mendoza and Dirzo 1999). However, a study of Lagunas de Montebello National Park in highland Chiapas showed somewhat higher rates of forest conservation than in a neighboring community (Johnson and Nelson 2004). The Maya Forest, as a trinational region between Mexico, Guatemala and Belize will be discussed in Section 4.3.

Evaluating all of these dynamics, it does not appear that Mexico as a whole is approaching a forest transition, but much more work remains to be done at identifying precise regions of forest loss and recovery. Nonetheless, a complex large-scale landscape dynamic can be glimpsed which suggests lower deforestation pressures and forest recovery, and the beginnings of a forest transition in some southern tropical areas, particularly in the Yucatan Peninsula in central Quintana Roo, the region around the Calakmul Biosphere Reserve, and the northern Lacandon forest. However, tropical deforestation likely continues along the Pacific and Atlantic coasts, particularly in dry tropical forests. At the same time, while most regions of the highlands may show forest cover stability in total area, particular areas are showing accelerating deforestation, becoming temperate zone deforestation hotspots. An important lacunae in forest cover dynamics is the apparent absence of studies of the Chimalapas, a montane tropical region of southern Oaxaca which is considered to be the third largest tropical forest mass in Mexico. Most research on regional forest cover in Mexico has focused on descriptive statistics on forest loss with only qualitative or few observations on the drivers of forest cover dynamics. An exception is Klooster's (2003) nuanced discussion of the peculiar dynamics of a regional forest transition in a context of continued underdevelopment in Michoacan.

### 5.4.2 Belize

The FAO figures suggest that the deforestation rate in Belize is 2.3% annually representing 36,000 ha/year and with almost 60% of the country still forested. Earlier reports suggest lower forest loss, with deforestation from 1989/1992 to 1994 placed at 25,000 ha annually (Chomitz and Gray 1996). In total, less than 10% of the land area has been converted to agriculture or settlements. Sixty-five percent is under broadleaf forest, close to the FAO figure cited above, and the remainder consists mostly of swamp, pine forest, and mangrove forest (Chomitz and Gray 1996).

Regional studies of deforestation in Belize have focused on the two largest blocks of forest in northern and southern Belize. Deforestation has been found to be heaviest in the north, with regional estimates for southern Belize less than 5,000 ha/year and for northern Belize, 13,000+ ha/year. A small area, focused on the private Rio Bravo conservation area in northwestern Belize, showed an average loss of 1,799 ha/year from 1993 to 1999 (Dushku et al. 2002). In the south, a study that focused on Toledo District, Belize from 1975 to 1999 found a total forest loss of almost 10% or 36,000 ha over the entire period. Deforestation expanded most notably in central Toledo District and along the border with Guatemalan border and the most densely forested areas were in the Maya Mountains in the northern part of the district (Emch et al. 2005).

Forest recovery, mostly because of abandoned pastures, is reported to be occurring in Belize, but it has not been quantified. Protected areas are the only forest use driver of importance in Belize. Coffee is produced in very small quantities in Belize and there is only one large certified forest, the Rio Bravo Conservation area with 104,888 ha, with other non-protected forest areas subject to logging concessions. Protected areas are thus by far the most substantial land use in Belize which contributes to forest conservation. An estimated 30% of Belizean territory is under some form of protection with 107 protected areas covering 12,604 km of national territory. It has been argued that protected areas in Belize have been effective at reducing deforestation (Chomitz and Gray 1996). In sum, data is limited on Belize but landscape dynamics appear to be primarily driven by the existence of the large northern and southern blocks of forest and an undocumented degree of pasture abandonment, with no systematic research on the drivers of these processes at any scale.

## *5.4.3 Guatemala*

### 5.4.3.1 The Continued Deforestation Pathway

The FAO report for Guatemala shows a 1.7% annual rate of forest loss in Guatemala from 1990 to 2000 with 26.3% forest cover (see Table 5.1). A recent study suggests that the net deforestation between 1991/1993 and 2001 was at a lower annual rate of 1.43%, with a net loss of 14% of the forest mass during that period. The study does not distinguish by forest type, but notes that 58% of municipalities in the central cordillera lost between 1% and 5% of their forests during this period while the forest losses in the municipalities of the Petén were between 15% and 25%, suggesting much higher rates of forest loss in the tropics than in temperate areas (UVG 2006). It is not evident if the intensive ongoing deforestation occurring in the highlands of neighboring Chiapas is occurring in highland Guatemala. In one regional study in the eastern highlands, a communally managed pine forest had declined in area 14.4% from 1954 to 1987 with significant degradation also taking place, with further forest cover declines from 1987 to 1996 (Holder 2004). Totonicapán, highlighted as a department with effective communal management institutions (see below), nonetheless had nearly 1% annual forest loss between 1991/1993 and 2001.

However, it is the forests of the lowland Petén which have shown the most intensive deforestation, with a rate of 1.81% and an annual loss of over 47,000 ha (UVG 2006). A study of the Maya Biosphere Reserve (MBR) for 1974–1997 showed 65.2% of the buffer zone lost during this period, and areas near roads showed increasing deforestation pressures in 1995–1997 (Hayes et al. 2002). Some of the parks that compose the MBR are also showing high rates of forest loss, losing 11.5% and 8.1% of forest cover between 1986 and 2004 with forest loss thought to be accelerating. However, the "multiple-use zone," dominated by community logging concessions, lost only 2.7% forest cover and allowed fewer illegal settlements (CONAP/WCS (2004). A study of a dry tropical forest region in eastern Guatemala (that compared it with an adjacent area in western Honduras, see below) found that it was about 70% forests and 30% with a dynamic flux of forest cover, but with deforestation being the strongest trend (Tucker et al. 2005). The Peten of Guatemala is one of the few areas in the entire region that is still undergoing intensive tropical colonization resulting in forest loss from subsistence agriculture, but the data are not sufficient to understand processes in the highlands.

#### 5.4.3.2 The Forest Recovery, Maintenance and Protection Pathway

There are indications of forest recovery in some regions of Guatemala. The most recent deforestation study showed that forests recovered by 3%, slightly offsetting the overall forest loss of 14%, and that this was concentrated in the northeastern departments. The causes of this forest recovery are not reported. The recovery dynamics varied between departments with, for example, Alta Verapaz gaining back about one-third of forests lost and in Huehuetenango closer to 50% forests recovered (UVG 2006). Coffee agroforestry has been an important forest use in maintaining modified forest cover in Guatemala, with an estimated 260,000 ha planted in coffee, most of it thought to be shade tree, but only around 5,000 ha of this is in organic coffee, which would give the coffee agroforest higher value (Varangis et al. 2003).

With respect to other forest use drivers, common property forests have maintained themselves in many of the traditional settlement areas of Guatemala, but the trends in these forests have not been well documented. The department of Totonicapán is reported to have strong communal management institutions and in the 1970s was documented as losing only 7% forest cover from 1954 to 1972 despite 80% population growth from the 1930s to the 1970s (Veblen 1978). However, forest loss of 0.95% annually has been reported more recently (UVG 2006). While below the national average it is also five times the rate of forest recovery in Totonicapán. The best documented region as to forest cover trends in community forests and protected areas, as in deforestation, has been the Petén and the Maya Biosphere Reserve (MBR). The MBR encompasses 1.6 million hectares and includes community forests embedded within a multiple use zone and protected areas. There are 12 community forest concessions (as well as two industrial concessions) which are dedicated to logging and the extraction of all other forest products. There are a total of 13 FSC certified forests in Guatemala with a total of 512,321 ha under certification, the second highest in the region after Mexico, all in the Petén. There are also 11 protected areas in various categories. The multiple use zone contains all of the community forest concessions, the industrial concessions, and both inhabited and uninhabited non-concessioned areas.

To what extent are both the community forests and the protected areas contributing to the forest use component of a forest transition? A recent study suggests a mixed picture. Some protected areas and some community forest concessions are showing relatively high rates of forest loss. Inhabited protected areas and recently inhabited community forest concessions have similar and relatively high deforestation rates. On the other hand, uninhabited protected areas and two long-inhabited community concessions and uninhabited community concessions all show very low deforestation rates. The differences can be attributed to distance from roads and population centers (although the long-inhabited concessions have both) but more broadly to remoteness from the colonization frontiers that are steadily advancing, especially on the western and central parts of the MBR (Bray et al. 2008). This same study is also one of the few that looks comparatively at the Maya Forest, and it found that long-inhabited forest ejidos and the Calakmul Biosphere Reserve, as a lightly inhabited protected area, have similarly low deforestation rates (Bray et al. 2008).

In another study that looked at the same forest mass across two countries, Tucker et al. (2005) look at tropical dry forest study sites in eastern Guatemala and western Honduras with comparable topography, climate, vegetation, and agriculture for the period 1987–1996. Landscapes in each country had about 70% of the area in stable forest and agriculture and about 30% of the area in various forms of flux. However, the Honduras side had larger patches and was less fragmented than the Guatemalan side. Further, Honduras had overall net forest regrowth, while in Guatemala the dominant trend was deforestation. There are two important trends present, (1) agricultural intensification that abandons more marginal agricultural lands to forest regeneration and (2) expansion of export crops at higher altitudes. Both processes are occurring in each country, but in Honduras the former dominates and in Guatemala the latter.

For Guatemala at the national level, deforestation is the dominant trend nationally, but rates of loss appear to be much higher in tropical over temperate areas, and some temperate regions are showing dynamic forest recovery. Regionally, subsistence agriculture through spontaneous colonization remains an important driver of deforestation in tropical regions of the Peten, with the cross-national study by Tucker et al. (2005) representing a notable level of sophistication that allows for an understanding of the cross-scale dynamics of forest cover processes.

Finally, there have been few efforts to study the contiguous trinational forest mass between Mexico, Guatemala, and Belize, the "Maya Forest" (Primack et al. 1998; Bray et al. 2008). Nonetheless, this review suggests strong tendencies towards forest conservation on the Mexican side, with the Belizean portion of the Maya Forest threatened to be bisected by deforestation in the central part of the country.

However, the most intense deforestation threats to the Maya Forest occur in the Peten of Guatemala with a colonization front advancing in several areas.

## *5.4.4 Honduras*

### 5.4.4.1 The Continued Deforestation Pathway

The FAO estimates deforestation in Honduras at 1% annually which implies an annual loss of 59,000 ha and a 2000 forest cover of 48.1% (see Table 5.1). Another study suggests that from 1965 to 1995 the pine forest landscape has been stable but that broadleaf forests lost about a third of their surface, around 1.3 million hectares over the 30 years (Pratt and Quijandría 1997). However, it is has also been noted that "trustworthy estimates of the rate of deforestation do not exist" (my translation) for Honduras and estimates from United Nations sources range from 62,000 ha annually to 108,000 ha (http://www.un.org/esa/agenda21/natlinfo/countr/honduras/natur.htm#forests; accessed 9/9/07; http://www.un.org/esa/agenda21/natlinfo/wssd/honduras.pdf; accessed 9/9/07). Despite this range of uncertainly, it is clear that the most dramatic losses have been in the forests of the Atlantic Coast, which have declined by 72.6% between 1962 and 1990, compared to only 30% loss for other broadleaf forests in the same period (Humphries 1998). This region also continues to suffer from spontaneous tropical colonization in the Mosquitia portion and parts of the Rio Platano Biosphere Reserve (see below). Tucker (1999) represents one of the few efforts to evaluate the relationship between forest tenure and deforestation anywhere in the region (but see Bray et al. 2008) and finds little difference between private property and common property forests in western Honduras. From most of the available data, one could conclude that forest cover trends in Honduras are entirely linear, but the few regional studies available suggest a more complex picture here also.

### 5.4.4.2 The Forest Recovery, Maintenance, and Protection Pathway

There is some evidence of forest recovery in Honduras, but at incipient levels. It has been suggested that fallow lands and secondary forests have expanded on 110,000 ha between 1974 and 1993, although the fate of these lands since then has not been documented (Kaimowitz 1996). Studies in western and central Honduras show that in highland areas dynamic landscape processes are present, with some local trends towards forest recovery. In western Honduras, forest recovery was noted in the 1987–1996 period but closer study found "a complex mosaic of land cover change processes that involve approximately equal amounts of reforestation and deforestation." Between 1987 and 1991, accessible areas showed greater deforestation and fragmentation, but this trend reversed in the following 5 years (1991–1996).

Deforestation increased in more distant and higher elevation areas because of government programs to promote coffee. In the more accessible regions, a logging ban and agricultural abandonment due to agricultural intensification elsewhere led to increased regrowth. At the same, there were cyclical vegetation oscillations due to swidden agriculture in higher areas near roads and regrowth occurring around edges of stable forest. All of these cover change processes occurred in relatively small areas in a matrix of stable forest and agricultural areas (Southworth and Tucker 2001; Nagendra et al. 2003; Munroe et al. 2005).

In the small La Lima watershed in the Yeguare river valley in central Honduras researchers found two distinct periods of forest cover dynamics. From 1955 to 1975 there was a linear period of agricultural expansion, with a deforestation rate of 1.2%/year. In the second 20-year period, 1975–1995, agricultural become more intensive and the deforestation rate declined to 0.6%/year. However, as in the case reported for Mexico by Klooster (2003), firewood extraction contributed to forest degradation even though forest cover remained (Kammerbauer and Ardon 1999, Kammerbauer et al. 2001). In these cases, forest cover is found to be stable or expanding slightly because of intensification of agriculture, not because of agricultural abandonment. There is also some evidence of forest recovery in Honduras using the technique of "repeat photography" (Bass 2004).

With respect to forest use drivers, coffee agroforestry has also been important in maintaining forest cover in Honduras with an estimated 237,000 ha. Of this, 95% of it shade tree (http://www.cafedehonduras.org/aboutus_esp.php; accessed 9/8/07) with small but growing exports of organic coffee beginning in 1998. Bass (2006) found in a mountain region of western Honduras that shade tree coffee largely accounted for a 17% increase in forest cover from 1954 to 1992. Forest management has gained new legal footing in Honduras, but it is unclear how much of it sustainable. Since the mid 1990s the Honduran forestry agency has required management plans for logging. As of 2001, there were 786 management plans authorized on 1.1 million hectares of land, which would suggest that about 20% of Honduras' 5.4 million hectares are under management. Data from 2002 suggests that 373 of these management plans are in the hands of communities with a total of 560,000 ha (www.cohdefor.hn/manejo_forestal/plan4.shtml,accessed9/10/07;AFE-COHDEFOR 2002). However, almost all of these management plans were in pine forests, with only 40 in tropical broadleaf forest ecosystems, all in national forests. In 2004 efforts by the World Wildlife Federation were underway to work with two Miskito Indian cooperatives on sustainable logging in the Mosquitia (Bray and Anderson 2005). However, as of 2007, there were still only three managed forests for a total of nearly 50,000 ha with FSC certification (Table 5.2). A recent decree promises to expand the possibility for community logging in Honduras (www.acicafoc.net; accessed 9/10/07).

Honduras currently has 99 protected areas covering nearly 29,761 sq km which constitute about 20% of the national territory (Table 5.2) but there has been little research on how effective they have been on conserving forest cover. However, one study suggests that Celaque National Park, established in 1987, has maintained forest cover while 25% of the surrounding area has been deforested for agriculture

(Southworth et al. 2004b). The largest park, Pico Bonito, at 107,300 ha has been reported to be under deforestation pressures (Humphries 1998) although it is now also the site of one of only eight Kyoto-certified forest carbon sequestration projects in the world. The largest protected region in Honduras is the Rio Platano Biosphere Reserve at 815,000 ha and is part of the Mosquitia Corridor that includes 5 million hectares of tropical forests between eastern Honduras and northern Nicaragua. However, the Rio Platano Biosphere Reserve is being threatened by agrarian reform projects and encroachments on its borders, with one zoned area losing 10% of its forest in a recent 6 year period (Hayes 2007). Hayes (2007) provides one of the few comparative studies of forest cover processes in and around protected areas in both the Honduran and Nicaraguan Mosquitia, and the Nicaraguan section will be discussed below. In sum, the Honduras forest landscape is likely characterized by relative stability in temperate areas with localized areas of fluxes in forest cover, but with continuing deforestation in tropical areas, with protection having variable impacts. No forest transition is close to taking place nationally, although it may be occurring in some very local areas. Several regional studies have captured multi-scale dynamics in forest cover and its drivers.

## 5.4.5 *El Salvador*

### 5.4.5.1 The Continued Deforestation Pathway

El Salvador is by far the most deforested country under study, with a mere 5.8% of the land area under forest cover for a total of 121,000 ha and ongoing forest losses of 7,000 ha/year (see Table 5.1). Most of the historical deforestation took place in earlier periods, although a 2.88% deforestation rate is reported for the 1990s. A substantial amount of the deforestation was for coffee in the first decades of the twentieth century, so coffee agroforests cover much larger areas in El Salvador. However, El Salvador is also the country where there exists the best documentation on forest recovery in any of the countries studied (Hecht et al. 2006; Hecht and Saatchi 2007). These studies suggest that remnant original forests constitute some 30,000–40,000 ha and coastal mangrove forests around 25,000 ha, well below the FAO figures.

### 5.4.5.2 The Forest Recovery, Maintenance and Protection Pathway

Agricultural abandonment occurred in El Salvador because of civil war and large-scale international migration. This has led to a rate of forest recovery estimated at 5.8% annually during the 1990s (Hecht and Saatchi 2007). Since one definition of a forest recovery is when the rate of new forest growth exceeds that of ongoing deforestation, it can be said that El Salvador is undergoing a forest transition, albeit at a point when natural forests had almost entirely disappeared. This recovery is also very partial, with an estimate of a 22% increase in areas with 30% forest cover and a 6.5% increase in higher-density forests since the early 1990s (Hecht and Saatchi 2007).

Coffee agroforests have been a crucial forest use in maintaining forest cover in the rural landscape of El Salvador with 161,000 ha in coffee, most of it thought to be shade tree coffee, with only around 7,000 ha of it being certified organic. As one measure of its significance for conservation in El Salvador, the area planted in coffee is ten times the area of protected areas (Varangis et al. 2003). Coffee in El Salvador grows in mountainous areas in the west, central and eastern regions, with about half in the western region. Ninety-five percent of it is shade grown with coffee farms in lowland areas having about 40% shade cover and in highland areas around 20%. A recent study has argued that shade tree coffee in El Salvador has provided a "bulwark against tree cover loss", particularly before 1990. In that that year, 51% of the noncoffee areas had no tree cover, while only 7% of the coffee regions had none. However, from 1990 to 2000 13% of the shade tree coffee area became deforested due to the fall in coffee prices, urbanization, and other factors (Blackman et al. 2006).

Due to the massive historical deforestation and the incipient nature of forest recovery in El Salvador, forest use for timber as a driver of forest maintenance is negligible in El Salvador and there are no FSC certified forests. With reference to protected areas, the World Database on Protected areas FAO figures reports 76 protected areas with 258 sq km. (Table 5.2). The largest of these is El Imposible at 5,000 ha and figures show that the protected areas remained stable in their forest cover in recent periods (Hecht and Saatchi 2007). Thus, the landscape of El Salvador may be said to be undergoing a very incipient forest transition, but at a point when native forests had almost entirely vanished, and with still modest processes of forest recovery. El Salvador has the most sophisticated analysis of the dynamics and drivers of forest recovery.

## 5.4.6 *Nicaragua*

### 5.4.6.1 The Continued Deforestation Pathway

There are wide variations in the estimate of total forest cover in Nicaragua, from 3.28 to 6 million hectares (see Table 5.1 and http://www.rlc.fao.org/proyecto/rla133ec/Guia%20Paises%20(1)/Nicaragua.PDF; accessed 9/13/07). Nicaragua is thought to have the second highest deforestation rate in the region at 3%, which implies a loss of 117,000 ha annually (Table 5.1). However, it has also been suggested that this high rate fell by the end of the 1990s and that the rate from 2000 to 2005 was more on the order of 1.3% (www.rainforests.mongabay.com, accessed 9/10/07). Nicaragua is also a region that is showing ongoing tropical forest loss through small-scale spontaneous colonization of the Atlantic Coast and the Mosquitia.

#### 5.4.6.2 The Forest Recovery, Maintenance, and Protection Pathway

Much of the historic deforestation in Nicaragua was due to the expansion of cattle ranching (and cotton farming), and when both began to decline in the 1980s, there was much pasture abandonment. In the 1990s it was estimated that Nicaragua had 1.1 million acres of scrub forest and another 900,000 ha in forest fallow, with the scrub forest in former cattle producing regions and the forest fallow in more humid areas (Kaimowitz 1996), but the fate of this forest landscape process has once again not been documented. Coffee agroforests as a forest use driver, on the other hand, cover less area than in the other Central America countries, with an estimated 108,000 ha of coffee in Nicaragua, with about 10% of that in specialty, organic and fair trade coffee. Much of the rest of it is likely to be shade tree coffee (Bacon 2005).

Sustainable forest management is in an incipient phase as forest maintenance driver. WWF is working with two communities on the Atlantic Coast of Nicaragua covering more than 40,000 ha to become FSC certified (Bray and Anderson 2005). Under new forest legislation passed in 1992 procedures for forest management plans were approved, with by 2000 a reported total of 519 management plans affecting a total area of 66,520, all on small and medium private landholdings in pine forests. In broadleaf tropical forests a total of 169 management plans were approved on 93,348 ha, for a total of 159,868 ha under management plans http://www.fao.org/docrep/008/j2628s/J2628S14.htm; accessed 19/13/07). However, few of these have FSC certification (only four forests with 20,766 ha) and there have been recent reports of many irregularities in the management plans on the Caribbean coast. (http://probidad.net/cs/index.php?option=com_content&task=view&id=693&Itemid=29, accessed 9/13/07). Nicaragua is reported to have 93 protected areas for a total of 29,405 sq km with little known about the status of most of them. The largest protected area is the Bosawas Biosphere Reserve created in 1991 and including some 2 million hectares in both the core and buffer zones. Some of the buffer zone areas of Bosawas occupied by mestizo farmers are being heavily deforested, with up to half of the forest gone. However, the indigenous occupied core zones of Bosawas are showing virtually no deforestation, with one such area having 97% forest cover in 2003 (Hayes 2007). In contrast, the Rio Platano Biosphere Reserve on the Honduran side of the Mosquitia is under great deforestation pressures because of failed efforts to centralize management in the government, while protection is much more effective in the Bosawas core area due to the decentralization of management in the hands of the indigenous inhabitants. In sum, the balance of forest cover dynamics in Nicaragua is not clear. The fate of the widespread incipient forest recovery of the 1990s has not been well-documented, nor has ongoing deforestation, although the multi-scale dynamics and ethnic and public policy drivers of forest cover in the Bosawas Biosphere Reserve have been well explicated. With respect to the binational Mosquitia, as mentioned, the pressures appear to be greatest on the Honduran side, although the areas outside the core of the Bosawas Biosphere Reserve area also under pressure.

## 5.4.7 *Costa Rica*

### 5.4.7.1 The Continued Deforestation Pathway

Costa Rica is the country where the best studies are available on deforestation and also on forest policy (Joyce and Sader 1988; Brockett and Gottfried 2002). The FAO reports that Costa Rica has the lowest current deforestation rate in the region at 0.8% implying an annual loss of 16,000 ha. The classic study of deforestation in Costa Rica to 1983 showed accelerating decline. From 1940 to 1977 forest cover dropped more than half, from 67% to 32%, dropping again by half in the following 6 years (1977–1983) from 32% to 17%. Tropical dry forest almost vanished entirely by 1961 (Janzen 1988). By 1983 only 17–20% of the country was forested (Joyce and Sader 1988). Much of this deforestation was driven by export cattle ranching, which underwent a serious contraction after 1989 (Arroyo-Mora et al. 2005). As recently as 1986–1991 the deforestation rate over half of Costa Rica's territory was estimated at 4.2% annually (Sanchez-Azofeifa et al. 2001). However, it is around this period when forest cover trends in Costa Rica began to turn around and after 1990 began on average to show a "markedly higher level" (Kleinn et al. 2004), with forest recovery occurring primarily in temperate zone forests.

### 5.4.7.2 The Forest Recovery, Maintenance, and Protection Pathway

Agricultural abandonment has led to trends of forest recovery in various regions of Costa Rica, but particularly in some of the dry tropical forest areas deforested the earliest, like the Chorotega region. In this region deforestation was high from 1960 to 1979 but followed by accelerating forest recovery from 1979 to 2000, reaching 4.91%/year in the last 14 year period (Arroyo-Mora et al. 2005). In the Tempisque Basin of Northwest Costa Rica, forest cover has increased significantly in the last 25 years due to agricultural intensification, plummeting beef prices, and conservation initiatives. However, agricultural intensification has also led to the loss of wetlands in lower parts of the watershed (Daniels, this volume). In some tropical areas agricultural abandonment does not yet appear to be taking place, although there are declines in the rate of deforestation. For example, in the Osa Peninsula, outside Corcovado National Park, forest cover declined over the entire period 1979–1997 although the rate went down from 1.5% annual deforestation rate in 1979–1987 to 0.83% from 1987 to 1997 (Sanchez-Azofeifa et al. 2001; Pfaff and Sánchez-Azofeifa 2004).

Coffee agroforestry is probably of least significance in Costa Rica. Although there are some 115,000 ha of coffee in Costa Rica, substantial areas of this are full sun coffee, and Costa Rica has very low rates of export of organic coffee, which is normally shade tree (Varangis et al. 2003). The sustainable forest management component in Costa Rica appears to be much less significant than protected areas. Despite the considerable focus on forests in Costa Rica only 19 forests with 76,547 ha under FSC certification and there have been many bureaucratic hurdles in the way of developing sustainable forestry among smallholders in Costa Rica (Brockett and Gottfried 2002).

Costa Rica is justifiably famous for its protected area networks, with an estimated 17,508 sq km in 183 protected areas, constituting 23.25% of the national territory (see Table 5.2). Studies of forest cover in and around these protected areas suggest that they are working within their boundaries but there are increasing problems outside their borders, making them in danger of becoming islands. A comprehensive study of deforestation in Costa Rica's park system found that deforestation inside protected areas was negligible for the period 1987–1997, and that 1-km buffer zones around the protected areas had a net forest gain for the same period. However, a 1% annual deforestation rate was found in a 10-km buffer zones, suggesting increased isolation of protected areas. (Sanchez-Azofeifa et al. 2003). The expansion of forest cover in the Chorotega region is attributed to the new conservation areas established during 1979–1986, Santa Rosa and Palo Verde National Parks, leading to the high restoration rates between 1986 and 1997 (Arroyo-Mora et al. 2005). Likewise, closer studies of Corcovado National Park show no deforestation inside its boundaries, but deforestation and forest degradation continue outside its boundaries, albeit at lower rates than in earlier periods (Sanchez-Azofeifa et al. 2002). Increased deforestation and fragmentation has also been found around Braulio Carillo National Park, another example of the island phenomenon (Schelhas and Sanchez-Azofeifa 2006). Private nature reserves are also important in Costa Rica, with a national association of private reserves counting 69 affiliates with more than 50,000 ha in the 1990s (Brockett and Gottfriend 2002).

Costa Rica is the only country where it has been well-documented where the important non-protected forest masses are, the Osa Peninsula, the Talamanca area and along the central volcanic range (Brockett and Gottfried 2002). The Talamanca region includes the third large forest block in Central America, after the Peten and the Mosquitia, in southern Costa Rica and northern Panama and is the largest unfragmented tract of montane tropical forest. It includes the La Amistad International Park and stretches across both Costa Rica and Panama including about 1.1 million hectares about equally divided on either side. On the Costa Rican side it includes 15 management units including a core zone, three national parks, two biological reserves, one forest reserve, seven Amerindian reserves, and one botanical garden (http://www.nmnh.si.edu/botany/projects/centres/amistad.htm; accessed 12/2/07). Although deforestation rates have been thought to be reduced on the Costa Rican side in recent years, unsustainable land uses were reported from some of the units in the 1990s (Kappelle and Juarez 1994). Thus, although Costa Rica historically suffered very high rates of deforestation, shifts in public policy and markets have led to reductions in deforestation in some regions and forest recovery in others. A more aggressive pursuit of these polices could lead to the beginnings of a forest transition in Costa Rica. Costa Rica has the highest number of regional studies, several of which document drivers, suggesting that the time is ripe for a more detailed metaanalysis of the multiscale dynamics and drivers of forest cover in Costa Rica.

Finally, the La Amistad International Park (PILA), the core component of the Amistad Biosphere bi-national World Heritage Site located in the Talamanca highlands of Costa Rica and Panama, is the only transnational forest mass in the region where international cooperation, in this case between Panama and Costa Rica has allowed for the formal recognition of the contiguous forest, so is an important model for the rest of the region.

## 5.4.8 *Panama*

### 5.4.8.1 The Continued Deforestation Pathway

The FAO figures show that Panama has 38.6% forest cover and is losing around 52,000 acres annually for a 1.6% annual deforestation rate. This represents an increase in deforestation in the recent period. Kaimowitz (1996) reported government statistics that deforestation fell from 46,000 ha annually 1970–1980 to 35,000 ha annually 1980–1987, but they may have risen again after that. The current estimates of forest loss cover an extremely wide range. A website that does not cite its source is the most optimistic, reporting an annual deforestation rate of 0.16% between 1990 and 2000 and decreasing and declining to 0.06% annually from 2000 to 2005 (http://rainforests.mongabay.com/deforestation/2000/Panama.htm, accessed 12/02/07) while other sources report 10% forest loss during the 1990s and 70,000 ha lost annually more recently (http://burica.wordpress.com/2007/11/03/estado-ambiental-de-panama-a-finales-del-siglo-xx/; accessed 12/02/07 Finally, the government reports that from 1992 to 2000 national forest reduced from 49% of the national territory to 45% (ANAM 2006a). In short, there is great unreliability in the figures. On the Panamanian side of the La Amistad Biosphere Reserve historically low deforestation rates have been increasing in recent years. Only 0.56% net forest loss occurred between 1987 and 1998, but deforestation accelerated to a loss of 2.34% of forest area between 1998 and 2001, with greatest pressure being felt in protected areas on the Caribbean side of the reserve (Forrestel and Peay 2006). The Darien is one of the few regions of Panama which has received specialized attention as to forest cover, with one study showing a decline of about 10% from 1987 to 1997 but with three-fourths of the province still forested (Harris et al. 2001).

### 5.4.8.2 The Forest Recovery, Maintenance, and Protection Pathway

Forest recovery in Panama is also poorly documented but there are press reports that it exists. It has been reported that the department of Los Santos showed 3.97% annual forest recovery from 1992 to 2000. Further, there are observational reports of unquantified forest recovery in many other areas of Panama, attributed to a change in government policies towards cattle grazing in the 1990s and consequent pasture abandonment and reforestation activities. (Cortes and Carrasco 2004). Figures on the extent of coffee agroforestry in Panama are hard to come by. The 1990 agricultural census of Panama reported 10,254 ha in over 40,000 farms, suggesting extremely small landholdings, while the 2001 agricultural census reported only around 30,000 farms but gave no figure on number of hectares. Sustainable forest management is very incipient with only eight FSC certified forests with 10,878 ha. Finally, Panama reports 34.3% of its national territory in 65 protected areas that have over 80% of the forest cover, the second highest in Central America after Belize, with a large percentage of this being in the indigenous territories in the Darien (ANAM 2006b). But these figures vary notably from the World Database on Protected Areas that shows 62 protected areas covering 24.59% of the national territory (Table 5.2). The Darien Peninsula continues to be protected because it serves as a barrier against hoof and mouth disease, with no cattle raising permitted within 25 km of the border and in the rest of the Darien only subsistence cattle raising is permitted. This has been further reinforced by Panamanian recognition of indigenous land rights over 700,000 ha of the Darien in the Kuna Yala and Embera *comarcas* (Kaimowitz 1996). Panama is the least documented of the countries in the region, with virtually no regional studies that look at multi-scale drivers and dynamics of forest cover, and no indications of formal recognition of the contiguous forest mass between Panama and Colombia.

## 5.5 Conclusions

This survey has attempted to synthesize a selection of the existing literature on forest cover dynamics in Mexico and Central America in order to bring into focus large-scale landscape dynamics of the forests of Mexico and Central America, and evaluate the extent to which a forest transition may be occurring at the country level. The survey does not pretend to comprehensiveness both because of the size of the area and the limitations of the published data. Nonetheless, this review makes it clear that our knowledge of forest cover dynamics in this world region is still quite rudimentary. There is wide disagreement or the absence of data on exact rates of deforestation and current states and dynamics of forest cover in many of the countries of the region. Local and regional studies are frequently more reliable and show trends which may be either similar to or varying from national tendencies, but the regional coverage is also quite limited. Of particular concern is the state of the four largest and transnational tropical forest masses in the region: The Maya Forest of Mexico, Guatemala, and Belize, the Mosquitia of Honduras and Nicaragua, the "La Amistad" forest between Costa Rica and Panama, and the Darien Peninsula forest, which also extends into Colombia. These "last frontier forests" have been the focus of efforts through the Mesoamerican Biological Corridor to conserve and link them through corridors, but efforts have foundered due to political problems and lack of participation by local communities (Bryant et al. 1997; Kaiser 2001).

There is also disagreement or inadequate data on many important forest uses in the region. The magnitude and status of forest recovery is very poorly documented, and there are important technical issues to be resolved in being able to monitor forest recovery (Castro et al. 2003). There is also little documentation on the precise extent of coffee agroforestry, the number of hectares under sustainable forest management, and even the number of hectares under protection (particularly differentiated by type of ecosystem and terrestrial vs marine), despite the attention focused on this last forest use. Further, few of the studies of forest cover dynamics in the region look explicitly at the issue of a forest transition. Only Klooster (2003) for a region of Mexico and Rudel (2005) at the level of Mexico and parts of Central America and the Caribbean specifically evaluate the reality and nature of forest transitions in the region. Thus, evidence for a forest transitions and its drivers must be gleaned from studies done without this perspective in mind.

Nonetheless, this review begins to show some of the general patterns of forest cover dynamics for some countries and regions, and where some of the most important lacunae are that must be filled to enhance our knowledge and as a foundation for management. A summary of the analysis by country of forest country dynamics is found in Table 5.3, and only some of the most notable findings from that table will be mentioned here. An intriguing suggestion from Mexico is some tropical hotspots are showing reduced deforestation and forest recovery, while some highland temperate areas, particularly in Chiapas and Michoacan, are emerging as "temperate deforestation hotspots". Mexico is the country where community forest management is an important driver in maintaining forests and reducing deforestation in both temperate and tropical areas, and most major tropical protected areas also appear to be fulfilling that function. In some temperate and tropical areas, forest recovery is occurring. In particular regions, such as central Quintana Roo, southern Campeche, and the northern Lacandon forest, forest transitions appear to be occurring. A temperate region of Michoacan is undergoing a forest transition in terms of forest cover, but with ongoing forest degradation, a significant variance from the pattern in industrial countries. While it is probable that Mexico as a whole is not undergoing a forest transition, inadequate information on apparent widespread agricultural abandonment makes it difficult to arrive at a definitive conclusion.

Belize is showing higher deforestation in its northern forest mass, which forms part of the trinational Maya Forest, while the southern mass is becoming increasingly isolated. Forest recovery is anecdotally said to be occurring, but has not been documented, and the only relevant forest use, which appears to be effective, is public protected areas. Information is inadequate to come to any conclusions about forest transitions in Belize. In Guatemala, accelerating deforestation is found in the Peten while temperate areas may be more stable, with some two temperate departments showing forest recovery. Guatemala is the only country with a national level study that evaluates both deforestation and forest recovery, showing deforestation outstripping forest recovery by a good margin. This review also suggests that the Maya Forest is stable or recovering in its Mexican portion, but under continued deforestation pressures in both Belize and especially Guatemala. Rates of deforestation in Honduras are not well-documented, but regional studies suggest that temperate forests may be relatively stable, but with dynamic local fluxes due to shifting markets. The Honduran Mosquitia appears to be under significant deforestation pressure, with protected areas in this region not showing much effectiveness.

El Salvador has the best-documented forest recovery processes in the region, but this occurred only when original forest cover had been reduced to very small

**Table 5.3** Summary of forest cover dynamics in Mexico and Central America

| | Continued deforestation | | Forest recovery, maintenance and protection | | | |
|---|---|---|---|---|---|---|
| Country | Tropical | Temperate | Agricultural abandonment | Coffee agroforestry | Sustainable forest management | Protected areas |
| Mexico | Tropical "hotspots" showing reduced deforestation | Some highland areas showing accelerating deforestation or forest transitions with degradation | Apparently widespread in some temperate and tropical areas but poorly documented | Important factor in maintaining forest cover in many highland and montane tropical regions | Important in maintaining forest cover in some temperate and tropical areas throughout country | Major ones in tropical areas appear to be reducing deforestation |
| Belize | Higher deforestation in northern Belize than southern | N/A | Pasture abandonment occurring but not well documented | Very minor | Confined to one large private reserve in northern Belize | Reportedly effective at reducing deforestation |
| Guatemala | Accelerating deforestation in Petén | Stable in some areas, but national trends not clear | Occurring in Alta Verapaz and Huehuetenango | Likely important in maintaining forest cover. | Occurring in some community concession in Peten and some communal forests in highlands | Under pressure in Peten |
| Honduras | Significant ongoing deforestation in Honduran Mosquitia | Temperate forests may be stable with dynamic fluxes in some regions | Some reportedly occurring but not well documented | Likely important in maintaining forest cover | Of minor significance | Rio Platano Biosphere Reserve losing forest rapidly |

| | | | | | | |
|---|---|---|---|---|---|---|
| El Salvador | Little remaining forest cover | Little remaining forest cover | Well-documented but still incipient | Very important in maintaining forest cover | N/A | Of minor significance |
| Nicaragua | High rates in tropical areas | Not well-documented | Reported widespread pasture abandonment but not well-documented | Important in some regions | Of minor significance | Core areas of Bosawas Biosphere Reserve effective |
| Costa Rica | Declining in recent years but still occurring | Not well-documented | Mot notable in some dry tropical forest areas | Of less significance because of widespread sun coffee plantings | Undocumented | Protected areas functioning but in danger of becoming islands |
| Panama | deforestation may be reduced in recent periods but mixed reports | N/A | Occurring in some areas but not well documented | Of probable minor significance but not well-documented | Undocumented | High percentage in protected areas; effectiveness not well documented except in Darien |

amounts, and the recovery is still incipient. Since conserving forests uses are still confined mostly to coffee, this forest recovery may also be regarded as precarious. Nicaragua has very poorly documented forest cover and forest use processes, so any generalizations are hazardous. The best-documented region is the Bosawas Biosphere Reserve, where it has been demonstrated that the indigenous-controlled core area is protected forests while buffers areas dominated by mestizo colonists are being rapidly deforested. Costa Rica, despite being well-studied in some respects, does not currently have national level studies that look at both deforestation and forest recovery that would allow judgments on a national forest transition, although it is clear that deforestation has been greatly reduced from earlier periods. Costa Rica does have the richest array of regional studies with several having a focus in deforestation in and around protected areas. These suggest that Costa Rican protected areas are in danger of becoming islands. Finally, generalizations in Panama are also hazardous because of a lack of information both nationally and regionally.

Overall, even though the prospects for forest transitions are not clear at the level of this large region, there are grounds for a cautious optimism. For much of the region, the worst periods of tropical deforestation occurred in the 1960s–early 1990s period, and these tendencies have now been much reduced or stabilized entirely. Some tropical areas, particularly in Mexico, may be undergoing a forest transition, while some tropical deforestation hotspots remain, particularly in the Guatemalan portion of the Maya Forest and the Honduran Mosquitia. Temperate forests, with the notable exception of the temperate forest hotspots in Mexico, may be more clearly stabilized or recovering than tropical forests.

Moving beyond description to theory, data is available for some regions that could begin to build the kind of hierarchical-panarchic framework proposed by Perz (2007; see also Perz and Almeyda, this volume). For example, agricultural abandonment is most commonly driven by rural–urban–international migration, best documented for Mexico. Migration chains express a socio-economic process that responds to large-scale movements in global economies, in this case, wage differentials between rural Mexico and industrial agriculture or urban economies in the US. However, each migration decision is taken first at the level of an individual household. Each individual male who decides to leave a household in a rural area usually implies a field or pasture left fallow for the indefinite future. As individuals from a village migrate, it makes it easier for other people from that village to migrate because of kinship–friendship relations, thus a particular village may begin to show a profile of increasing amounts of secondary vegetation due to out-migration and agricultural abandonment. Entire regions may become characterized by having a high percentage of emigrant households, leading to a regionally discernible recovery of forests from agricultural abandonment. Scaling up to a national level in a cumulative process, an entire country may begin to show signs of forest recovery because of processes that originate both in large-scale economic processes and the decisions of individual actors and households. This is a "cascade of causation" (Perz 2007) where global economics and individual households interact to create multi-scale changes in forest cover.

A panarchic framework calls for examination of both social and biophysical factors with dynamics occurring at varying temporal or spatial scales (Perz 2007). Tropical colonization, leading to extensive deforestation, occurred in Mexico over a decades-time frame from the 1950s to 1990s, depending on the region. This process has now slowed considerably or even been reversed in Mexico. On the other hand, temperate zone deforestation accelerated rapidly in Michoacan in recent years due to the opening of US markets to Mexican avocados (Barsimantov and Navia Antezana 2008). Thus, in an earlier period, demands for agricultural land by landless peoples drove tropical deforestation. More recently, demands for labor and avocados in the US has driven both forest recovery and new waves of deforestation, sometimes on very short time scales. Put another way, over the last several decades, there has been a transition from linear deforestation processes to more complex nonlinear patterns of deforestation, forest recovery, and forest maintenance in spite of proximate deforestation pressures.

A more detailed mapping of these hierarchical-panarchic dynamic interactions between forest cover and broader socio-economic processes could help build both theory and policy. A greater awareness of the short and long-term impacts of particular policies or processes could help create more sustainable relationships between forests and the local and global communities that depend on them. The current degree of uncertainty about basic facts of forest cover dynamics is becoming increasingly critical as the world begins to face the degree of the challenge represented by our current rate of carbon emissions and associated global warming. According to Stern (2007), it is currently estimated that global deforestation account for more than 18% of global greenhouse gas emissions, more than the entire transportation sector. Stern (2007) reviews "key messages" on deforestation that apply well to the situation in Mexico and Central America. Some of these messages are that

1. Curbing deforestation is a highly cost-effective way of reducing greenhouse gas emissions and has the potential to offer significant reductions fairly quickly.
2. Policies on deforestation should be shaped and led by the nation where the forests stand but there should be strong help from the international community, which benefits from their actions.
3. At a national level, establishing and enforcing clear property rights to forestland, and determining the rights and responsibilities of landowners, communities and loggers, is key to effective forest management. This should involve local communities, and take account of their interests and social structures, work with development goals and reinforce the process of protecting the forests.
4. Compensation from the international community should be provided and take account of the opportunity costs of alternative uses of the land, the costs of administering and enforcing protection, and managing the transition.
5. Action to preserve the remaining areas of natural forest is urgent. Large-scale pilot schemes are required to explore effective approaches to combining national action and international.

With respect to the last point, this review makes clear that actions to preserve remaining areas of natural forest are already occurring. Forest uses such as shade tree coffee, sustainable management of forests for timber by local communities, and protected areas are all being effective in the region in maintaining and protecting forests. In addition, the increasing amounts of secondary forest in many regions need to be valued and managed. International scientific bodies should be established that will establish protocols and provide financial assistance to standardize forest cover and forest use monitoring at the national and regional level, and encouraging bi or multinational management institutions for important cross-boundary forest masses where they exist. This are important steps towards the kind of large-scale but locally empowered management of landscape processes that can move us towards a more democratic and sustainable "earth system engineering and management" (Allenby 2005). Mexico and Central America and their connected forests and forest patches as a whole are not yet clearly moving towards the "Great Restoration" suggested by Victor and Ausubel (2000) but foundations are being laid, and visionary policymakers with good scientific support and wide grassroots participation could make it a reality in the first half of the twenty-first century.

**Acknowledgments** Thanks to Daniel Gann, Florida International University GIS Laboratory for help in image processing and to Danna Planas for research assistance.

# References

AFE-COHDEFOR (2002) Informe sobre la funcionalidad global e institucional del sector forestal. Adminstración forestal del Estado AFE-COHDEFOR, Tegucigalpa, Honduras

Allenby B (2005) Reconstructing earth: technology and environment in the age of humans. Island Press, Washington, DC

ANAM (2006a) Indicatores Ambientales de la República de Panama. 2006. Autoridad Nacional del Ambiente

ANAM (2006b) Informe: El Sistema Nacional de Áreas Protegidas. Autoridad Nacional del Ambiente: Dirección Nacional de Áreas Protegidas y Vida Silvestre

Arroyo-Mora J, Calvo J, Janzen D, Rivard B, Sanchez-Azofeifa G (2005) Dynamics in landscape structure and composition for the Chorotega region, Costa Rica from 1960 to 2000. Agric Ecosyst Environ 106:27–39

Bacon C (2005) Confronting the coffee crisis: Can Fair Trade, organic, and specialty coffees reduce small-scale farmer vulnerability in Northern Nicaragua? World Dev 33(3):497–511

Barsimantov J, Navia Antezana J (2008) Land Use and Land Tenure Change in Mexico's Avocado Production Region: Can Community Forestry Reduce Incentives to Deforest for High Value Crops?. Presented at "governing shared resources: connecting local experience to global challenges," 12th Biennial Conference of the International Association for the Study of Commons, Cheltenham, England, July 14–18, 2008

Bass JO (2004) More trees in the tropics. Area 36:19–32

Bass JO (2006) Forty years and more trees: land cover change and coffee production in Honduras. Southeast Geogr 46:51–65

Bartra A (1996) El México bárbaro: plantaciones y monterías del sureste durante el Porfiriato. El Atajo Ediciones, Mexico, DF

Blackman A, Ávalos-Sartorio B, Chow J (2006) Tree cover loss in El Salvador's shade coffee areas. Discussion Paper 07-32. Resources for the Future, Washington, DC

Blackman A, Albers HJ, Ávalos-Sartorio B, Crooks L (2007) Land cover in a managed forest ecosystem: Mexican shade coffee. Discussion Paper 07-30. Resources for the Future, Washington, DC

Bray DB, Ellis EA, Armijo N, Beck CT (2004) The institutional drivers of sustainable landscapes: a case study of the 'Mayan Zone' in Quintana Roo, Mexico. Land Use Policy 21:333–346

Bray DB, Klepeis P (2005) Deforestation, forest transitions, and institutions for sustainability in Southeastern Mexico, 1900–2000. Environ Hist 11:195–223

Bray DB, Anderson AB (2005) Global Conservation Non-Governmental Organizations & Local Communities: Perspectives on Programs and Project Implementation in Latin America. Working Paper No. 1. Conservation and Development Series. Institute for Sustainability Science in Latin America and the Caribbea, Florida International University, Miami, FL

Bray DB, Duran E, Anta S, Martin GJ, Mondragón F (2008) A new conservation and development frontier: community protected areas in Oaxaca, Mexico. Curr Conserv. 2(2):7–9

Bray DB, Duran E, Ramos VH, Mas J-F, Velázquez A, Barry D, Radachowsky J (2008) Tropical deforestation, community forests and protected areas in the Maya forest. Ecol Soc 13(2):56 [online] URL: http://www.ecologyandsociety.org/vol13/iss2/art56

Brockett CD, Gottfried RR (2002) State policies and the preservation of forest cover: lessons from contrasting public-policy regimes in Costa Rica. Lat Am Res Rev 37(1):7–40

Brower LP, Castilleja G, Peralta A, López-Garcia J, Bojorquez-Tapia L, Diaz S, Melgarejo D, Missrie M (2002) Quantitative changes in forest quality in a principal overwintering area of the monarch butterfly in Mexico, 1971–1999. Conserv Biol 16(2):346–359

Bryant D, Nielsen D, Tangley L (1997) The last frontier forests: ecosystems and economics on the edge. World Resources Institute, Washington, DC

Calo M, Wise TA (2005) Revaluing peasant coffee production: organic and fair trade markets in Mexico. Global Development and Environment Institute, Medford, MA

Castro KL, Sanchez-Afofeifa GA, Rivard B (2003) Monitoring secondary tropical forests using space-borne data: implications for Central America. Int J Rem Sens 24:1853–1894

Cayuela L, Echeverría C, Rey J (2006) Clearance and fragmentation of tropical montane forests in the Highlands of Chiapas, Mexico (1975–2000). For Ecol Manag 226:208–218

Chomitz KM, Gray DA (1996) Roads, land use, and deforestation: a spatial model applied to Belize. World Bank Econ Rev 10(3):487–512

Chowdhury RR, Schneider C (2004) Land cover and land use: classification and change analysis. In: Turner BL II, Geoghegan J, Foster DR (eds) Integrated land-change science and tropical deforestation in the Southern Yucatán. Oxford University Press, New York

Consejo Nacional de Areas Protegidas/Wildlife Conservation Society (CONAP/WCS) (2004) Deforestation estimation in the Maya Biosphere Reserve, 2003–2004. Report. U.S. Agency for International Development. Washington, DC

Cortes A, Carrasco L (2004) Positiva recuperación de bosques. La Prensa. October 3

de Jong BHJ, Ochoa-Gaona S, Castillo-Santiago MA, Ramírez-Marcial N, Cairns MA (2000) Carbon flux and patterns of land-use/land-cover change in the Selva Lacandona, Mexico. Ambio 29:504–511

Duran E, Mas J-F, Velázquez A (2005) Land-use cover change in community-based forest management regions and protected areas in Mexico. In: Bray DB, Merino-Pérez L, Barry D (eds) The community forests of Mexico. University of Texas Press, Austin, TX

Dushku A, Brown S, Hall M (2002) Modeling the deforestation and carbon emissions baseline in the Rio Bravo Conservation and Management Area Climate Action Project, Belize, 1993–2035. Cooperative Agreement between Winrock International and the EPA ID# CR 827293-01-0. Product 5. Report of the application of GEOMOD to the Rio Bravo, Belize, project area. Winrock International, Arlington, VA

Emch M, Quinn JW, Peterson M, Alexander M (2005) Forest cover change in the Toledo District, Belize from 1975 to 1999: a remote sensing approach Prof Geogr 57(2):256–267

Food and Agriculture Organization (2005) State of the World's Forests. FAO, Rome

Forrestel A, Peay KG (2006) Deforestation in a complex landscape: La Amistad Biosphere Reserve. J Sustain For 22(1/2):49–71

Geist HJ, Lambin EF (2001) What drives tropical deforestation? a meta-analysis of proximate and underlying causes of deforestation based on subnational case study evidence. LUCC International Project Office, Louvain

Gómez-Mendoza L, Vega Penya E, Isabel Ramírez M, Palacio-Prieto JL, Galicia L (2006) Projecting land-use change processes in the Sierra Norte of Oaxaca, Mexico. Appl Geogr 26:276–290
Grandia L (2006) Unsettling: land dispossession and enduring inequity for the Q'eqchi' Maya in the Guatemalan and Belizean frontier colonization process. Ph.D. thesis. Department of Anthropology, University of California-Berkeley, Berkeley, CA
Hayes DJ, Sader SA, Schwartz NB (2002) Analyzing a forest conversion history database to explore the spatial and temporal characteristics of land cover change in Guatemala's Maya Biosphere Reserve. Landsc Ecol 17:299–314
Hayes TM (2007) Does tenure matter? A comparative analysis of agricultural expansion in the Mosquitia forest corridor. Hum Ecol 35:733–747
Hecht SB, Kandel S, Cuellar IG, Rosa H (2006) Globalization, forest resurgence, and environmental politics in El Salvador. World Dev 34:308–323
Hecht SB, Saatchi SS (2007) Globalization and forest resurgence: changes in forest cover in El Salvador. BioScience 57(8):663–672
Holder CC (2004) Changes in structure and cover of a common property pine forest in Guatemala 1954–1996. Environ Conserv 31(1):22–29
Humphries S (1998) Milk cows, migrants, and land markets: unraveling the complexities of forest-to-pasture conversion in Northern Honduras. Econ Dev Cult Change 47(1):95–124
Janzen DH (1988) Tropical dry forests: the most endangered major tropical ecosystem. In: Wilson EO (ed) Biodiversity. National Academic Press, Washington, DC
Johnson KA, Nelson KC (2004) Common property and conservation: the potential for effective communal forest management within a national park in Mexico. Hum Ecol 32(6):703–733
Joyce A, Sader S (1988) Deforestation rates and trends in Costa Rica, 1940 to 1983. Biotropica 20(1):11–19
Kaimowitz D (1996) Livestock and deforestation, Central America in the 1980s and 1990s: a policy perspective. Center for International Forestry Research, Bogor, Indonesia
Kaiser J (2001) Bold corridor project confronts political reality. Science 293:2196–2199
Kammerbauer J, Cordoba B, Escolán R, Flores S, Ramirez V, Zeledón J (2001) Identification of development indicators in tropical mountainous regions and some implications for natural resource policy designs: an integrated community case study. Ecol Econ 36:45–60
Kammerbauer J, Ardon C (1999) Land use dynamics and landscape change pattern in a typical watershed in the hillside region of central Honduras. Agric Ecosyst Environ 75(1–2):93–100
Kappelle M, Juarez ME (1994) The Los Santos forest reserve: a buffer zone vital for the Costa Rican La Amistad Biosphere Reserve. Environ Conserv 21(2):166–169
Kleinn C, Corrales L, Morales D (2004) Forest area in Costa Rica: a comparative study of tropical forest cover estimates over time. Environ Monit Assess 73:1
Klooster D (2003) Forest transitions in Mexico: institutions and forests in a globalized countryside. Prof Geogr 55(2):227–237
Latifovic R, Zhu Z-L, Cihlar J, Giri C (2002) Land cover of North America 2000. Natural Resources Canada, Canada Center for Remote Sensing, US Geological Service EROS Data Center
Mathews AS (2003) Supressing Fire and Memory: Environmental Degradation and Political Restoration in the Sierra Juarez of Oaxaca, 1887–2001. Env Hist 8(1):4–25
Mather A (1992) The forest transition. Area 24:367–379
Mendoza E, Dirzo R (1999) Deforestation in Lacandonia (southeast Mexico): evidence for the declaration of the northernmost tropical hot-spot. Biodivers Conserv 8:1621–1641
Moguel P, Toledo V (1999) Biodiversity conservation in traditional coffee systems of Mexico. Conserv Biol 13(1):1–21
Munroe DK, Southworth J, Tucker CM (2005) The dynamics of land-cover change in western Honduras: exploring spatial and temporal complexity. Agric Econ 27(3):355–369
Nagendra H, Southworth J, Tucker CM (2003) Accessibility as a determinant of landscape transformation in western Honduras: linking pattern and process. Landsc Ecol 18:141–158
National Geographic Society/Cultural Survival (2002) Map: the coexistence of indigenous peoples and the natural environment, National Geographic Society: Washington, DC

National Geographic Society/Center for the Support of Native Lands (2002) Map: indigenous peoples and natural ecosystems in Central America and Southern Mexico. National Geographic Society, Washington, DC
Ochoa-Gaona S, González-Espinosa M (2000) Land use and deforestation in the highlands of Chiapas, Mexico. Appl Geogr 20:17–42
Perfecto IR, Rice A, Greenberg R, Van der Voort ME (1996) Shade coffee: a disappearing refuge for biodiversity. BioScience 46(8):598–608
Perz S (2007) Grand theory and context-specificity in the study of forest dynamics: forest transition theory and other directions. Prof Geogr 59(1):105–114
Perz SG, Skole DL (2003) Secondary forest expansion in the Brazilian Amazon and the refinement of forest transition theory. Soc Nat Resour 16(1):277–294
Pfaff ASP, Sanchez-Azofeifa GA (2004) Deforestation pressure and biological reserve planning: a conceptual approach and an illustrative application for Costa Rica. Resour Energy Econ 26:237–254
Pratt L, Quijandría G (1997) Sectoral forestal en Honduras: análisis de sostenibilidad. INCAE, Costa Rica
Primack RB, Bray D, Galletti HA, Ponciano I (1998) Timber, tourists, and temples: conservation and development in the Maya Forest of Belize, Guatemala, and Mexico. Island Press, Washington, DC
Rudel TK (2005) Tropical forests: regional paths of destruction and regeneration in the late twentieth century. Columbia University Press, New York
Rudel TK, Coomes OT, Moran E, Achard F, Angelsen A, Xu J, Lambin E (2005) Forest transitions: towards a global understanding of land use change. Glob Environ Change 15:23–31
Sanchez-Azofeifa GA, Harriss RC, Skole DL (2001) Deforestation in Costa Rica: a quantitative analysis using remote sensing imagery. Biotropica 33(3):378–384
Sanchez-Azofeifa GA, Rivard B, Calvo J, Moorthy I (2002) Dynamics of tropical deforestation around national parks. Mt Res Dev 22(4):352–358
Sanchez-Azofeifa GA, Daily GC, Pfaff ASP, Busch C (2003) Integrity and isolation of Costa Rica's national parks and biological reserves: examining the dynamics of land-cover change. Biol Conserv 109:123–135
Schelhas J, Sanchez-Azofeifa GA (2006) Post-Frontier forest change adjacent to Braulio Carillo National Park, Costa Rica. Hum Ecol 34(3):407–420
Southworth J, Tucker C (2001) The influence of accessibility, local institutions, and socioeconomic factors on forest cover change in the mountains of Western Honduras. Mt Res Dev 21(3):276–283
Southworth J, Nagendra H, Carlson L, Tucker C (2004a) Assessing the impact of Celaque National Park on forest fragmentation in western Honduras. Appl Geogr 24:303–322
Southworth J, Munroe D, Nagendra H (2004b) Land cover change and landscape fragmentation – comparing the utility of continuous and discrete analyses for a western Honduras region. Agric Ecosyst Environ 101:185–205
Stern N (2007) Economics of climate change: the Stern review. Cambridge University Press, Cambridge
Trejo I, Dirzo R (2000) Deforestation of seasonally dry tropical forest a national and local analysis in Mexico. Biol Conserv 94(1):133–142
Tucker CM (1999) Private versus common property forests: forest conditions and tenure in a Honduran community. Hum Ecol 27:201–230
Tucker C, Munroe D, Nagendra H, Southworth J (2005) Comparative spatial analyses of forest conservation and change in Honduras and Guatemala. Conserv Soc 3(1):174–200
Tucker RP (2000) Insatiable appetite: the United States and the ecological degradation of the third world. University of California Press, Berkeley, CA
UVG (2006) Dinámica de la cobertura forestal de Guatemala durante los años 1991,1996 y 2001 y mapa de cobertura forestal 2001. *Fase II:Dinámica de la cobertura forestal*. Universidad del Valle de Guatemala, Instituto Nacional de Bosques, Consejo Nacional de Áreas Protegidas: Guatemala, Guatemala
Varangis P, Siegel P, Giovannucci D, Lewin B (2003) Dealing with the coffee crisis in Central America: impacts and strategies. Policy Research Working Paper 2993. Development Research Group, Rural Development. World Bank, Washington, DC

Veblen TT (1978) Forest preservation in the western Highlands of Guatemala. Geogr Rev 68(4):417–434

Velázquez A, Duran E, Ramírez I, Mas J-F, Bocco G, Ramírez G, Palacio JL (2003) Land use-cover change processes in highly biodiverse areas: the case of Oaxaca, Mexico. Glob Environ Change 13:175–184

Velázquez A, Durán E, Mas J-F, Bray D, Bocco G (2005) Situación Actual y perspectiva del cambio de la cubierta vegetal y usos de suelo en México. Ed. Elena Zúñiga Herrera. México, antes los desafíos de desarrollo del milenio. Consejo Nacional de Población, Mexico City, Mexico

Victor DG, Ausubel JH (2000) Restoring the forests. Foreign Aff 79(6):127–144

Vreugdenhil D, Meerman J, Meyrat A, Gómez LD, Graham DJ (2002) Map of the ecosystems of Central America: final report. World Bank, Washington, DC

# Chapter 6
# Reforestation in Central and Eastern Europe After the Breakdown of Socialism

**Gregory N. Taff, Daniel Müller, Tobias Kuemmerle, Esra Ozdeneral, and Stephen J. Walsh**

G.N. Taff (✉) and E. Ozdeneral
Department of Earth Sciences, University of Memphis, TN, USA
e-mail: gntaff@memphis.edu

D. Müller
Leibniz Institute of Agricultural Development in Central and Eastern Europe (IAMO), Halle (Saale), Germany and Department of Geography, Humboldt-University in Berlin, Berlin, Germany

T. Kuemmerle
Department of Geography, Humboldt-University in Berlin, Berlin, Germany and Department of Forest and Wildlife Ecology, University of Wisconsin-Madison, WI, USA

S.J. Walsh
Department of Geography & Carolina Population Center, University of North Carolina, NC, USA

H. Nagendra and J. Southworth (eds.), *Reforesting Landscapes: Linking Pattern and Process*, Landscape Series 10,
DOI 10.1007/978-1-4020-9656-3_6, © Springer Science+Business Media B.V. 2010

# 6.1 Overview

## *6.1.1 Introduction*

Eastern Europe experienced a period of rapid and radical changes of its political, institutional, demographic, and socioeconomic structures after the fall of the Iron Curtain in 1989 and the breakdown of the Soviet Union in 1991. These events triggered widespread land use change, most notably the abandonment of vast areas of cropland. Some of this land has experienced secondary forest succession. While this trend is acknowledged at a macro-level, the rates and spatial patterns at which it is occurring locally and regionally remain poorly understood. Little is known about differences in reforestation trends and the pattern-process relations of land use/land cover change among different regions in Eastern Europe. Furthermore, the environmental and societal causes and consequences of reforestation have not been widely analyzed.

The goals of this chapter are to provide an overview of reforestation trends, including the drivers of reforestation, in Eastern Europe after the breakdown of socialism. Because recent forest cover trends cannot be fully understood without considering land use legacies from pre-socialist and socialist times, we begin with a brief review of forest use and reforestation trends through the nineteenth and twentieth centuries. We then assess reforestation in Eastern Europe by comparing national-level forest resource data, and by discussing three detailed case studies that are illustrative of the transformations seen throughout Eastern Europe. These case studies, drawn from research conducted in Latvia, Romania, and Albania, span Eastern Europe from North to South (Fig. 6.1), and represent geographic places in differing stages of economic development. The three cases thus illustrate the variety of environmental, institutional, economic, and demographic factors that have shaped the rates and spatial patterns of post-socialist reforestation in the region.

In this chapter we define reforestation as a change in land cover from non-forest cover to forest cover, including forest expansion due to natural succession and forest planting. Forest land includes forest stands and disturbed areas (e.g., forest clear-cuts expected to grow back), whereas forest cover refers to forested stands only. The term Eastern Europe refers to all former socialist countries and Soviet Republics in Central and Eastern Europe, including European Russia.

## *6.1.2 Historic Forest Cover Trends*

According to Zerbe and Brande (2003), forest cover in Central Europe reached a minimum throughout the region near the end of the eighteenth century. By this point, much of Central Europe's forests had been transformed to open landscapes. This stage in Central Europe conforms to the second stage in the *forest transition* theory (Mather 1992). According to this theory, also called the forest transition phenomenon, there is generally an initial period (stage 1) of relative stability of

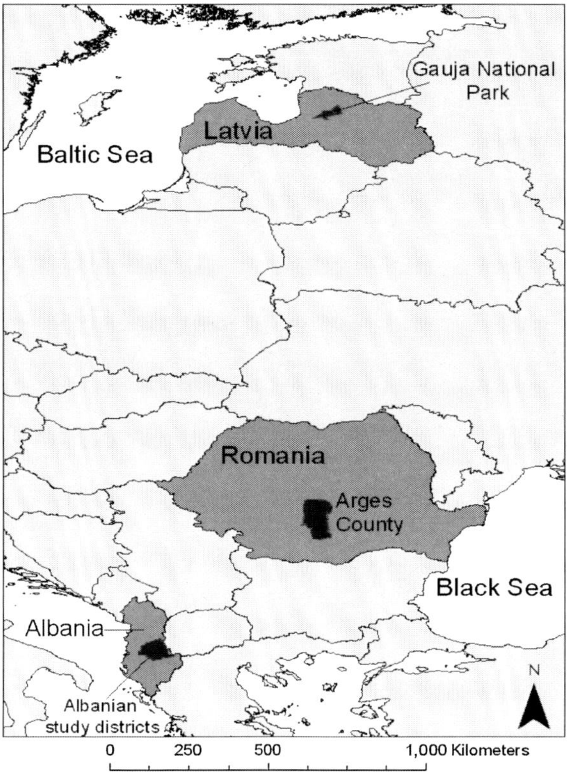


**Fig. 6.1** The three study sites within Eastern Europe

forest cover in a particular region, followed by a drastic decrease in forest cover (stage 2) associated with economic development (late eighteenth century in Central Europe), and finally, an increase in forest cover (stage 3) associated with an even higher level of economic development (Mather 1992, Rudel 1998, Rudel et al. 2005). The underlying processes leading to the forest transition may be referred to as the 'economic development path', coined by Rudel et al. (2005). In the nineteenth and twentieth centuries, forests recovered in many areas worldwide due to urbanization, industrialization, and land use extensification (Baldock et al. 1996; MacDonald et al. 2000; Kauppi et al. 2006). This third stage of the forest transition occurred in several areas in Eastern Europe during the late nineteenth and early twentieth centuries, often through planting of coniferous forest (*Pinus sylvestris* in lowlands and *Picea abies* in mountainous regions, Zerbe and Brande 2003).

Available statistics that describe forest cover trends in Eastern Europe during the period of socialism are often of unknown quality. Satellite-derived image products for characterizing land use/land cover change are generally only available beginning in the early 1970s through the Landsat program. However, major trends in land use during this era are relatively well-understood. During this period, natural resources, including forests, were seen as an engine of economic growth in all socialist countries, particularly in the former Soviet Union. This frequently resulted in unsustainable forest practices (Pryde 1991; Peterson 1995; Oldfield 2000). In addition, great efforts were undertaken to industrialize and collectivize the agricultural sector throughout the region, and agricultural lands were greatly expanded into forests, even into marginal areas (Turnock 2002). Due to these practices, the third stage of the forest transition was delayed across much of Eastern Europe during the socialist period, and forest cover increased only in a few areas, mostly in remote settings and along the borderlands of Western Europe (Augustyn 2004). Few old growth forests were left untouched, and there was a substantial shift in the age distribution of forests towards younger stands during this period. Zerbe and Brande (2003) reports that soil eutrophication occurred throughout Central Europe since the early 1980s, promoting quicker forest growth in areas where regrowth did occur. The increase in soil nutrients also stimulated forest regrowth rates in the post-socialist era. Zerbe finds that the eutrophication occurred throughout the region with differential rates, depending on liming and fertilizer application, as well as differential rates of atmospheric nitrogen deposition.

### 6.1.3 Forest Cover Trends in the Post-socialist Period

The transition from the command-driven to the market economy dramatically altered land use and the agricultural conditions of the region. During the transition, prices were liberalized, fertilizer and fuel prices increased dramatically, most communal farms were dissolved, and privatization and land reforms often led to unclear land tenure. Farmers faced new global competition (including from heavily subsidized agriculture, particularly in Western Europe) and the Council for Mutual Economic Assistance (COMECON) disappeared. These factors led to a sudden and substantial decrease in the economic attractiveness of farming as compared to other household livelihood strategies. As a consequence, there were unprecedented rates of farmland abandonment that was coupled with out-migration from rural to urban areas, and the human impacts on the region's rural landscapes substantially decreased (Bicik et al. 2001; Ioffe and Nefedova 2004; DLG 2005). In particular, this population movement from rural to urban places commonly led to natural forest succession processes on abandoned farmland.

Forest management changed drastically during the post-socialist transition. Though each country chose different means to manage their forests, an increase in forest timber extraction, including substantial illegal logging, became common throughout the region. The illegal logging was particularly evident during the early transition years when poverty was at its peak and institutional oversight of forests was at its weakest. Documented cases of such illegal logging include Latvia and

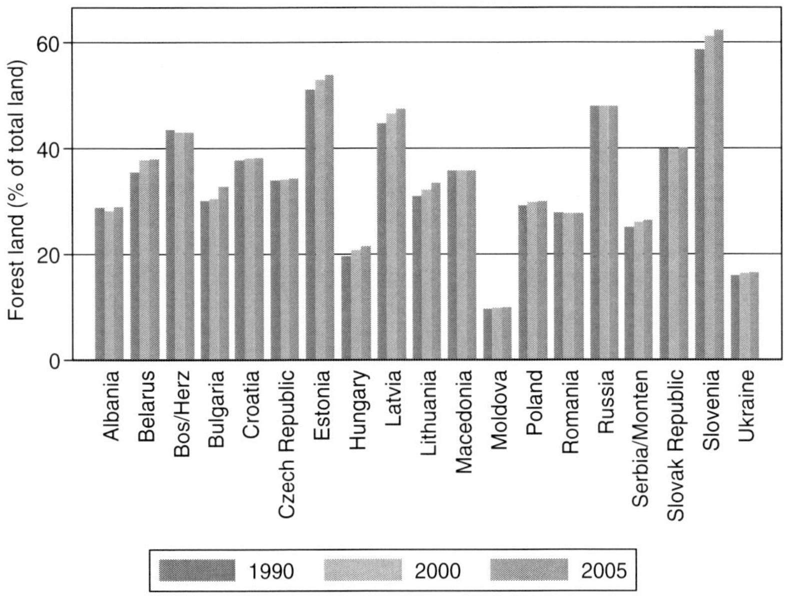


**Fig. 6.2** Percent forested land in East European countries in years 1990, 2000, 2005 (*Source*: FAO 2006)

Romania (this chapter), Bulgaria (Staddon 2001), and particularly high illegal logging rates in Albania (this chapter, and Bouriaud 2005), Armenia (Schmithüsen et al. 2001), Ukraine (Nijnik and van Kooten 2000), and Estonia and Slovenia (Bouriaud 2005). In addition, during this period, forest planting virtually stopped in many parts of Eastern Europe (Buksha et al. 2003). In spite of the increased timber extraction and decreased forest planting during the early transition period, almost all Eastern European countries gained forest cover between 1990 and 2005, with exceptions in Bosnia/Herzegovina, Macedonia, Romania, and Russia, where forest cover remained virtually stable (Fig. 6.2). The observed overall increase in forest throughout most of this region was mostly due to reforestation on agricultural lands, primarily as natural forest succession.

Major differences in reforestation patterns and other land use/land cover changes have been occurring, however, between the countries of Eastern Europe since the demise of socialism. Much of these differences depend on the method of land privatization that the country chose to pursue (Swinnen 2001). In addition, a positive association is found between the increase in wealth of the countries in the region and the increase in forest cover. A significant positive correlation (correlation coefficient=0.5, significant at the 5% level) exists between changes in GDP per capita and changes in percent forest cover among all the countries of Eastern Europe in the period of 1990–2005 (Fig. 6.3). This association is present among the three case studies explored in this chapter. Albania is one of the very poorest countries

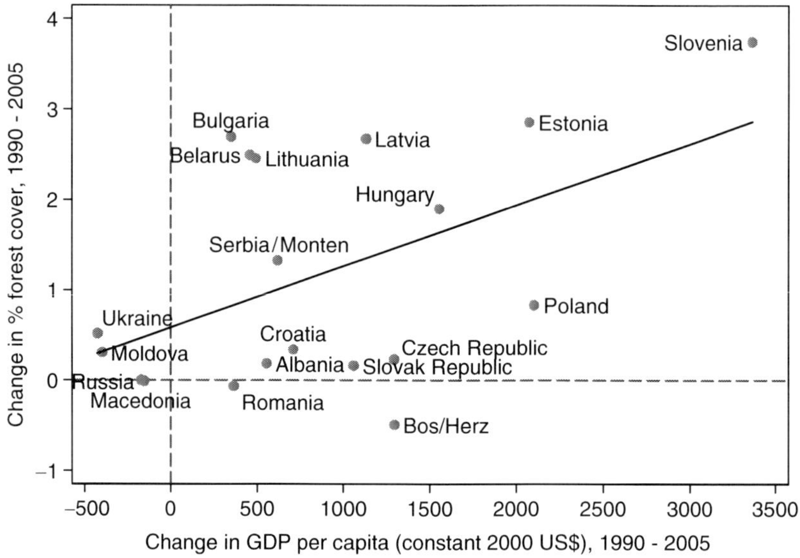


**Fig. 6.3** Plot of change in GDP per capita vs. change in percent forest cover between 1990 and 2005 for Eastern European countries
(*Source*: World Bank 2007, *Note*: the forest cover data used here by the World Bank was taken from the FAO [2006])

in Eastern Europe, and since the collapse of socialism has experienced more economic pressure to make use of timber resources than other Eastern European countries. Latvia, one of the wealthiest countries within Eastern Europe since the collapse of socialism, has experienced forest cover increase. Romanian's economic situation since the collapse of socialism has been between the levels of Albania and Latvia, and Romania has shown no significant change in forest cover during this period. While there is ample evidence that many factors affect the amount of forest change in each country, the general economic condition of the country may be a key factor.

To date there has been very little research on reforestation in Eastern Europe, yet there are important implications of reforestation in this region. Benefits of reforestation particular to this region begin with a list provided by the former Lithuanian Minister of Environment, Arnas Kundrotas: increased forest cover improves air quality and groundwater resources, prevents pollution runoff from industrial and agricultural sources into water bodies, reduces soil erosion, and provides for recreational opportunities (Kundrotas 2002). Moreover, forest expansion increases carbon sequestration in forests and offers countries international negotiation power in terms of carbon-releasing industrial development. In addition to these functions, forests and reforestation can serve societies in this region through promoting cultural values, amenity values, spiritual values, aesthetics, and wood products (Koch and Skovsgaard 1999), as well as providing income for rural livelihoods (Chazdon 2008). Also, many non-timber goods are often extracted from forests in Eastern Europe, such as mushrooms, berries, herbs, tree resin, and peat, and forests offer hunting and tourism opportunities.

In the post-socialist period, however, there has been concern over the loss of cultural landscapes and the accompanying biota due to reforestation on abandoned agricultural fields and meadows throughout the region (Baldock 1999; The World Conservation Union 2004; Kobler et al. 2005; Taff 2005; Baur et al. 2006; Kozak et al. 2007b; Kuemmerle et al. 2009). While definitions of cultural landscapes differ by region and are often subjective, a common thread among most definitions of the cultural landscapes throughout Eastern Europe is the existence of a fine-grained matrix of small agricultural fields and wooded lands (Bunkše 2000; Angelstam et al. 2003, Herzon and O'Hara 2007). If national reforestation policies do not take the protection of cultural landscapes into consideration, extensive reforestation on abandoned agricultural lands will lead to losses or degradation of this matrix in Eastern Europe (Baldock 1999, The World Conservation Union 2004). The perceived threat of reforestation in the region applies not only to the persistence of cultural landscapes but also to the cultural practices associated with them; examples are documented in Latvia (Bunkše 2000; Schwartz 2001; Taff 2005), Poland (Angelstam et al. 2003; Kozak et al. 2007b), the Czech Republic (Lipsky et al. 1999) and Slovenia (Kobler et al. 2005).

Reforestation on abandoned agricultural fields and meadows can result in the loss of some biodiversity; this loss is primarily due to the disappearance of meadow habitats and edge habitats common in cultural landscapes (Watkins 1993; Herzon and O'Hara 2007; Kozak et al. 2007a; Pil ts 2007). Such cases of biodiversity loss have been reported from Lithuania (Lazdinis et al. 2005), Latvia (Taff 2005), Northeast Germany (Dabbert 1995), Poland (Angelstam et al. 2003), Romania (Cremene et al. 2005; Baur et al. 2006) and Ukraine (Elbakidze and Angelstam 2007). The IUCN recommends conserving non-forest areas of high ecological value which are part of traditional landscapes, particularly through increasing incentives for landowners to keep up traditional land uses (The World Conservation Union 2004).

### *6.1.4 National Level Forest Resource Statistics*

UNECE & FAO (2000) compiled forest area change statistics for all Eastern European countries before and after the system change. Different 'before' and 'after' reference dates were used in this dataset, depending on data availability for each country, with 'before' reference dates ranging from 1957 to 1990, and 'after' reference dates ranging from 1990 to 1998. Based on these data, all Eastern European countries were found to experience an increase in forest area during this time period, except for Albania (with an average annual forest area decrease of 7,800 ha between 1957 and 1995) and the former Yugoslavia (with an average annual forest area decrease of 1,450 ha between 1979 and 1995). In a separate analysis by the FAO (2006), the percentages of forest cover in 1990, 2000, and

2005 were compiled (see Fig. 6.2) for the Eastern European countries; this analysis uses the FAO definition of forest, which includes areas with canopy cover greater than 10% and tree height of at least 5 m, including areas currently undergoing reforestation or expected to regenerate. There has, however, been considerable concern about the reliability of official forest resource statistics, especially from the socialist period. Therefore, comparing these data across time and between countries is challenging. Moreover, forest resource data often do not consider forest degradation and illegal logging, and rarely account for reforestation on abandoned farmland. It is for these reasons that in-depth case studies, such as the ones in this chapter that use remote sensing to map changes in land cover and forest patterns, in addition to using qualitative analyses, are crucial to understanding forest trends and reforestation processes in this region.

## 6.2 Case Studies

### *6.2.1 Case Study I: Reforestation and Forest Cover Change in Latvia's Gauja National Park*

#### 6.2.1.1 Study Site and National Historical Context

In 2005, forests covered 47.4% of the land area of Latvia, compared to 44.7% in 1990 (FAO 2006). These statistics include cleared forest lands expected to undergo reforestation, so the amounts of standing forest cover are less due to recent forest clear-cuts. The amount of land area affected by forestry activities each year (which includes both forest clear-cut areas and selective harvesting) was in the range of 1,540 $km^2$ (~ 5% of Latvia's forest area) to 2,320 $km^2$ (~ 8% of Latvia's forest area) in the period of 1995–2000 (Central Statistical Bureau of Latvia 2003).

National parks were first created in the Soviet Union in the 1970s. Nature reserves (zapovedniki), however, preceded national parks in the Soviet Union, and were created as early as 1916 in Russia exclusively to preserve nature. Few people (only scientists and park rangers) were allowed (and are currently allowed) to enter these reserves. Soviet national parks were later created as multi-purpose parks, allowing for recreational activities, limited economic activities, and living (dwelling) space. Gauja National Park (GNP) is Latvia's oldest and largest national park (917.5 $km^2$, or 1.5% of Latvia's land area), established in 1973 as the second Soviet national park, and is located in north central Latvia (see Fig. 6.1). GNP contains forests, agricultural fields, meadows, wetlands, water bodies, cultural-historical sites, villages, and rural homes. Approximately 870 plant, 166 bird, and 52 mammal species inhabit the Park (Pilāts 2007). Since the fall of the Soviet Union, land restitution was implemented in Latvia to give land back to former owners (as of 1940) and their descendents. Eighty percent of GNP has been privatized since 1990 through post-socialist land restitution, though Park-related land use restrictions have accompanied this privatization (Gauja National Park Management 1998; Strautnieks 2002).

#### 6.2.1.2 Post-socialist Forest Change Patterns and Processes

Land cover was assessed in GNP between the late Soviet era and 2002. Four Landsat Thematic Mapper (TM) satellite images from the summers of 1985, 1994, 1999, and 2002 were classified and post-classification change analyses were performed. The Landsat images were each georectified and then classified using a hybrid supervised/unsupervised approach (Jensen 2005). In the summer of 2001, the geographic locations for two separate sets of ground control sites for each of 16 landcover types were documented in the field for the purposes of training the supervised portion of the classification and for accuracy assessment. Orthophotos from 1997 were used in conjunction with the 2001 field data to verify landcover classes for classification training and accuracy assessment of the 1999 image. The accuracy assessment for the 1999 land cover map yielded an overall accuracy of nearly 96%, with a Kappa statistic of 0.94. Kappa statistics over 0.8 represent strong agreement according to Congalton (2004). To classify the other three image dates, training data were selected by determining landcover through comparison to the 1999 image via visual interpretation and spectral profile analysis. Just three of the land cover classes are considered here: forest, shrubs, and cropland/grassland.

Between 1985 and 2002, overall forest cover in GNP increased by 3% or 13.6 km$^2$ (Fig. 6.4). The timing of forest losses and gains in GNP tell interesting stories about the causes of reforestation and land cover change. In the first period of analysis (1985–1994), GNP experienced an overall loss of 22 km$^2$ of forest. This period straddles Latvian independence (1991). During Soviet times, very little forest cutting, and practically no clear-cutting, took place in the Park (M. Sēstulis 2002, personal communication, O. Nikodēmus 2003, personal communication). Therefore, most of the forest loss seen during this period is assumed to have taken place between 1991 and 1994. Although land restitution in Latvia began in 1990, only about 3% of land parcels were restituted by 1994 (Balsevics 2004), so the reduction in forest area is not likely the result of land cover modifications induced by returnees to their land. The loss of forest is likely due to cutting by profiteers taking advantage of a chaotic legal system during the early independence years, including criminal illegal cutting. In addition, the GNP Administration exerted (and continues to exert) managed forest cuts on state lands in the Park (the GNP Administration has been forced to fund its budget in large part through these means since Latvian independence due to lack of full financial support from the state) (Gauja National Park 2001).

Between 1994 and 1999, GNP experienced a forest area increase of 45.8 km$^2$ (Fig. 6.4). It is evident from satellite image post-classification change analyses that much of this increase in forest area was due to regrowth of forests cut in the previous period (just after independence). Much of the increase in forest area also comes

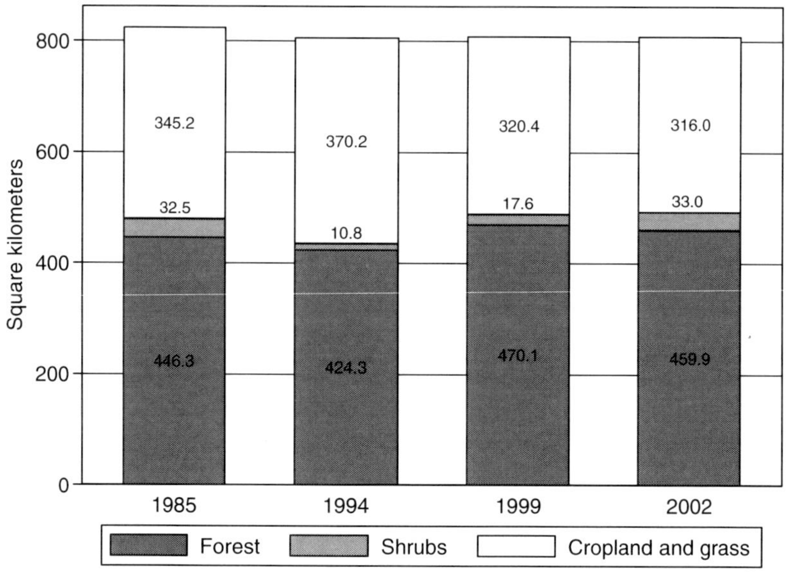


**Fig. 6.4** Changes in land cover type in GNP, Latvia, 1985–2002
(*Source*: Authors)

from natural succession on abandoned agricultural fields. Abandonment of agriculture began in the very early 1990s in response to the severe economic depression and the breakup of the state and communal farms at this time. During this time, Latvia's Gross National Product per capita fell from 3,909 USD in 1990 to 2,135 USD in 1993 (calculated in 2000 US dollars) (Central Statistical Bureau of Latvia 2002). Furthermore, as new nature protection policies were being developed and implemented throughout Eastern Europe during this period, a new set of laws for the protection and management of GNP was set forth by the Ministry of Environmental Protection and Regional Development in 1994, replacing former Soviet era laws. The new laws reflected policies that were more appropriate for the mixed ownership land that ensued in the Park after independence, and were focused on preserving core areas of the Park (Petersen 1999). The core areas, most of which are forested, include 40% of the land area of the Park, and are a mix of privately and publicly owned lands; the increased protection of these lands led to overall forest area increase within the Park. The new policies reflect Latvia's commitments to international environmental treaties and conventions, and European Union (EU) Directives (Latvia became an EU member in 2004).

Between 1999 and 2002, GNP experienced an overall forest loss of 10.2 $km^2$ (Fig. 6.4). The bulk of the land restitution process in Latvia took place in the late 1990s. Although land restitution in Latvia began in 1990 and was completed (except for a small number of cases) in 1999, only 40% of land restitution claims had been resolved by January of 1997 (Freedom House 1998; Balsevics 2004). Since 60% of Latvian restitution claims were settled between 1997 and 1999, the majority of the loss in forest between 1999 and 2002 is likely due to cutting activities of recent returnees to their land. Some forest loss can also be traced back to the GNP Administration's continued managed forest cuts on state lands to support the Administration's budget.

A key service provided by the protection of forests is to protect biodiversity. Lazdinis et al. (2005) note that the spatial pattern of forests in Lithuania, in addition to the sheer quantity of reforestation, is important from a biodiversity perspective. They use quantitative landscape indices to measure the effects of forest patterns on habitats, particularly for rare birds. Similarly, we measured the spatial patterns of forest fragmentation in GNP to assess changes in potential habitat for two rare indicator species that are sensitive to forest fragmentation. The Black Stork (*Ciconia nigre*) and the Lesser Spotted Eagle (*Aquila pomarina*) prefer core forest area, generally nesting more than 250 m from the forest edge (the Lesser Spotted Eagle prefers to nest at least 500 m, yet less than 1 km, from the forest edge) (Auninš 2002, Pilāts 2002). The FRAGSTATS software (McGarigal and Marks 1995) was used to assess the total area in GNP satisfying these nesting criteria (labeled as "core forest area") for each image date. Analyses showed that the amount of core forest area in GNP was 2.7, 1.6, 2.1, and 3.4 $km^2$, in 1985, 1994, 1999, and 2002, respectively. The aforementioned intense forest-cutting that took place soon after independence was a likely cause of the reduction in core forest area between 1985 and 1994 (1.1 $km^2$). The drastic increase in core forest area in the latter two periods, between 1994 and 2002, totaled 1.8 $km^2$. This shows that an Administration policy focus on preserving core zones of the Park not only increased the total number of hectares of forest but also more than doubled the amount of remote forest areas important for maintaining key species habitats.

### 6.2.2 Case Study II: Reforestation Potential and Forest Cover Change in Argeş County, Romania

#### 6.2.2.1 Study Site and National Historical Context

The Carpathian Mountains are Europe's largest mountain range and most extensive continuous temperate forest ecosystem. Centuries of land use history in the Carpathians have created today's typical land mosaics of farmland, forests, and villages (Baur et al. 2006; Elbakidze and Angelstam 2007). Yet the region has remained relatively undisturbed compared to Western Europe, and is a biodiversity hotspot that harbors many rare and endangered species as well as large populations of several flagship species (Webster et al. 2001; Ioras 2003).

Since the early twentieth century, reforestation has resulted in increasing forest cover in the Carpathians (Kozak et al. 2007b; see also Chapter 11). However, the transition from command to market-oriented economies after 1989 brought about substantial land use changes, including forest expansion on abandoned cropland (Hrivnak and Ujhazy 2005, Kuemmerle et al. 2008) as well as increased, sometimes illegal, logging (Nijnik and Van Kooten 2000; Kuemmerle et al. 2007).

The effects of these two land use processes on net forest cover change in the Carpathians remain poorly understood.

Romania encompasses the majority of the Carpathian Mountains within its boundaries, including almost the entire southern Carpathians. Moreover, more than 92% of Romania's forests (~ 5.7 million ha) are found in the mountains and foothills of the Carpathians (Ioras and Abrudan 2006). These forests occur in three zones of potential natural vegetation: a montane zone (1,500–2,200 m above sea level) with coniferous forest (e.g., mountain pine, Norway spruce, and silver fir); a foothill zone (250–1,500 m) with mixed and broadleaved forest (e.g., European beech, silver fir, and hornbeam); and the plains (< 250 m) dominated by oak forest (Enescu 1996, Mihai et al. 2007). Forestry has traditionally played an important role for Romania's rural economy (Ioras and Abrudan 2006). During socialism, all Romanian forests were nationalized and managed by the state. Forests were exploited at high, sometimes unsustainable, rates, resulting in progressively younger and thinner stands (Turnock 2002). After the fall of the Iron Curtain, forests were restituted in three phases (restitution laws in 1991, 2000, and 2005), targeting about 65% of Romania's forests in total (Ioras and Abrudan 2006). These ownership changes and the economic hardships of the transition period raised considerable concern about increased forest harvesting and illegal logging in the region (Nijnik and Van Kooten 2000; Turnock 2002; Strimbu et al. 2005).

Romania also harbors vast areas of farmland. In the Carpathians, animal husbandry and small-scale crop farming are important agricultural activities in the mountain valleys, while large-scale crop farming dominates the plains. Whereas agricultural production had been expanded during socialism, after 1989 there was a substantial decline in agriculture due to both its diminishing profitability (Turnock 1998; Trzeciak-Duval 1999) and migration from rural to urban areas (Ioffe and Nefedova 2001). Moreover, post-socialist land reforms resulted in a restructuring of Romania's agricultural sector and land use patterns. All farmland was either restituted or distributed among the former workers of the cooperatives (Parliament of Romania 2000), but ownership rights were often unclear. These processes resulted in widespread farmland abandonment (DLG 2005; Baur et al. 2006; Kuemmerle et al. 2009).

#### 6.2.2.2 Post-socialist Forest Change Patterns and Processes

Although these land use trends are acknowledged at a general level, little is known about the rates and spatial patterns of post-socialist forest disturbance, reforestation, and farmland abandonment at the local level. This is unfortunate, because forest cover changes have widespread effects on Carpathian biodiversity and ecosystem services (Baur et al. 2006). Moreover, the effects of the collapse of the socialist system on overall forest trends in the region are still in question. Therefore, changes in post-socialist forest cover and farmland abandonment were examined in Argeş County in southern Romania (Fig. 6.1) using Landsat Thematic Mapper (TM) and Enhanced Thematic Mapper Plus (ETM+) images. Argeş County was selected for study because it covers a wide range of environmental conditions (elevation ranges from 100 m to above 2,544 m) and has a socio-economic setting typical of many rural areas in Romania.

To quantify land cover change in the study region, individual land cover maps were derived for the years 1990, 1995, 2000, and 2005. A total of nine mid-summer Landsat images were necessary to cover the full study region for all years. The analysis involved the masking of all settlements, roads, water bodies, and rivers based on topographic maps. The remaining areas were stratified into "forest", "cropland", and "grassland" (permanent grassland including shrubland) based on a hybrid classification approach (Bauer et al. 1994; Kuemmerle et al. 2006). The four land cover maps all had overall accuracies between 90.6% and 92.5% (Kappa statistics between 0.85–0.89) based on 765 ground control points mapped in the field and from very high resolution Quickbird images. Post-classification map comparison was used to quantify land cover changes for the time periods 1990–1995, 1995–2000, and 2000–2005. Forest fragmentation was assessed using Riitters' fragmentation measures (Riitters et al. 2002; Kuemmerle et al. 2006; Kuemmerle et al. 2007). A detailed description of the image processing is provided in Kuemmerle et al. (2009).

Conversions from cropland to grassland were the dominant land cover changes between 1990 and 2005 in Argeş County. In 1990, just after the system change (1989), cropland amounted to 2,400 km², whereas cropland amounted to 1,885 km² in 2005 (Fig. 6.5). Thus, 21% of the cropland cultivated during the last years of socialism was abandoned by 2005 (17.3%, 3.9%, and 1.0% for the periods 1990–1995, 1995–2000, and 2000–2005, respectively). Conversion rates were highest in mountain valleys, but in terms of area the majority of the conversions occurred in the foothill zone.

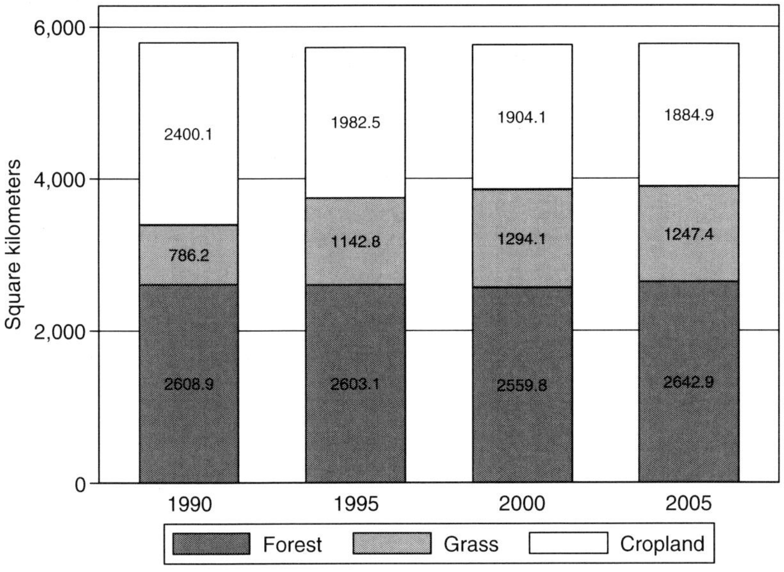


**Fig. 6.5** Changes in land cover type in Argeş County, Romania, 1990–2005 (*Source*: Authors)

Forest cover remained remarkably stable throughout the transition period and increased only marginally (about 1%) between 1990 and 2005. No evidence of extensive large-scale forest disturbance was found, and only a very small proportion of former croplands had already reverted back to forests (roughly 1% by 2005). Similarly, forest fragmentation in Argeş County did not change substantially in the time period studied: a consistent proportion of about 37% of consolidated (non-fragmented) forest persisted.

Contrary to initial concerns (Webster et al. 2001; Turnock 2002), large-scale ownership transfers, economic difficulties in the 1990s, and a lower level of administrative control after the breakdown of socialism did not trigger large-scale forest clear-cutting or an increase in forest fragmentation in Argeş County. Thus our analyses support findings by Strimbu et al. (2005) who found only marginal changes in forest canopy based on analyses of six southern Romanian forest inventory plots. Nonetheless, our results were surprising, because harvesting rates increased after the system change in a number of other Carpathian countries (Kuemmerle et al. 2007). Romania's long silvicultural tradition based on selective logging, annually revised state plans for harvesting levels, forest certification, and new forest legislation targeting multifunctional forestry may all have been factors preventing large-scale logging on restituted land (Ioras and Abrudan 2006; WWF 2007). However, forest restitution in Romania is far from being complete and much of Romania's forests were only recently privatized (Ioras and Abrudan 2006). Large-scale changes in forest cover may therefore lie ahead. Furthermore, while extensive clear-cutting did not occur between 1990 and 2005, forest degradation and fine-scale illegal logging are challenging to assess using Landsat's 30×30 m pixel resolution. Such disturbances may still have been substantial and are of major concern in the Romanian Carpathians (Turnock 2002; Bouriaud 2005; Strimbu et al. 2005). Although cropland abandonment was widespread during the transition period in Argeş County, reforestation on former cropland areas has not been extensive. Reforestation in the Romanian Carpathians occurs via natural succession, and a 15 year time period may be too short for forests to develop in this region. Moreover, field visits confirmed that at least some of the abandoned areas are occasionally grazed at low intensities, which may retard forest regrowth.

### *6.2.3 Case Study III: Reforestation and Forest Cover Change in Southeast Albania*

#### 6.2.3.1 Study Site and National Historical Context

During socialism Albania was the only country in Central and Eastern Europe that nationalized all its agricultural land. After the collapse of the socialist regime, Albania embarked on a comprehensive land distribution program in 1991. Based on equity principles, the majority of the farm land was distributed equally among rural dwellers, which led to a fragmented, small-scale agricultural structure (Müller and

Munroe 2008). The strategy could not have been more different for forest resources. Ninety-four percent of Albania's forested areas were already in state ownership at the start of the collectivization. The remaining private forests were nationalized in 1944 (Meta 1993, de Waal 2004). Contrary to the agricultural land reform policies, the Albanian government started to restitute forest land to pre-1944 owners in 1996 (Ministry of Agriculture and Food 2002). However, this affected only the 6% of forested area that was in private ownership before the nationalization of these lands (de Waal 2004). The majority of the forested areas remained in state ownership.

Between 1960 and 1980, state-organized deforestation aimed at expanding agricultural areas that served as a pillar of the socialist economy. This led to a decrease in forest cover by 10% to 20% despite approaches aimed at sustainable forest management (Meta 1993; UNECE 2001). Forest cover loss slowed following the demise of socialism, but is still cited as one major environmental problem, particularly in the more accessible forested areas (UNECE 2001).

In a quest to improve forest management, the government started to devolve forest land to local communities in 1996. In 2006, the goal of transferring use rights of 40% of the forested land and 60% of the pasture land to local individuals and communities was almost completed (Ministry of Agriculture and Food 2002, World Bank 2006). In particular, use rights to forests located in close proximity of villages were transferred to local authorities (World Bank 2006). Decision makers have started to think about turning the use rights into ownership rights, but this debate was ongoing at the time of writing (J. Stahl, personal communication). Forests that remained in state ownership continued to be under the jurisdiction of the General Directorate of Forests and Pastures (the General Directorate of Forests and Pastures was part of the Ministry of Agriculture and Food until the establishment of the Ministry of Environment in 2005, which then took over the jurisdiction over forests and pasture lands).

Since the collapse of socialism, Albania's dense forests decreased from approximately 31% of the total land surface in 1991 to 25% in 2001 (Jansen et al. 2006). Dominant changes were from broadleaved forest to broadleaved woodland and to herbaceous cropland (Jansen et al. 2006). In addition, forest degradation was widespread as tree densities decreased. These forest cover modifications and conversions reflect, to a large extent, the high levels of forest product extraction. But changes in forest extent and quality did not proceed linearly during the post-socialist transition. Two periods of rapid forest product extraction include the period following the collapse of socialism in 1991 due to the state of lawlessness, and the period following 1996, when the collapse of the pyramid investment schemes turned out to be detrimental to forest integrity (Hashi and Xhillari 1999, UNECE 2001). During these crises, forests virtually became an open-access resource. Fuelwood, timber, and non-timber products were rapidly extracted and hoarded in these periods of quasi-anarchy.

Forest use is still a central aspect of household livelihood strategies in rural Albania today and the majority of rural households rely predominantly on fuelwood for heating and cooking (Stahl 2007). Other important benefits derived from forests include the grazing of animals, the collection of non-timber forest products such as medicinal herbs and pine resin, and the sale of timber in local markets (Meta 1993,

Stahl 2007). Stahl's (2007) qualitative study showed that in accessible forested areas, the economic rents from forests may have been significantly higher than from agricultural production.

#### 6.2.3.2 Post-socialist Forest Change Patterns and Processes

In the following, empirical evidence is presented for the four districts of Elbasan, Librazhd, Gramsh, and Pogradec in the Southeast of Albania (see Fig. 6.1). The data collection strategies (for details see Müller and Sikor 2006, Müller and Munroe 2008) include a socioeconomic survey in 100 villages, the interpretation of Landsat satellite images, and the collection of land use relevant GIS variables. In addition to the quantitative components, qualitative in-depth case studies provided evidence on underlying patterns and processes within villages. Changes in land cover were assessed from Landsat images for the years 1988 (TM) and 2003 (TM and the Terra Advanced Spaceborne Thermal Emission and Reflection Radiometer [ASTER]). On-screen interpretation was conducted on a scale not smaller than 1:40,000 (see Müller and Munroe 2008, for details of the image analysis). Before and during interpretation in 2004, more than 300 reference points on the ground were compared to the derived land cover classes and cross-checked with secondary statistics. Overlays of the two output land cover maps produced land cover change maps that indicate the extent and locations of changes between 1988 and 2003.

Total forested areas (combining closed forest with more than 30% crown cover and open forest with a crown cover between 10% and 30%) remained stable at 39% during the study period – from near the end of the socialist period (1988) to 2003 (Fig. 6.6; note that only relevant land cover types are reported in this figure). However, we observe the occurrence of forest degradation, with a decrease in closed forest of 1.4% of the total area, in part giving way to open forest that increased by 1.6% of the total area. The largest change was the decrease in cropland from 935 $km^2$ (26%) to 677 $km^2$ (19%), equivalent to a decline of 27.6% of the cropland under production in 1988.

Forest degradation was dominated by the conversion of closed forest cover into open forests and shrubs (Table 6.1). Deforestation, the complete removal of tree cover, was observed on 202 $km^2$ of forest land, or on 13.5% of all forest in 1988. At the same time, 142 $km^2$ of cropland, shrubland, and open forest in 1988 regenerated into closed forest by 2003.

As expected, most forest cover, and in particular closed forest, is concentrated at higher altitudes and on more undulating terrain (Müller and Sikor 2006). Forest cover changes also depended markedly upon elevation. Deforestation and reforestation were predominantly observed at medium and high elevation terciles (Fig. 6.7). Modifications within the forest cover category (forest degradation, i.e., a change from closed to open forest; and forest regeneration, i.e., a change from open to closed forest) were modest compared to the magnitudes of forest conversions. Forest regeneration encompasses 32 $km^2$ in total with small variations across elevation groups. Forest degradation increases with elevation, and amounts to 42 $km^2$ in the highest elevation tercile.

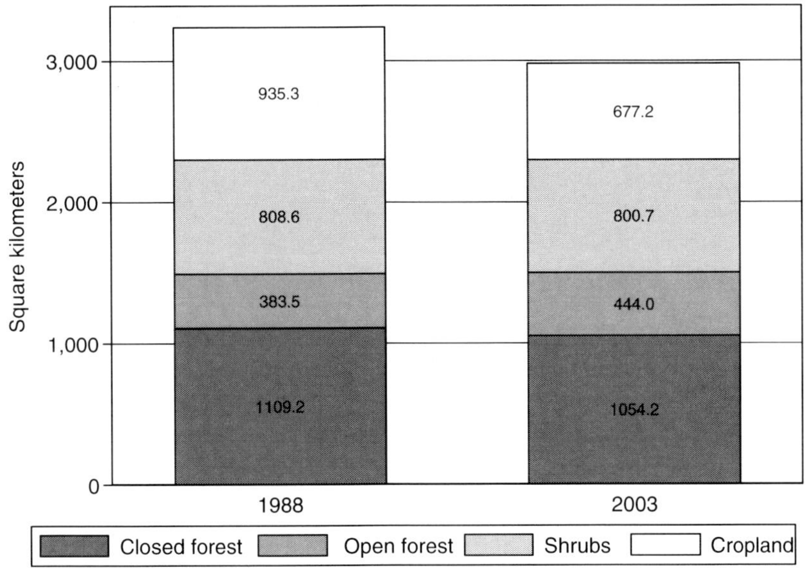


**Fig. 6.6** Changes in land cover type in Elbasan, Librazhd, Gramsh, and Pogradec, Albania, 1988–2003
(*Source*: Authors)

**Table 6.1** Forest cover change matrix (in km$^2$)

| Observed forest cover, 1988 | Observed forest cover, 2003 | | | | |
|---|---|---|---|---|---|
| | Closed forest | Open forest | Shrubs | All other | Total |
| Closed forest | **912.2** | 67.2 | 55.9 | 73.8 | 1,109.2 |
| Open forest | 32.0 | **279.6** | 29.8 | 42.1 | 383.5 |
| Shrubs | 30.1 | 50.3 | **613.3** | 114.9 | 808.6 |
| All other | 79.9 | 46.9 | 101.6 | **1,280.5** | 1,509.0 |
| Total | 1,054.2 | 444.0 | 800.7 | 1,511.4 | **3,810.2** |

Two land-change processes are central to forest degradation in Albania: first, fuelwood extraction mainly for subsistence purposes and, second, more commercially-oriented illegal logging (Müller and Sikor 2006, Müller and Munroe 2008). Survey results for the four districts showed that 94% of the rural households relied on fuelwood as their main source of energy in 2004. This large demand led to an active local firewood trade that is arguably the motivation for considerable forest extraction. Deforestation and forest degradation are therefore connected with small-scale fuelwood extraction, which represents an important livelihood strategy in many forested areas (Müller and Sikor 2006). Figure 6.7 also suggests that most of these dynamics take place at medium and higher elevations where forest cover is higher. However, no reliable secondary data exist on the amount of wood extracted, as most of the logging activities are illegal. Nevertheless, estimates of average household fuelwood consumption imply that the remote sensing data may

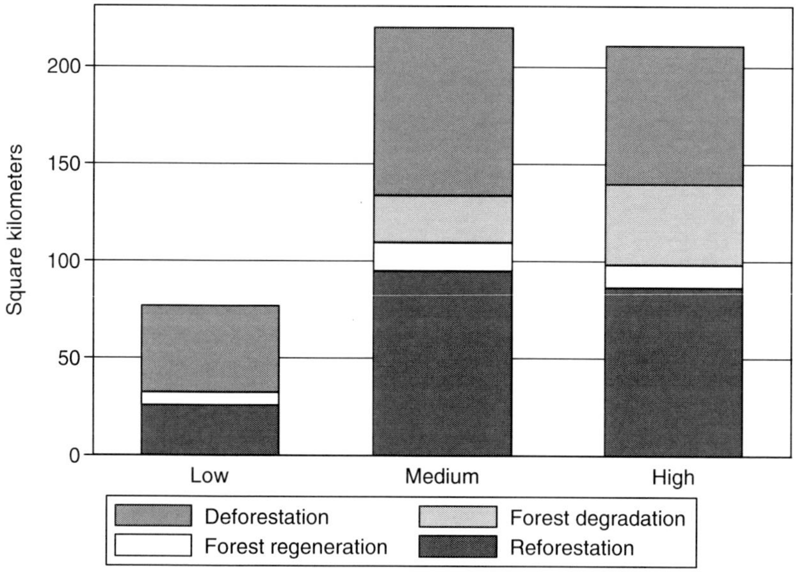


**Fig. 6.7** Forest cover conversions and modifications between 1988 and 2003 by elevation tercile (*Source*: Authors)

significantly underestimate the degree of forest degradation (Müller and Sikor 2006). While a complete removal of tree cover above the minimum mapping unit can be readily derived from Landsat data, it is challenging to assess forest degradation. More detailed data may be needed to adequately monitor forest degradation caused by selective timber extraction and the removal of understory vegetation. The hypothesis of forest degradation not observed by the satellite imagery is corroborated by the survey interviews where most villagers perceived rapid forest degradation around their village territories. Similarly, in-depth case studies involving participatory mapping exercises and qualitative research revealed much higher rates of forest resource extraction around the investigated villages (Stahl 2007).

## 6.3 Discussion

### *6.3.1 Drivers of Reforestation in Post-socialist Eastern Europe*

Three primary mechanisms drive reforestation around the globe: spontaneous reforestation, the growing of forest plantations, and the spreading of agro-forests (Rudel, Chapter 3 of this volume). The primary reforestation process in post-socialist Eastern Europe is spontaneous reforestation in the form of natural succession on abandoned agricultural lands. However, Forest Services in some countries have been planting on traditional forest lands and some abandoned farmlands, for example in Russia, Hungary, Bulgaria, Latvia, Lithuania, and Croatia (World Resources Institute 2007). These patterns tend to occur most on marginal lands, such as less fertile lands with adverse accessibility, and often on topographically rugged terrain (cf. MacDonald et al. 2000).

Much of the Eastern European countryside has experienced substantial depopulation throughout most of the twentieth century, and particularly in the last 20 years. This decrease of rural population was frequently associated with agricultural abandonment (MacDonald et al. 2000; Nikodēmus et al. 2005). In the Northern Carpathian Mountains, for instance, Kozak et al. (2007b) found that the overall proportion of forests increased from 5% in the 1930s to 31% in the 1990s as a result of post-World War II resettlement and depopulation of rural areas in favor of cities. Kozak et al. (2007b) also found that change in forest cover was inversely related to population. Elbakidze and Angelstam (2007) connect land abandonment to rural depopulation in the Ukrainian Carpathians from 1970 to 2005.

Interestingly, some researchers have found that population *increase* is occasionally associated with nearby land abandonment (and increased forest area) in Eastern and Central Europe (Kozak 2003; Başkent and Kadioğullari 2007). Such increases of forest cover are not likely to be a direct result of the population changes, but instead are related to the forest transition phenomenon. This association is found primarily in recent periods during increased economic development that pulls part of the labor force out of agricultural employment due to higher wages paid in the service and industrial sectors (Rudel et al. 2005). Contributing to this trend, across Eastern Europe the ratio of output to input prices decreased substantially due to the post-socialist transition (Rozelle and Swinnen 2004). Latvia is a good example of such a transition pathway, as its agricultural sector shrank from 23% of GDP in 1991 to a mere 4% in 2005 (World Bank 2007). Higher wages in the other sectors of the economy absorbed agricultural laborers, as agricultural output prices failed to rise sufficiently to support similar wage increases within agriculture (Strijker 2005). Deteriorating terms of trade in agriculture and concurrently rising opportunity costs of labor led to widespread cropland abandonment. The results from our case studies show that the decreasing economic importance of agricultural production relative to other opportunities led to agricultural land abandonment in the Romanian Carpathian Mountains and in Gauja National Park, Latvia.

Several factors directly related to the fall of socialism are shown to have fostered land abandonment. The post-socialist land reforms and related breakup of the collective farms have been shown to lead to fragmentation of landholdings resulting in parcels too small, and sometimes too fragmented, to enable competitive incomes from agriculture in Lithuania (Kundrotas 2002), Albania (World Bank 2006), the Ukrainian Carpathian Mountains (Elbakidze and Angelstam 2007), Argeş County, Romania (Müller et al. 2009), and in Gauja National Park, Latvia (Taff 2007). Furthermore, restitution as the dominant land reform strategy in Eastern Europe often resulted in landowners living far from their land and often lacking interest and experience in farming, such as in Lithuania (Kundrotas 2002), Latvia (Taff 2007), and the Carpathians (Webster et al. 2001, Kuemmerle et al. 2008). We suggest that these factors likely led to a surge in reforestation rates during the early post-socialist period, similar to reforestation trends that occurred after major political and socio-economic events in the past, such as World War II (Augustyn 2004; Kozak et al. 2007b) or the Bubonic Plague (Yeloff and van Geel 2007). Interestingly, rates of farmland abandonment in Poland were found to be strongly associated with the type of farmland ownership that existed during socialism (i.e., state owned land vs. privately owned land) (Kuemmerle et al. 2008).

Government land use policies in the region have been important drivers of both reforestation and deforestation. In an early example, the Austrian government made the first attempt to protect young forests to maturity in 1894, which led to reforestation in the (current) Ukrainian Carpathians (Elbakidze and Angelstam 2007). In another example in Gauja National Park, Latvia, post-socialist nature protection policies have helped to increase forest cover, and particularly core forest area important for biodiversity conservation. In addition to policies protecting isolated nature areas, the Soviet Union instituted spruce timber production during the Soviet era, and at the same time dictated that forests be cleared to increase the area for collective agriculture and pasture (Elbakidze and Angelstam 2007, Kuemmerle et al. 2007). These examples suggest that land use policies in each country are likely to heavily influence national forest cover area and patterns.

Finally, we note that some regions experience overall forest cover increase, but simultaneously experience a degradation of existing forests. In this chapter, results show that household fuelwood extraction and small-scale timber extraction (predominantly illegal logging) have resulted in forest degradation in Albania, even though overall forest area has increased in the country since the collapse of socialism. Başkent and Kadioğullari (2007) and Kozak et al. (2007a) have similarly found increases in forest area coinciding with forest degradation along the Eagan Sea coast of Turkey and the Northern Carpathian Mountains, respectively.

### *6.3.2 Outlook: Future Reforestation in the Region*

With the plethora of drivers of land cover change in the region, projections of countrywide and regional forest cover change patterns are difficult to establish. However, abandonment and subsequent reforestation of agricultural lands (primarily due to natural forest succession) is likely to continue throughout most of the region, particularly in more marginal areas (i.e. high altitudes, areas on steeper slopes, and less accessible areas), which may possibly lead to an overall increase in forest area in the coming decades. Continued economic development and structural changes may further raise non-farm incomes and, therefore, the opportunity costs of agricultural labor, leading to additional agricultural abandonment and subsequent reforestation. This process may become particularly noticeable in economies that still have large agricultural populations (such as Albania and Romania).

The current low regional birth rates and ongoing emigration from rural areas will lower population densities further. In addition to population decreases, changing age structures already have substantial effects on rural landscapes. It is mostly the younger and economically active that leave rural areas in search of better employment opportunities, leaving an aging rural population structure in much of rural Eastern Europe (Nikodēmus et al. 2005; Müller and Munroe 2008). This reduces the demand for land and may foster abandonment of agriculture, which would, in turn, lead to more reforestation.

In addition, forests are likely to re-grow and be preserved through the increased designation of protected or conservation forests in the region. Furthermore, markets for carbon credits such as those set up by the Clean Development Mechanisms (CDMs) may make reforestation profitable. CDMs, established by the Kyoto Protocol, offer industrialized countries opportunities to gain carbon credits by funding reforestation projects in developing and transitional countries.

Some other regional processes may decrease the amount of future reforestation in Eastern Europe. The increasing demand for bio-fuels and rising food prices may elevate the profitability of agriculture, spurring reutilization of abandoned agricultural lands. The continuation of payments under the Common Agricultural Policy (CAP) and EU rural development support to countries in Eastern Europe may also encourage agricultural production in the region, particularly where good farming conditions prevail. Land prices in Eastern Europe have increased dramatically and much land has been acquired by speculators, especially in the new accession countries. Agricultural land use could expand again if speculators begin to sell these lands to agricultural companies.

Although many of the aforementioned processes will affect future reforestation in Eastern Europe, the stage of a country's economic development in the post-socialist period seems to be a good indicator of the amount of reforestation occurring within its borders. Similarly, economic growth seems to be associated with the rate of forest regrowth. We speculate that the countries in Eastern Europe are at various phases between the second stage of the forest transition (i.e., decrease in forest cover associated with economic development) and the third stage of the forest transition (i.e., increase in forest cover associated with an even higher level of economic development). For instance, Poland (Kozak, Chapter 11 this volume) and Latvia (this chapter) may already be in the midst of the third stage of the forest transition. The Polish Carpathian Mountains (Kozak, Chapter 11 this volume) and Latvia's Gauja National Park have experienced high levels of reforestation, yet this may also be partly due to the mountainous character of the Carpathians (i.e., less suitable for agriculture) and the protected status of Gauja National Park. Romania, with a GDP per capita less than half that of Latvia, may be on the brink of entering the third forest transition stage, while Albania, the poorest of the three countries presented in this chapter, has not experienced large-scale reforestation, and may still be at the end of the second stage of the forest transition.

Increasing GDP, ongoing structural economic changes, rural to urban migration, the economic shift away from agriculture towards the industrial sector and, in particular, towards the service sector, may soon push Romania and Albania into the third stage of the forest transition. The strong decrease in rural population and high emigration rates in Albania may already signal this approaching transition. Still, the share of rural population remains quite high in Albania relative to other Eastern European countries (World Bank 2007). In addition, the current intensive reliance on traditional biomass in Albania may prohibit significant increases in forest cover. As long as forest resources provide high and competitive land rents, an efficient protection of forests will be difficult to achieve.

## 6.4 Conclusions

Policy makers and land managers urgently need multi-scale and up-to-date information on the spatial patterns of reforestation, deforestation, and other land use changes. But small-scale forest cuts and forest degradation are difficult to monitor with medium and low resolution satellite imagery, which suggests the need for research to incorporate high resolution satellite imagery, aerial photography, and ground-based investigations including qualitative data collection. Such studies can provide more detailed and comprehensive pictures of land use change, and they can frame these changes within relevant local and regional socio-political contexts. Such studies can help land managers to balance the trade-offs and opportunities offered by the multitude of potential land uses for economic prosperity, ecosystem service provision, cultural heritage preservation, carbon sequestration, and biodiversity conservation in Eastern Europe.

These case studies and previous research have shown that widespread abandonment of agricultural lands throughout much of Eastern Europe has already led to reforestation in some areas, and has a large potential for reforestation in other areas. Agricultural land abandonment is shown to be occurring as a result of the interactions between decreasing agricultural profitability, changes in the sectoral compositions of the economies, rural population decrease, and an aging rural population structure. Fragmentation and degradation of forests are also common in Eastern Europe, both in areas with overall forest cover increase (e.g., Gauja National Park, Latvia) and overall forest cover loss (e.g., Albania), often due to illegal forest cutting and extensive use of forest products. While a plethora of benefits are widely recognized regarding increased forest cover in the region, there is also a concern over the loss of cultural landscapes and the accompanying biota due to reforestation on abandoned agricultural fields and meadows.

This chapter illustrates the significant diversity in local patterns and processes of reforestation and forest dynamics in Eastern Europe. Yet it shows how local forest dynamics are conditioned or mediated by the broad scale, regional context and the timing of historical events (i.e., the demise of socialism in Eastern Europe, economic development, and the eastward expansion of the European Union). The variations in forest dynamics between countries are likely associated with economic development levels, differences in rural demographic patterns, diverse land use histories, and the national-level strategy of post-socialist land privatization. As this research has demonstrated, only through the integrated study of people and the environment can we effectively describe land use/land cover change patterns and give rationale and explanation to the revealed trajectories and processes of change.

## References


Angelstam P, Boresjö-Bronge L, Mikusiński SU, Wästfelt A (2003) Assessing village authenticity with satellite images: a method to identify intact cultural landscapes in Europe. Ambio: J Hum Environ 32:594–604

Augustyn M (2004) Anthropogenic changes in the environmental parameters of the Bieszczady Mountains. Biosph Conserv 6:43–53

Auninš A (2002) Zoologist Project Manager, Latvian Fund for Nature, interview with Gregory N. Taff. Riga, Latvia

Balsevics E (2004) Report Based on Exchange Programme Documents Provided by Edvins Balsevics. Celk Center, Budapest, Hungary

Baldock D (1999) Farming in a living countryside. Eur Nat 3:15

Baldock D, Beaufoy G, Brouwer F, Godeschalk F (1996) Farming at the margins: Abandonment or redeployment of agricultural land in Europe. Institute for European and Environmental Policy and Agricultural Economics Research Institute, London

Başkent E, Kadioğullari A (2007) Spatial and temporal dynamics of land use pattern in Turkey: a case study in İnegöl. Landsc Urban Plan 81:316–327

Bauer M, Burk T, Ek A, Coppin P, Lime S, Walsh T, Walters D, Befort W, Heinzen D (1994) Satellite inventory of Minnesota forest resources. Photogramm Eng Rem S 60:287–298

Baur B, Cremene C, Groza G, Rakosy L, Schileyko AA, Baur A, Stoll P, Erhardt A (2006) Effects of abandonment of subalpine hay meadows on plant and invertebrate diversity in Transylvania, Romania. Biol Conserv 132:261–273

Bicik I, Jelecek L, Stepanek V (2001) Land-use changes and their social driving forces in Czechia in the 19th and 20th centuries. Land Use Policy 18:65–73

Bouriaud L (2005) Causes of illegal logging in Central and Eastern Europe. Small-Scale Forest Econ Manag Policy 4:269–292

Buksha I, Pasternak V, Romanovsky V (2003) Forest and forest products country profile Ukraine. In: UNECE/FAO Timber and forest discussion papers. UNECE/FAO, Geneva

Bunkše E (2000) Reality of rural landscape symbolism in the formation of a post-Soviet, postmodern Latvian identity. Nor Geogr Tidsskr 53:121–138

Central Statistical Bureau of Latvia (2003) Statistical yearbook of Latvia 2003. Central Statistical Bureau of Latvia, Riga, Latvia

Central Statistical Bureau of Latvia (2002) Statistical yearbook of Latvia 2002. Central Statistical Bureau of Latvia, Riga, Latvia

Chazdon R (2008) Beyond deforestation: restoring forests and ecosystem services on degraded lands. Science 320:1458–1460

Congalton R (2004) Putting the map back in map accuracy assessment. In: Lunetta R (ed) Remote sensing and GIS accuracy assessment. CRC Press, Boca Raton, FL

Cremene C, Groza G, Rakosy L, Schileyko A, Baur A, Erhardt A, Baur B (2005) Alterations of steppe-like grasslands in Eastern Europe: a threat to regional biodiversity hotspots. Conserv Biol 19:1606–1618

Dabbert S (1995) Agroforestry and land-use change in industrialized nations: a case study from northeastern Germany. Agrofor Syst 31:157–168

De Waal C (2004) Post-socialist property rights and wrongs in Albania: An ethnography of agrarian change. Conserv Soc 2:19–50

DLG (2005) Land abandonment, biodiversity, and the CAP. Land abandonment and biodiversity in relation to the 1st and 2nd pillars of the EU's Common Agricultural Policy; Outcome of an international seminar in Sigulda, Latvia, 7–8 October, 2004. Government Service for Land and Water Management of the Netherlands (DLG), Utrecht, The Netherlands
Elbakidze M, Angelstam P (2007) Implementing sustainable forest management in Ukraine's Carpathian mountains: The role of traditional village systems. Forest Ecol Manag 249:28–38
Enescu V (1996) Forest genetic resources conservation in Romania. In: Food and Agriculture Organization of the United Nations (ed) Forest genetic resources N.24. Food and Agriculture Organization of the United Nations (FAO), Rome
Food and Agriculture Organization of the United Nations (2006) Global forest Resources Assessment 2005: Progress Towards Sustainable Forest Management, FAO Forestry Paper 147. FAO, Rome
Freedom House (1998) http://www.freedomhouse.org/nit98/latvia.html. Cited 2005
Gauja National Park Management (1998) Legislation Relating to the Gauja National Park, a compilation. Gauja National Park Management, Sigulda, Latvia
Gauja National Park Management (2001) Gauja National Park yearbook. Sigulda, Latvia
Hashi I, Xhillari L (1999) Privatization and transition in Albania. Post-Comm Econ 11:99–125
Herzon I, O'Hara R (2007) Effects of landscape complexity on farmland birds in the Baltic states. Agric Ecosys Environ 118:297–306
Hrivnak R, Ujhazy K (2005) Changes of the mountain grassland vegetation after abandonment and colonization by Norway spruce. Ekol Bratislava 24:231–253
Ioffe G, Nefedova T (2001) Land use changes in the environs of Moscow. Area 33:273–286
Ioffe G, Nefedova T (2004) Marginal farmland in European Russia. Eurasian Geogr Econ 45:45–59
Ioras F (2003) Trends in Romanian biodiversity conservation policy. Biodivers Conserv 12:9–23
Ioras F, Abrudan IV (2006) The Romanian forestry sector: privatisation facts. Int Forest Rev 8:361–367
Jansen L, Carrai G, Morandini L, Cerutti P, Spisni A (2006) Analysis of the spatio-temporal and semantic aspects of land cover/use change dynamics 1991–2001 in Albania at national and district levels. Environ Monit Assess 119:107–136
Jensen J (2005) Introductory digital image processing, 3rd edn. Prentice Hall, Upper Saddle River, NJ
Kauppi P, Ausubel J, Fang J, Mather A, Sedjo R, Waggoner P (2006) Returning forests analyzed with the forest identity. Proc Natl Acad Sci U S A 103:17574–17579
Kobler A, Tomaž C, Pirnat J (2005) Modelling spontaneous afforestation in Postojna area, Slovenia. J Nat Conserv 13:127–135
Koch N, Skovsgaard J (1999) Sustainable management of planted forests: some comparisons between Central Europe and the United States. New Forest 17:11–22
Kozak J (2003) Forest cover change in the Western Carpathians in the past 180 years; a case study in the Orawa region in Poland. Mt Res Dev 23:369–375
Kozak J, Estreguil C, Vogt P (2007a) Forest cover and pattern changes in the Carpathians over the last decades. Eur J Forest Res 126:77–90
Kozak J, Estreguil C, Troll M (2007b) Forest cover changes in the northern Carpathians in the 20th century: a slow transition. J Land Use Sci 2:127–149
Kuemmerle T, Hostert P, Perzanowski K, Radeloff VC (2006) Cross-border comparison of land cover and landscape pattern in Eastern Europe using a hybrid classification technique. Rem Sens Environ 103:449–464
Kuemmerle T, Hostert P, Radeloff VC, Perzanowski K, Kruhlov I (2007) Post-socialist forest disturbance in the Carpathian border region of Poland, Slovakia, and Ukraine. Ecol Appl 17: 1279–1295
Kuemmerle T, Hostert P, Radeloff VC, van der Linden S, Perzanowski K, Kruhlov I (2008) Cross-border comparison of post-socialist farmland abandonment in the Carpathians. Ecosystems 11:614–628

Kuemmerle T, Müller M, Griffiths P, Rusu M (2009) Land use change in southern Romania after the collapse of socialism. Regional Environ Change 9:1–12
Kundrotas A (2002) The role of afforestation in the Lithuanian environmental protection policy. Balt For 8:70–71
Lazdinis M, Roberge J, Kurlavičius P, Mozgeris G, Angelstam P (2005) Afforestation Planning and biodiversity Conservation: Predicting Effects on Habitat Functionality in Lithuania. J Environ Plan Manag 48:331–348
Lipsky Z, Kopecky M, Kvapil D (1999) Present land use changes in the Czech cultural landscape. Ecology (Bratislava) 18:31–38
MacDonald D, Crabtree JR, Wiesinger G, Dax T, Stamou N, Fleury P, Gutierrez Lazpita J, Gibon A (2000) Agricultural abandonment in mountain areas of Europe: Environmental consequences and policy response. J Environ Manag 59:47–69
McGarigal K, Marks B (1995) FRAGSTATS: Spatial Pattern Analysis Program for Quantifying Landscape Structure, General Technical Report PNW-351. USDA Forest Service
Mather AS (1992) The forest transition. Area 24:367–379
Meta M (1993) Forests in Albania. Forest J 91:27–28
Mihai B, Savulescu I, Sandric I (2007) Change detection analysis (1986–2002) of vegetation cover in Romania. A study of alpine, subalpine, and forest landscapes in the Iezer Mountains, Southern Carpathians. Mt Res Dev 27:250–258
Ministry of Agriculture and Food (2002) Annual report 2002. Ministry of Agriculture and Food, Tirana
Müller D, Kuemmerle T, Rusu M, Griffiths P (2009) Lost in transition: determinants of post-socialist cropland abandonment in Romania. J Land Use Sci 4:109–129
Müller D, Munroe DK (2008) Changing rural landscapes in Albania: Cropland abandonment and forest clearing in the postsocialist transition. Ann Assoc Am Geogr 98:855–876
Müller D, Sikor T (2006) Effects of postsocialist reforms on land cover and land use in South-Eastern Albania. Appl Geogr 26:175–191
Nijnik M, Van Kooten GC (2000) Forestry in the Ukraine: the road ahead? Forest Policy Econ 1:139–151
Nikodēmus O (2003) Professor of Geography & Landscape Scientist, University of Latvia, interview with Gregory N. Taff. Riga, Latvia
Nikodēmus O, Bell S, Grine I, Liepins I (2005) The impact of economic, social and political factors on the landscape structure of the Vidzeme Uplands in Latvia. Landsc Urban Plann 70:57–67
Oldfield JD (2000) Structural economic change and the natural environment in the Russian federation. Post-Comm Econ 12:77–90
Parliament of Romania (2000) Legislative survey: Law on juridical circulation of land (Law No. 54/1998). http://www.cdep.ro/legislatie/eng/vol48eng.pdf. Cited 10 Jan 2008
Petersen V (1999) Gauja National Park nature protection plan. Gauja National Park Management, Sigulda, Latvia
Peterson DJ (1995) Russia environment and natural resources in light of economic regionalization. Post-Sov Geogr 36:291–309
Pilāts V (2002) Director of Research, Gauja National Park, interview with Gregory N. Taff. Sigulda, Latvia
Pilāts V (ed) (2007) Biodiversity in Gauja National Park. Gauja National Park Management, Sigulda, Latvia
Pryde P (1991) Environmental management in the Soviet Union. Cambridge University Press, Cambridge, UK
Riitters K, Wickham J, O'Neill R, Jones K, Smith E, Coulston J, Wade T, Smith J (2002) Fragmentation of continental United States forests. Ecosystems 5:815–822
Rozelle S, Swinnen J (2004) Success and failure of reform: Insights from the transition of agriculture. J Econ Lit 42:404–456
Rudel T (1998) Is there a forest transition? Deforestation, reforestation, and devleopment. Rural Sociol 63:533–552
Rudel TK, Coomes OT, Moran E, Achard F, Angelsen A, Xu J, Lambin E (2005) Forest transitions: towards a global understanding of land use change. Glob Environ Change A 15:23–31

Schmithüsen F, Iselin G, Le Master D (2001) Experiences with new forest and environmental laws in European countries with economies in transition. Proceedings of the International Symposium jointly organized by the IUFRO Research Group 6.13.00 and the Austrian Federal Ministry of Agriculture and Forestry, Ossiach, Austria

Schwartz K (2001) Wild horses and great trees: National identity and the global politics of nature in Latvia. PhD dissertation in political science for The University of Wisconsin, Madison, WI

Sēstulis M (2002) Head Forester of the Cēsis Region of the State Forest Service, interview with Gregory N. Taff. Cēsis, Latvia

Staddon C (2001) Local forest-dependence in postcommunist Bulgaria: A case study. GeoJ 55:517–528

Stahl J (2007) The political ecology of postsocialist land use change: Case studies from three villages in Southeastern Albania. Doctoral dissertation, Humboldt-Universität zu Berlin, Berlin

Strautnieks J (2002) Director of Gauja National Park, interview with Gregory N. Taff. Sigulda, Latvia

Strijker D (2005) Marginal lands in Europe - causes of decline. Basic Appl Ecol 6:99–106

Strimbu BM, Hickey GM, Strimbu VG (2005) Forest conditions and management under rapid legislation change in Romania. Forest Chron 81:350–358

Swinnen J (2001) Transition from collective farms to individual tenures in Central and Eastern Europe. In: De Janvry A, De Gordillo G, Platteau J, Sadoulet E (eds) Access to land, rural poverty, and public action, a study prepared for development economics research of the United Nations University. Oxford University Press, Oxford

Taff G (2005) Conflict between global and local land-use values in Latvia's Gauja National Park. Landsc Res 30:415–430

Taff G (2007) Competing stakeholder interests in Latvia's Gauja National Park. In: Taff G Influence of Post-Soviet land reform on the patterns of land use dynamics in Gauja National Park, Latvia. Ph.D. dissertation for the Department of Geography at the University of North Carolina, Chapel Hill, NC

Trzeciak-Duval A (1999) A decade of transition in central and Eastern European agriculture. Eur Rev Agric Econ 26:283–304

Turnock D (1998) Introduction. In: Turnock D (ed) Privatization in rural Eastern Europe. The process of restitution and restructuring. Edward Elgar, Cheltenham

Turnock D (2002) Ecoregion-based conservation in the Carpathians and the land use implications. Land Use Policy 19:47–63

UNECE (2001) Environmental performance review of Albania. United Nations Economic Commission for Europe, Geneva

UNECE & FAO (2000) Forest resources of Europe, CIS, North America, Australia, Japan and New Zealand, Geneva Timber and Forest Study Paper No. 17 (ECE/TIM/SP/17). UNECE, New York and Geneva

Watkins C (1993) Forest expansion and nature conservation. In: Watkins C (ed) Ecological effects of afforestation: Studies in the history and ecology of afforestation in Western Europe. CAB International, Wallingford, UK

Webster R, Holt S, Avis C (eds) (2001) The status of the Carpathians. A report developed as a part of The Carpathian Ecoregion Initiative. WWF, Vienna

World Bank (2006) Status of the land reform and real property markets in Albania. The World Bank Office, Tirana

World Bank (2007) World development indicators 2007. The World Bank, Washington, DC

World Conservation Union (2004) Afforestation and reforestation for climate change mitigation: Potentials for Pan-European action. IUCN Programme Office for Central Europe, Warsaw

World Resources Institute WRI (2007) Earth trends: the environmental portal – forest extent – forest plantations area, net change. http://earthtrends.wri.org/searchable_db/index.php?theme=9&variable_ID=305&action=select_countries. Cited 20 Aug 2008

World Wildlife Fund [WWF] (2007) Romania leads in sustainable forestry – tax incentives and certification for FSC wood. http://www.panda.org/about_wwf/where_we_work/europe/what_we_do/danube_carpathian/index.cfm?uNewsID=89140. Cited 26 Sep 2007

Yeloff D, van Geel B (2007) Abandonment of farmland and vegetation succession following the Eurasian plague pandemic of AD 1347–52. J Biogeogr 34:575–582

Zerbe S, Brande A (2003) Woodland degradation and regeneration in Central Europe during the last 1, 000 years – a case study in NE Germany. Phytocoenologia 33:683–700

# Chapter 7
# Reforestation and Regrowth in the Human Dominated Landscapes of South Asia

**Harini Nagendra**

H. Nagendra (✉)
Indiana University, Bloomington, IN, USA and
Ashoka Trust for Research in Ecology and the Environment (ATREE), Bangalore, India
e-mail: nagendra@indiana.edu

H. Nagendra and J. Southworth (eds.), *Reforesting Landscapes: Linking Pattern and Process*, Landscape Series 10,
DOI 10.1007/978-1-4020-9656-3_7, © Springer Science+Business Media LLC 2010

## 7.1 Introduction

The densely populated landscapes of South Asia pose a particular challenge to researchers, practitioners and others interested in understanding and managing the dynamics of forest change. In most parts of South Asia (here defined as the countries of Bangladesh, Bhutan, India, Nepal, Pakistan and Sri Lanka), forests and humans have coexisted in close proximity for centuries. Much of this region is dominated by rural agricultural settlements, highly dependent on forest products for subsistence as well as for their livelihoods. These are landscapes of contrast, where it is common to find long standing traditional institutions of forest management coexisting with large scale forest logging, or government protected areas located alongside clearings for large infrastructure projects.

The countries in this region have all experienced (and continue to experience) large scale forest clearing. Yet, while primary forests continue to be threatened across South Asia, the countries located in this region have dealt with challenges of deforestation by using a range of different approaches. While some countries such as Nepal have experimented with innovative, successful policies of community forestry, others such as Bhutan have strictly controlled visitor flows to reduce pressures on the natural environment.

These approaches have had varying results, with differences in the direction and extent of forest change observed in adjacent countries (FAO 2005; Global Forest Resources Assessment 2005). As Table 7.1 indicates, Nepal, Pakistan and Sri Lanka experienced overall declines in forest cover between 1990 and 2000. Yet, during the same time period, Bhutan largely maintained its forest cover, while Bangladesh and India have noted slight increases in forest cover. Further, although Bangladesh has recorded a slight decrease in growing stock density during the same time period, forests in Bhutan and India increased in density. Forests in Nepal, though decreasing in area, recorded a significant increase in density (Kauppi et al. 2006). Of the six countries in this region, only two – Pakistan and Sri Lanka – have experienced decreases in forest cover (deforestation), as well as declines in forest density (degradation). The other four countries – Bangladesh, Bhutan, India and Nepal – have experienced increases in forest area (reforestation), and/or increases in forest density (regrowth).

Neither this overall trend towards reforestation and/or regrowth, nor differences observed between countries can be directly attributed to commonly identified drivers of forest change. Population growth, affluence, technology and development are frequently identified as primary underlying drivers of deforestation in the developing tropics (Allen and Barnes 1985; Ehrlich and Ehrlich 1990; Mather and Needle, 2000). Yet, Table 7.1 also indicates that population and development – two factors commonly implicated as driving deforestation in multiple studies – cannot explain the variations in the direction and rates of forest change observed at the country scale. Among the six countries located in this region, Bangladesh and India have the highest population densities, but both countries record increases in forest area. Although Sri Lanka, Pakistan and India have the highest per capita GDP levels in

**Table 7.1** Information on countries and forests of South Asia (data from FAO 2005;Global Forest Resources Assessment 2005)

| | Bangladesh | Bhutan | India | Nepal | Pakistan | Sri Lanka | World |
|---|---|---|---|---|---|---|---|
| Population density – 2003 (/km$^2$) | 1,127.3 | 48.0 | 358.4 | 176.0 | 199.2 | 295.0 | 48.2 |
| Rural population % – 2003 | 75.8 | 91.5 | 71.7 | 85.0 | 65.9 | 79.0 | 51.7 |
| GDP per capita – 2003 | 385 | 303 | 555 | 233 | 498 | 913 | NA |
| Forest area – 2000 (% of total land area) | 10.2 | 64.2 | 21.6 | 27.3 | 3.1 | 30.0 | 29.6 |
| Forest plantations – 2000 (‘000 ha) | 625 | 21 | 32,578 | 133 | 980 | 316 | 186,733 |
| Forest area per capita – 2000 (ha) | ~0 | 1.5 | 0.1 | 0.2 | ~0 | 0.1 | 0.6 |
| Consumption of wood fuel – 2000 (‘000 tons) | 27,763 | 4,348 | 300,564 | 12,728 | 25,013 | 5,774 | 1,795,496 |
| Annual change in forest cover, 1990–2000 (‘000 ha) | 17 | ~0 | 38 | –78 | –39 | –35 | –9391 |
| Forest designated for production (%) | 31.7 | 15.9 | 21.2 | 5.1 | 32.0 | 8.8 | 34.1 |
| Forest designated for protection or conservation (%) | 28.7 | 73.0 | 36.5 | 33.5 | 11.4 | 29.9 | 20.5 |
| Annual rate of change in forest cover, 1999–2000 (%) | 1.3 | ~0 | 0.1 | –1.8 | –1.5 | –1.6 | –0.2 |
| Annual change in growing stock per hectare ($m^3$/ha/year) | –0.38 | 1.22 | 0.08 | 8.16 | –0.36 | –0.16 | NA |

the region, Sri Lanka and Pakistan record forest decreases, while India shows a net increase in forest cover and density. While much previous research on land use/ land cover change has relied on explanations that incorporate population and affluence as main drivers, an exclusive focus on these factors ignores several other critical drivers of change that can modify and even reverse trajectories of forest decline (see also Angelsen and Kaimowitz 1999; Geist and Lambin 2001; Ostrom and Nagendra 2006).

The Environmental Kuznets curve and forest transition "theory" has been extensively used to explain the observation that many countries invest in forest conservation and encourage regrowth once they have reached certain levels of affluence, and industrialization (e.g. Bhattarai and Hammig 2001). Such explanations, while adequately describing much change in countries located in the economically developed world, from Scotland to France to the USA, fail to account for the turnarounds in forest cover observed in economically developing countries such as India and Bangladesh, the protection of forest cover observed in economically poor Bhutan, or the large increases in forest density observed in Nepal (see also Mather 2007). The lack of adequate discussion on the drivers of reforestation and regrowth in low to medium economies can in part be attributed to the general reluctance of the land cover change community to focus on these processes. In recent years this has changed, however, with several recent studies highlighting the presence of large scale reforestation in many developing and transition economies, as well as discussing and developing new frameworks that can help explain these turnarounds.

In general, three major pathways have been introduced to explain increases in forest area and density (Rudel et al. 2005; Rudel this volume). Simplified, the first of these states that forest regrowth occurs in rural areas which have been abandoned at a large scale, because of migration to cities. Labor scarcities in these areas mean that people no longer can work their farms, and these abandoned farming areas then revert to woodlands or forests. This reforestation is largely referred to as "natural" or "spontaneous", meaning that the land is left to regenerate naturally, and there is little or no planting of trees, or specific government policies that encourage this trend.

The second explanation is that governments and local communities, facing critical shortages of wood and other forest products, engage in protection of forests through the creation of community forestry or national parks. They often engage in extensive plantation of degraded areas. Sometimes, degraded forest areas are also handed over to private agencies for large-scale plantation to meet industrial needs. The third pathway is somewhat linked to the second, and describes the planting of trees by rural farmers on individual farmlands, creating mixed areas of agroforestry, to supplement the needs of local households.

Given the significance of these pathways, it is surprising that there is such little research that examines their applicability in actual field contexts. Indeed, it almost appears that these discussions of pathways have been developed in parallel with field studies of reforestation, rather than being developed from a foundation of field work. While some of the papers that discuss pathways of reforestation cite specific field studies (e.g., Rudel et al. 2005), almost none of the field studies of reforestation in South Asia discuss drivers of reforestation and their applicability in the developing country context. Neither are there any large scale studies that integrate information on the drivers of reforestation across a large number of study sites in a region, unlike the many such meta-analyses now conducted on deforestation and other forms of land cover change (e.g. Geist and Lambin 2001, 2002; Keys and McConnell 2005; Rudel 2005). Yet, given the increasing number of case studies of reforestation reported from around the globe, it should now be possible to incorporate findings from several of these studies into a larger analysis. Such an approach will enable an evaluation of the suitability of these pathways to explain the mechanisms that drive reforestation and forest regrowth in a range of locations and contexts, and enable their further development.

Such meta-analyses have proved extremely helpful in other contexts, enabling the identification of major drivers of specific types of land cover change (Geist and Lambin 2001; Keys and McConnell 2005; Rudel 2005). They can benefit greatly, however, from drawing on detailed field studies of specific locations to contextualize their findings. While meta-analyses provide breadth, case studies bring context, both critical for an in-depth understanding of the drivers of reforestation in developing countries. Unfortunately, integrating findings from field studies on individual landscapes is often challenging, since these studies are conducted by investigators working at different spatial scales, in different cultural, institutional and socio-economic contexts, using different methods, and coming at the problem from different disciplinary viewpoints (Keys and McConnell 2005; Young et al. 2006; Poteete and Ostrom 2008).

This chapter studies the drivers of reforestation and regrowth in South Asian, highly populated forested landscapes, using a combination of case studies and a larger meta-analysis. First, a meta-analysis of 24 papers from South Asia was conducted to evaluate whether current explanations of the drivers of reforestation were adequate to explain the pathways of forest change in these contexts. Subsequently, four case studies of reforestation and regrowth in landscapes located in India and Nepal are discussed in detail, enabling the description of the human drivers of reforestation and regrowth in these landscapes to a level of detail not possible in the meta-analysis, and to differentiate between drivers of increase in forest area (reforestation), and those of increase in forest density (regrowth). Integrating findings from both these approaches provides an enriched understanding of the drivers of reforestation and forest regrowth in the densely populated, biodiversity rich landscapes of South Asia.

## 7.2 Meta-Analysis

### *7.2.1 Methods*

A number of academic databases, including Web of Science, Scirus and Google Scholar, were used to locate peer reviewed journal articles that provided case studies of reforestation and forest regrowth located in South Asia. Papers were selected if they discussed reforestation, forest regrowth, or increase in tree density in agroforestry areas over time – and if they also contained information on the human drivers and pathways that led to an increase in forest cover or tree density. Twenty-four case studies were selected, ranging in spatial scale from the local (village), to that of an entire country. Of these, two studies were located in Bangladesh, six in India, 14 in Nepal (reflecting the greater intensity of research on forest change in this country), one in Pakistan, and one in Sri Lanka. No peer reviewed published literature on reforestation or regrowth in Bhutan could be located, reflecting the relative paucity of land cover change research in this country (see Meyfroidt and Lambin this volume, for more information on reforestation in Bhutan).

From these papers, information was gathered on the drivers of reforestation and forest regrowth, both proximate and underlying; the nature of reforestation and regrowth, whether natural or planted; the types of actors involved in reforestation and regrowth, from local communities to aid agencies to governmental agencies; and the threats to reforestation and regrowth noted by the authors, if any. In some papers, multiple pathways to reforestation and regrowth were identified – in such cases, information on each pathway was recorded separately.

### *7.2.2 Results*

As shown in Table 7.2, the case studies of reforestation and regrowth range in scale from that of entire countries, ranging over thousands of square kilometers, to villages, of the scale of 1–10 km$^2$. Interestingly, though, the pathways to reforestation and regrowth that are reported here do not appear to differ with the spatial scale of examination. They do, however, differ across location. Often, within a location, multiple drivers and pathways to reforestation and regrowth were described. These have been recorded separately.

In all, 35 pathways to reforestation and regrowth were described in 24 case studies. Following Rudel (this volume) and Rudel et al. 2005, these were categorized into three different types – natural regeneration (also described as "spontaneous" regeneration by Rudel (2005)), where no explicit protection or planting of trees takes place; regeneration through planting or protection; and farm agroforestry. Of these, regeneration through planting or protection was the most common, with 20 of the 35 recorded pathways (57%) belonging to this category. 10 cases of agroforestry (29%) were recorded, while only five of the pathways described (14%) were described as occurring naturally.

The dominant pathway to reforestation in South Asia thus appears to be that of protection of degraded forests, in conjunction with encouraging regrowth through the planting of trees. Why are such protection or plantation efforts taking place? Of the 20 cases studied, ten (50%) cited an increase in conservation awareness, 12 (60%) cited increasing scarcity of forest products, while ten (50%) cited government initiatives towards decentralizing forest management and handing over control to forest communities. In addition, one case in Nepal (Fox 1993) stated that alterations

**Table 7.2** Drivers of reforestation in various landscapes

| Sl. No. | Country | Reference | Spatial scale | Pathway | Proximate drivers | Underlying drivers | Type of reforestation | Actors involved | Threats |
|---|---|---|---|---|---|---|---|---|---|
| 1 | Bangladesh | Giri et al. 2008 | Country | Protection and planting | Community forestry | Increased conservation awareness | Natural regeneration and plantation | Local communities, national government | Agricultural expansion, aquaculture |
| 2 | Bangladesh | Safa 2006 | Country | Protection and planting | Social forestry | Scarcity of forest products | Plantation | Farmers, local communities, national government, non-governmental organizations | None mentioned |
| 3a | India | Bhat et al. 2001 | Country ($10^6$–$10^7$ $km^2$) | Protection and planting | Participatory forestry through Joint Forest Management | Increased conservation awareness, decentralization of forest management to strengthen local institutions, scarcity of wood products | Plantation, sometimes followed by natural regeneration | Local communities, national government | Agricultural expansion, grazing, fire, extraction of forest products |
| 3b | | | | Protection and planting | Social forestry on degraded village commons | Scarcity of wood products | Plantation, sometimes followed by natural regeneration | National government | Agricultural expansion, grazing, fire, extraction of forest products |
| 4 | India | Foster and Rosenzweig 2003 | Country ($10^6$–$10^7$ $km^2$) | Protection and planting | Participatory forestry through Joint Forest Management | Decentralization of forest management to strengthen local institutions, scarcity of wood products | Not available | Local communities, national government | Not mentioned |
| 5 | India | Ghate and Nagendra 2005 | Villages ($10^0$– $10^1$ $km^2$) | Protection and planting | Participatory forestry through Joint Forest Management | Decentralization of forest management to strengthen local institutions, scarcity of wood products | Plantation and natural regeneration | Local communities, national government, non-governmental organizations | Grazing, fire, extraction of forest products |

(continued)

**Table 7.2** (continued)

| Sl. No. | Country | Reference | Spatial scale | Pathway | Proximate drivers | Underlying drivers | Type of reforestation | Actors involved | Threats |
|---|---|---|---|---|---|---|---|---|---|
| 6 | India | Nagendra et al. 2006 | Protected area($10^2$–$10^3$ $km^2$) | Protection and planting | Protection by national government | Increased conservation awareness | Natural regeneration | National government | Grazing, harvest of forest products |
| 7 | India | Ostrom and Nagendra 2006 | Protected area($10^2$–$10^3$ $km^2$) | Protection and planting | Protection by national government | Increased conservation awareness | Natural regeneration | National government | Grazing, harvest of forest products |
| 8 | India | Robbins et al. 2007 | Protected area($10^2$–$10^3$ $km^2$) | Protection and planting | Protection by national government | Increased conservation awareness | Plantation and natural regeneration | National government | Grazing, extraction of forest products |
| 9a | Nepal | Awasthi et al. 2002 | Watershed ($10^2$ – $10^3$ $km^2$) | Protection and planting | Community forestry | Increased conservation awareness, scarcity of forest products | Natural regeneration | Local communities, national government, international donor agencies | Agricultural expansion, timber harvesting |
| 9b | | | | Protection and planting | Private plantations | Increased national awareness, scarcity of forest products | Plantation | Landowners, national governments, international donor agencies | Agricultural expansion, timber harvesting |
| 10 a | Nepal | Bhandari and Grant 2007 | Watershed ($10^2$–$10^3$ $km^2$) | Natural regeneration | Abandonment of unproductive marginal agriculture | Unprofitability of farming, outmigration of labour to city | Natural regeneration | Farmers | Not mentioned |
| 10b | | | | Protection and planting | Community forestry | Increased conservation awareness, decentralization of forest management to strengthen local institutions, scarcity of forest products | Plantation | Local communities, national government, international donor agencies | Extraction of forest products, conflicts due to wildlife attacks from reforested areas |

| | | | | | | | | | |
|---|---|---|---|---|---|---|---|---|---|
| 11a | Nepal | Bhattarai and Conway 2008 | District ($10^3$–$10^4$ km$^2$) | Protection and planting | Community forestry | Increased conservation awareness, decentralization of forest management to strengthen local institutions, scarcity of forest products | Plantation and natural regeneration | Local communities, national government | Grazing, extraction of forest products |
| 11b | | | | Farm agro-forestry | Tree planting on farmland | Increased conservation awareness, scarcity of forest products, scarcity of labour | Plantation | Farmers, national government | Timber harvesting |
| 12a | Nepal | Brown and Shrestha 2000 | Watershed ($10^2$–$10^3$ km$^2$) | Protection and planting | Community forestry | Scarcity of forest products | Plantation and natural regeneration | Local communities, national government, international donor agencies | Extraction of forest products |
| 12b | | | | Farm agroforestry | Tree planting on farmland | Scarcity of forest products | Plantation | Farmers | Not mentioned |
| 13 | Nepal | Carter and Gilmour 1989 | Watershed ($10^2$–$10^3$ km$^2$) | Farm agroforestry | Tree planting on farmland | Scarcity of forest products, scarcity of labor | Plantation | Farmers | Not mentioned |
| 14 | Nepal | Carter 1992 | Wards ($10^2$–$10^3$ km$^2$) | Farm agroforestry | Tree planting on farmland | Scarcity of forest products, erosion control | Plantation and natural regeneration | Farmers | Timber and fuelwood harvesting |
| 15a | Nepal | Fox 1993 | Village ($10^0$–$10^1$ km$^2$) | Protection and planting | Community forestry | Decentralization of forest management to strengthen local institutions, scarcity of forest products, alteration in agricultural and livestock husbandry practices leading to decreased requirement for forest products | Plantation followed by natural regeneration | Local communities, national government, international donor agencies | Conflicts due to wildlife attacks from reforested areas, extraction of forest products, increasing livestock density |
| 15b | | | | Farm agroforestry | Tree planting on farmland | Scarcity of forest products | Not mentioned | Farmers | Not mentioned |

(continued)

**Table 7.2** (continued)

| Sl. No. | Country | Reference | Spatial scale | Pathway | Proximate drivers | Underlying drivers | Type of reforestation | Actors involved | Threats |
|---|---|---|---|---|---|---|---|---|---|
| 16a | Nepal | Gautam et al. 2004 | Watershed ($10^2$–$10^3$ km$^2$) | Natural regeneration | Abandonment of unproductive marginal agriculture | Unprofitability of farming, outmigration of labour to city | Plantation and natural regeneration | Farmers | Extraction of forest products |
| 16b | | | | Protection and planting | Community forestry | Scarcity of forest products | Plantation and natural regeneration | Local communities, national government | Extraction of forest products |
| 17 | Nepal | Gilmour 1988 | District ($10^3$–$10^4$ km$^2$) | Farm agroforestry | Tree planting on farmland | Scarcity of forest products | Plantation and natural regeneration | Farmers | Not mentioned |
| 18a | Nepal | Gilmour and Nurse 1991 | Watershed ($10^2$–$10^3$ km$^2$) | Protection and planting | Community forestry | Decentralization of forest management to strengthen local institutions | Not mentioned | Local communities | Not mentioned |
| 18b. | | | | Natural regeneration | Reforestation in degraded government forests | Alteration in agricultural and livestock husbandry practices leading to decreased requirement for forest products | Natural regeneration | Farmers | Not mentioned |
| 18c. | | | | Farm agroforestry | Tree planting on farmland | Scarcity of labor for collection of products from forest | Not mentioned | Farmers | Not mentioned |
| 19a. | Nepal | Jackson et al. 1998 | District ($10^3$–$10^4$ km$^2$) | Protection and planting | Community forestry | Decentralization of forest management to strengthen local institutions | Plantation and natural regeneration | Local communities | Grazing, extraction of forest products |
| 19b. | | | | Natural regeneration | Abandonment of unproductive marginal agriculture | Scarcity of farm labour | Natural regeneration | Farmers | Grazing, extraction of forest products |
| 19c. | | | | Farm agroforestry | Tree planting on farmland | Scarcity of forest products | Plantation and natural regeneration | Farmers | Not mentioned |

| | | | | | | | | | |
|---|---|---|---|---|---|---|---|---|---|
| 20. | Nepal | Nagendra et al. 2004 | Landscape ($10^2$–$10^3$ $km^2$) | Protection and planting | Community forestry | Increased conservation awareness, decentralization of forest management to strengthen local institutions, revenue from wildlife tourism | Plantation and natural regeneration | Local communities, national government, international donor agencies | Lack of decision making power provided to local communities, withdrawal of national and international support |
| 21. | Nepal | Nagendra 2007 | Country | Protection and planting | Community forestry | Decentralization of forest management to strengthen local institutions, revenue from wildlife tourism | Plantation and natural regeneration | Local communities, national government | Lack of decision making power provided to local communities |
| 22. | Nepal | Virgo and Subba 1994 | Landscape ($10^2$–$10^3$ $km^2$) | Farm agroforestry | Tree planting on farmland | Scarcity of forest products | Plantation | Farmers | Not mentioned |
| 23. | Pakistan | Dove 1993 | Country | Farm agroforestry | Tree planting on farmland | Scarcity of forest products | Plantation | Farmers | Not mentioned |
| 24. | Sri Lanka | Jayatissa et al. 2002 | Lagoon ($10^0$–$10^1$ $km^2$) | Natural regeneration | Expansion of mangrove into shallow areas of lagoon | Decrease in lagoon salinity due to upstream irrigation project | Natural regeneration | Not mentioned | Agricultural expansion |

in agricultural and livestock practices of local communities have led to decreased demand for forest products, making it easier for communities to protect forests. Another case, also from Nepal (Nagendra et al. 2005), discussed the importance of revenues from wildlife tourism that help support community efforts to protect local forests. Of course, several cases cited the congruence of multiple factors.

Some proponents of the Environmental Kuznets curve argue that nations which achieve a level of economic development are often motivated to reforest areas because of an increased national conservation awareness, and desire to live in proximity to nature. Such explanations have not been used to explain reforestation or regrowth in less wealthy countries. In developing country contexts, it is largely the scarcity of forest products that is believed to drive increases in forest area and density. Such a focus on scarcity as the major driver of reforestation in developing countries, disregards the capacity of poor, rural, forest dependent households and developing country governments to be motivated about conservation. The capacity to care about one's environment is not restricted only to wealthy countries. This analysis shows clear evidence that that poor communities can and often do care strongly enough about forest conservation enough to protect forests, even at the cost of temporarily reduced access to forest products. The fact that 50% of the cases discussed an increased awareness of conservation as driving reforestation and regrowth bears testimony to its significance.

Clearly, the increased global trend towards decentralization of forest management (Agrawal et al. 2008) has also helped significantly in enabling local communities to reforest their landscapes. Although this analysis confirms the criticality of market dynamics, and in particular the impact of increasing scarcity of forest products on encouraging reforestation and regrowth, this does not appear sufficient in itself to promote reforestation in a majority of cases. In many cases, reforestation has required the additional commitment of local communities and national governments towards protection, as well as being facilitated by the devolution of forest management to local communities who often engage in tree planting activities to encourage regrowth (Poffenberger et al. 1996).

Providing further confirmation, reforestation and regrowth appeared to be conducted and assisted by a range of actors including national governments, local communities, private landowners and national and international aid organizations. National governments and local communities were involved in an overwhelming majority of the cases (18 and 16, or 90% and 80% respectively). Nine cases (45%) recorded additional involvement of local and international NGOs and aid agencies. Although some international agencies have called for an increase in private ownership of forests as a way to combat deforestation, only one of the cases of reforestation or regrowth recorded involvement of private landowners in plantation forestry (Awasthi et al. 2002).

Of the 20 cases of protection, two did not provide information on whether trees were planted or allowed to regenerate naturally, following protection from lopping, cutting, fire and grazing. Of the remaining 18 cases, 12 (67%) involved both plantation and natural regeneration – three cases only mentioned natural regeneration, and three cases only mentioned planting (one of these being the single instance of private forestry discussed in Awasthi et al. (2002)). Thus, unlike previously discussed, this pathway to reforestation is not dominated by tree planting alone, and appeared to be equally composed of planted and naturally regenerated trees following protection.

Farm agroforestry constituted the second most commonly described pathway to reforestation. As previous syntheses indicate, this pathway was predominantly driven by perceived scarcities of forest products (nine out of ten cases, or 90%). In all cases, farmers were the main actors involved – only in one of the cases (Bhattarai and Conway 2008) was the role of the national government mentioned, through the provision of incentives for farm plantations. Scarcity of labor was also mentioned in three cases (30%), along with a single mention each of conservation awareness and soil erosion control. Farm agroforestry appears to be primarily a self-initiated, local scale effort that is playing an increasingly significant role across South Asia. Contrary to previous discussions, farmers not only planted trees, they also allowed natural in-growth of trees (Gilmour 1988; Carter 1992; Jackson et al. 1998; Bhattarai and Conway 2008). This helped farmers to maintain an increased diversity of tree species on their farmland, and enabled easy availability of timber and fuelwood without the expenditure of much time, effort or money.

The natural regeneration pathway to reforestation appears to be less common in the densely populated landscapes of South Asia. While three of the five studies recorded the expected path of forest regrowth on abandoned farmlands following the unprofitability of marginal agriculture, and out migration to cities, one study (Gilmour and Nurse 1991) indicated regrowth in government forests following changes in agricultural and animal husbandry practices that reduced the need for biomass inputs from the forest. A shift from organic to chemical inputs reduced the need to maintain large numbers of animals for manure, thus reducing the pressure on the forest. A shift in cattle rearing practices from forest grazing to stall feeding, further reduced pressure on the forest. Finally, one study in Sri Lanka (Jayatissa et al. 2002) recorded the expansion of mangrove forests into a shallow lagoon due to changes in lagoon salinity following an upstream irrigation project. Clearly, reforestation is not always a positive event, and even trees can play an invasive role!

While these case studies provide some insights into the human drivers of reforestation and regrowth, they do not provide in-depth discussions of the processes and pathways that cause forests to increase in area or in density. What these kinds of studies provide in breadth, they lack in depth (Poteete and Ostrom 2008). In the next part of this chapter, four case studies of landscapes in India and Nepal are discussed, to provide insights into the human drivers and processes that act to increase forest area and density at the human–forest interface.

## 7.3 Detailed Case Studies

### *7.3.1 Methods*

Remote sensing provides a particularly effective tool for studying patterns, processes and drivers of land cover change. Satellite image analysis is perhaps the most frequently used technique used to obtain quantitative information about the extent, spatial location and configuration of land cover and its changes over time. Satellites provide the examination of spatial land cover data from multiple time points starting from the early 1970s, allowing the creation of land cover maps over landscape-scale spatial extents and over frequent time steps (Tucker et al. 2004). These instruments enable evaluations of rates of land cover change in a relatively unbiased manner compared to expensive and detailed assessments based on field data collection and interpretation.

The distribution of the four landscapes in Nepal and India is depicted in Fig. 7.1. The basic approach, methods, spatial scale and time period of analysis have been maintained across all four landscapes, thus enabling an in-depth discussion of the processes and patterns of reforestation and regrowth in a consistent manner. For each landscape, a land cover change analysis was conducted using Landsat Thematic Mapper (TM) image from around 1989/1990 and an Enhanced Thematic Mapper (ETM+) image from around 2000/2001. (Specifically, for the Chitwan and Kabhrepalanchowk landscapes of Nepal, images of 1989 and 2000 were used; for the Mahananda landscape of India, images of 1990 and 2000 were used; and for Tadoba Andhari Tiger Reserve, satellite images of 1989 and 2001 were used). Images were georegistered to topographic survey sheets, and subjected to atmospheric correction, radiometric calibration and radiometric rectification procedures

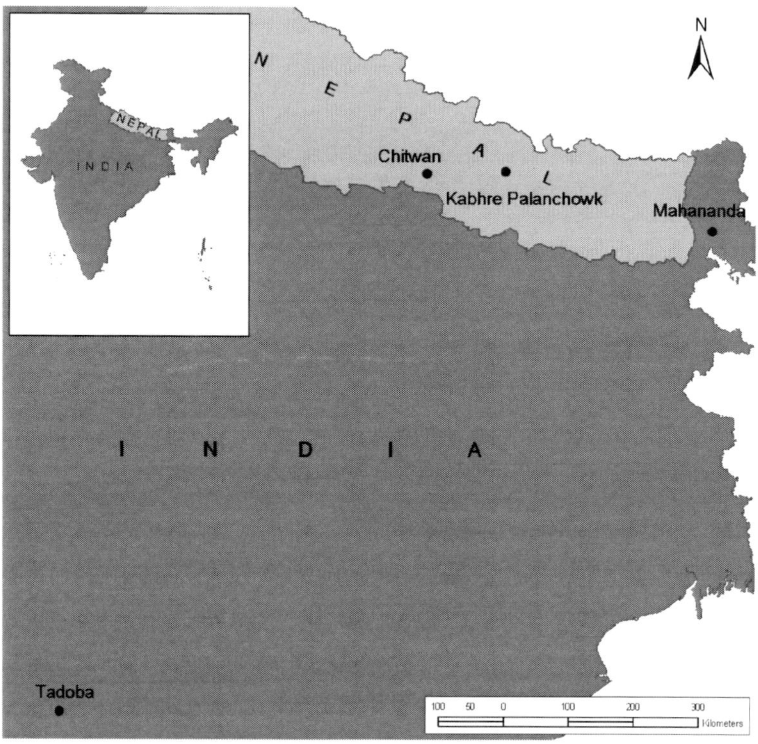


**Fig. 7.1** Location of four case study landscapes

to facilitate comparability across dates (Jensen 2000). Root mean square errors of registration were maintained at levels below 0.5 pixels and registration was verified visually by overlaying and swiping registered images, to ensure that sliver areas of misregistration did not bias estimates of land cover change (Jensen 2000).

Based on extensive collection of field training data, supervised classifications of images of both time periods were conducted into open forest, dense forest, and nonforest categories. Evaluations of classification accuracy using independent sets of test points not utilized for the classification confirmed that the accuracy of classification was high, with producer's accuracies of over 85% for all landscapes.

For assessing land cover change, individual classifications for both image dates were combined using ARC/INFO™ software to provide a change image that identifies sequences of land cover classes for both observation dates (Petit et al. 2001). Since there were three land cover classes in each date, this recoding resulted in a total of nine change classes. Since the focus of this analysis was on stability and increase in forest cover, these pixels were grouped into four change categories depending on the nature of change in forest cover that they represented.

Pixels forested both in the 1989/1990 image and in the 2000 (forest-forest) represented a "stable forest" category. Land cover *conversion* occurs when a land cover type is completely replaced by another, completely different type (Turner and Meyer Turner and Meyer 1994). Thus, "Reforestation" comprised of pixels that were nonforest in 1989/1990 and changed to either open or dense forest in 2000/2001. Finally, land cover *modification* consists of changes that affect the quality or density of forest cover without changing the nature of the land cover class (forest). Accordingly, pixels that changed from open forest in 1989/1990 to dense forest in 2000/2001 were categorized as "Regrowth". All other pixels were grouped as stable nonforest/deforestation/degradation, and were not considered further for this analysis. This approach enables an enriched understanding of increase in forest cover and density. Using information from interviews with local communities enables us to differentiate the human drivers of reforestation, for instance occurring through on formerly logged areas, from those associated with regrowth, such as the increase in forest density following protection of degraded community forests (Nagendra et al. 2008).

### 7.3.2 Landscape 1 – Chitwan District, Nepal Terai Plains

Located along the Nepal–India border, this study landscape is located in the Terai plains, to the south of the foothills and valleys below the Himalayan mountains. In the early 1950s, a malaria eradication program in the Terai opened the way for this densely forested area to be occupied, giving rise to a period of intense deforestation, and converting the region to a densely populated mix of agriculture and forests. Following concerns about rapid deforestation in this region, two protected areas were established in this landscape – Nepal's first protected area, Chitwan National Park, bordered by Parsa Wildlife Reserve. The pressure on forests however continued to be high, with hundreds of villages surrounding the park which depend on the surrounding forests to a significant extent.

Following the success of community forestry programs in the Nepal middle hills, the Government of Nepal launched community forestry programs in the Terai plains in the early 1990s. These programs enable local communities to protect patches of forest, and to earn incomes from the sale of forest products including timber. Alongside this program, in areas bordering Chitwan National Park, the park buffer zone program was established as a variant of the community forestry program, where local communities protect forest areas bordering the park, and earn revenues from the sale of forest products as well as from ecotourism.

Figure 7.2 depicts the areas of stable forest cover, reforestation and regrowth in the landscape between 1989 and 2000. Clear differences can be observed across management regimes. Community forests contain a substantial proportion of stable forest cover, along with some forest regrowth, while the buffer zone forests are dominated by areas of reforestation, along with some forest regrowth. In contrast, there is hardly any reforestation taking place in the peripheral areas of the park (park periphery), although there is some forest regrowth.

These distributions have been shaped by land use, land tenure and differences in biophysical location (Nagendra et al. 2004, 2008). Since community forests are largely located at higher elevations and steeper slopes, they have been protected in part by their relative inaccessibility, giving rise to the relatively higher proportions of stable forest cover in these areas. However, while these forests have not been cleared for agriculture or other land uses, they have been subjected to significant levels of extraction of forest products, and to grazing, resulting in a condition of relative degradation in the early 1990s. Following their protection through the community forestry program, forest density has increased, resulting in the regrowth

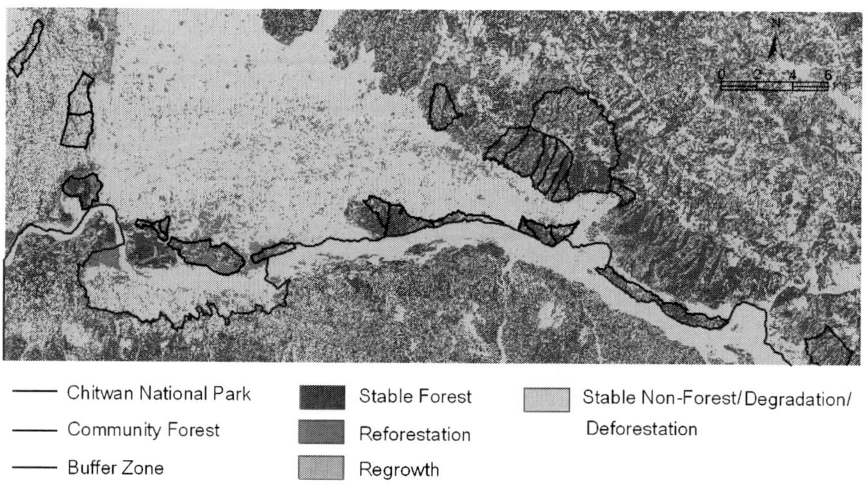


**Fig. 7.2** Map of forest change in Landscape 1 – Chitwan district, Nepal *Terai* plains (*see Color Plates*)

observed in the land cover change analysis. In contrast, several areas in the buffer zone experienced significant land conversion to agriculture and urban uses prior to 1990. After being brought under protection through the park buffer zone program in the early 1990s, these areas were extensively planted with trees as well as protected from grazing and harvest of forest products, resulting in rapid regrowth of forest cover in this fertile alluvial area.

In the peripheral areas of the Chitwan National Park, there is frequent and extensive flooding from the Rapti River, as well as burning for traditional harvest of thatch grass during specific times of the year, leading to forest degradation. Since the spatial location of flooding and thatch grass burning shifts from year to year, some regrowth of tree cover is observed in areas that were subject to these activities. Finally, in the surrounding landscape which is largely comprised of areas in the fertile Chitwan agricultural valley, small scattered patches of regrowth are visible corresponding to small private plantations and to tree planting programs conducted by the East-Rapti Irrigation Project of the Department of Irrigation along irrigation canals.

### *7.3.3 Landscape 2 – Kabrepalanchowk District, Nepal Middle Hills*

The middle hills of Nepal, located between the Terai plains to the south, and the high Himalayan ranges to the north, have been densely populated for centuries. The Kabrepalanchowk district, where this study landscape was located, has the distinction of being the oldest district in Nepal to initiate experiments with community forestry. This landscape constitutes a mix of upland forested and terraced agricultural areas, and plantations and irrigated agriculture in river valleys. As with most parts of Nepal, this landscape is characterized by a dense mix of forests, agriculture and settlements. Local communities depend on the forest for timber, fuelwood, fodder and a variety of non-timber forest products.

Forests in this landscape are largely managed under three different tenure regimes: community forests, leasehold forests and national forests. The boundaries of 11 community forests, ten leasehold forests and 11 national forests were mapped and forest cover change studied using satellite images of 1990 and 2000 (Fig. 7.3). Under the community forestry program, community forests are managed by local communities organized into user groups. These forests are largely protected from harvest and use, but communities are permitted to sell forest products including timber at specified intervals of time, and to utilize the income for developmental activities (Gautam et al. 2004). In this district, the first to implement community forestry, international aid agencies – in particular the Australian Agency for International Development – have provided technical and financial support to the community forestry program since its inception. As Fig. 7.3 indicates, most community forests appear well protected with large patches of stable forest. This tenure regime also contains the greatest proportion of area under forest regrowth, indicating the impact of the protection provided to these forests by local communities.

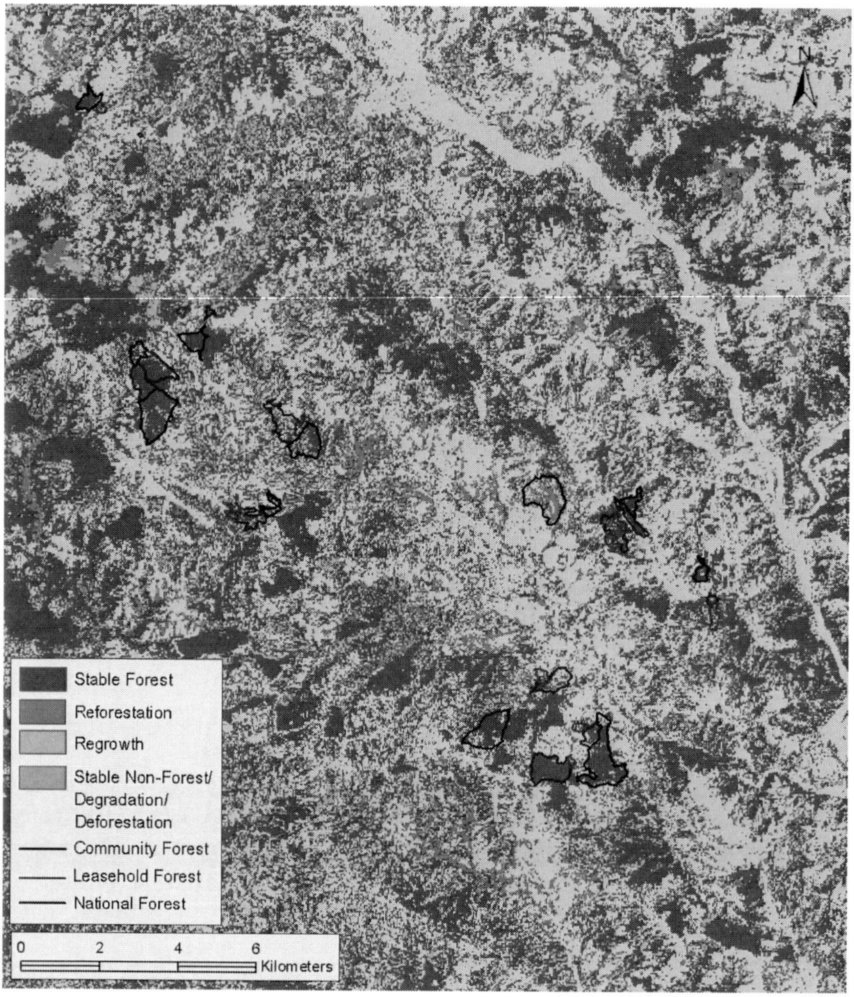


**Fig. 7.3** Map of forest change in Landscape 2 – Kabrepalanchowk district, Nepal middle hills (*see Color Plates*)

Smaller patches of highly degraded forests are handed over to small groups of very poor families under the leasehold forestry program, for conducting agroforestry and horticultural forestry (Nagendra et al. 2005). This program received initial support from the Food and Agriculture Organization of the United Nations, as well as the International Fund for Agricultural Development. However, leasehold forest users have faced significant opposition from other forest users who have been excluded from this program, and often find it hard to protect the boundaries of their forests from encroachment by other users. The leasehold forest patches are much smaller than community forests or national forests, with very little forest cover compared to the other forest tenure categories. These forests show only a small amount of forest regrowth, indicating the limited success of this program. Finally, most remaining patches of forest in the district that are not under community forestry or under leasehold forestry, are managed as national forests. Technically under the protection of the Forest Department, in practice some of these forests are under the de facto protection of local communities, showing some forest regrowth, while others are accessed regularly by nearby settlements.

### *7.3.4 Landscape 3 – Mahananda Wildlife Sanctuary, Eastern Indian Foothills*

The Mahananda Wildlife Sanctuary (MWS) extends between an elevation of 350–1,500 m above sea level, and is located along the foothills of the Himalayas. There are moderate to steep slopes and high ridges to the north with mixed evergreen and deciduous forest cover, sloping down to gentle, almost flat stretches of the Terai dominated by *Shorea robusta* (sal) and grassy alluvial plains, to the south (Wildlife Circle 1997; Shankar 2001). The forested habitat within the sanctuary harbors a rich diversity of flora and fauna, and forms the largest compact block of forest at the western end of the elephant migration route in this region. Plantations of various economically valuable tree species are also interspersed throughout the forest.

Substantial portions of the park, especially in the riverine plains and foothills, were maintained for commercial timber extraction until the early 1990s. After this period, various forest regulations including the National Forest Policy of 1988, and the subsequently passed Tree Felling Act and Supreme Court Orders of 1996 curtailed tree felling and fuelwood extraction within the sanctuary (IIFM 2001).

An analysis of land cover change was conducted using Landsat TM and ETM+ satellite images of 1990 and 2000 (Ostrom and Nagendra 2006). As Fig. 7.4 indicates, clearly, the overall landscape is experiencing a trend towards reforestation and regrowth. There is some regrowth in the surrounding landscape, much of which appears to have taken place in the areas of shade grown tea gardens surrounding the MWS. Some of these tea gardens have been abandoned, and several are no longer cultivated as extensively as they used to be. Since tea in these areas is shade grown and interspersed with trees, these areas have consequently witnessed substantial increases in tree canopy cover.

Within the MWS, large patches of stable forest to the north indicate the impact of natural protection through the steep topography of the northern section. Substantial regrowth has taken place in the less hilly southern sections where commercial timber extraction was discontinued by the Government of India during the early to mid-1990s, enabling the forest to regenerate. This is especially observed in areas of low elevation located towards the center of the protected areas, where human impacts are relatively lower.

There is still substantial clearing towards the peripheral areas of the MWS which are connected to illegal timber markets by road and railway networks. On all sides except the northern boundary, the sanctuary is surrounded by tea garden settlements

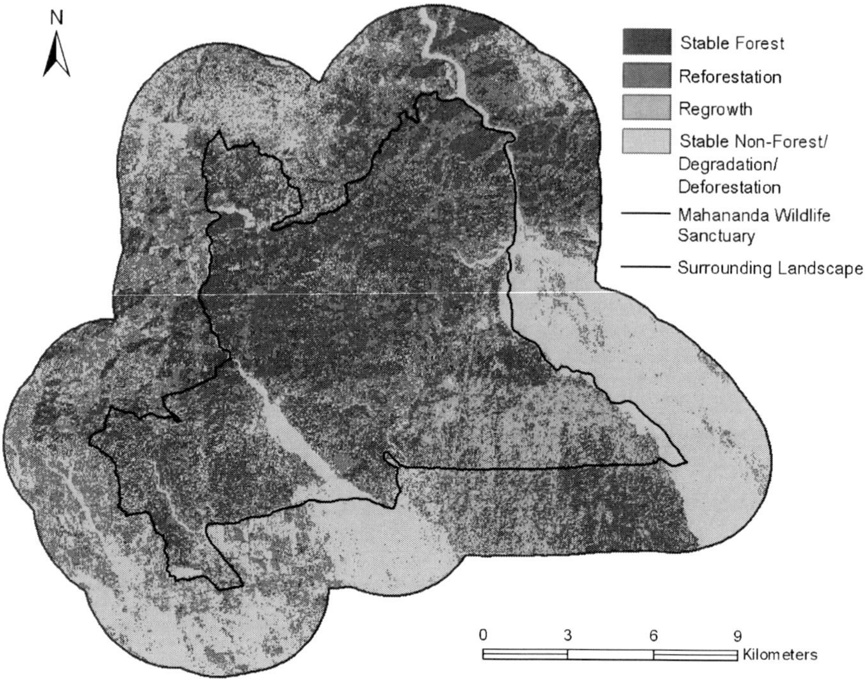


**Fig. 7.4** Map of forest change in Landscape 3 – Mahananda Wildlife Sanctuary, eastern Indian foothills (*see Color Plates*)

and villages which depend on the forest for fuelwood, timber and other forest products (Das and Guha 2003). Despite regular patrolling by Forest Department guards armed with guns, and a network of electric fences in important wildlife habitat areas, there is considerable pressure on the park from these settlements (Ostrom and Nagendra 2006). There is extensive transport of illegally extracted timber outside the sanctuary through a dense network of smaller roads just outside the MWS. The Forest Department, ill-staffed due to financial constraints, nevertheless continues frequent monitoring of the MWS, and has seized significant volumes of illegal timber from individual and large-scale poachers. The pressure on the park from the surrounding settlements is significant, though, and makes it difficult for them to completely protect the forest.

Development initiatives in the region, while providing a much-needed impetus for local economies, pose a significant threat to the future extent and connectivity of forest cover in this landscape. The Indian Railway's plans to convert the existing meter-gauge railway line in this area into broad-gauge is likely to significantly increase rail traffic, and to adversely affect the protected forests in this region. Timber smuggling activities also continue to take place at an organized scale, some even with support from other agencies of the Indian Government, despite efforts by the Forest Department to limit these activities.

### *7.3.5 Landscape 4 – Tadoba Andhari Tiger Reserve, Central Indian Plains*

The Tadoba Andhari Tiger Reserve (TATR) is situated in the central plains of India, in the eastern part of Maharashtra state. The reserve consists of a national park and wildlife sanctuary covering a 625 $km^2$ landscape that is an interspersion of grasslands, water bodies and dry tropical deciduous forests along with patches of riparian forest alongside streams. Six villages within the boundaries of the park, and 53 villages located on the periphery, depend on the protected area for fuel, fodder, timber and other non-timber forest produce requirements. Most of the peripheral villages are located on the northern and western boundary, and the road network in the northern part of the reserve is quite well developed. On the southern and eastern sides, the TATR is surrounded by Government controlled Reserve Forest and Protected Forest areas.

Analysis of forest cover change between 1989 and 2001 (Fig. 7.5) indicates little forest area where the density of village settlements and road networks is greatest, to the northeast and northwest sides (Nagendra et al. 2006). Towards the south, where the surrounding categories of State owned protected forest buffer the TATR from human influence, forest cover is relatively well preserved. Forest cover in the park interior also appears to be comparatively protected.

While there is hardly any increase in forest cover within the park – possibly a consequence of this area having already been largely forested by 1989 – the analysis of change trajectories shows that there is substantial increase in density, or regrowth, within formerly open, degraded forest. This can be observed both within the park towards the northern boundary, as well towards the southern end both within the park and in the adjacent Protected and Reserve Forest areas. In 1993, the Maharashtra State Government designated the area within the Tadoba National Park and Andhari Wildlife Sanctuary as a Tiger Reserve. However, rights regarding collection of minor forest products were suspended in 1992, even before the formal declaration of the Tiger Reserve. It seems as though efforts by the park managers to limit forest extraction and to protect the area at the periphery seem to be paying off in terms of limiting forest clearing and encouraging regrowth.

Such restrictions have not come without socioeconomic cost. Villages within and outside the protected area have faced substantial difficulties after the loss of their traditional rights over forest harvest. Many continue to collect forest produce and to graze cattle within the park, despite efforts from forest guards to protect the TATR. This analysis indicates that the park itself appears well protected despite these problems, with forest cover maintained and even increasing slightly during the period of observation. However, protection by guards may not be sustainable in the long run unless local communities are involved. Previous experience in this region indicates that, if the support of local communities can be obtained, the park can be much better protected against wildlife poachers and timber loggers, ensuring its long term survival and maintenance.

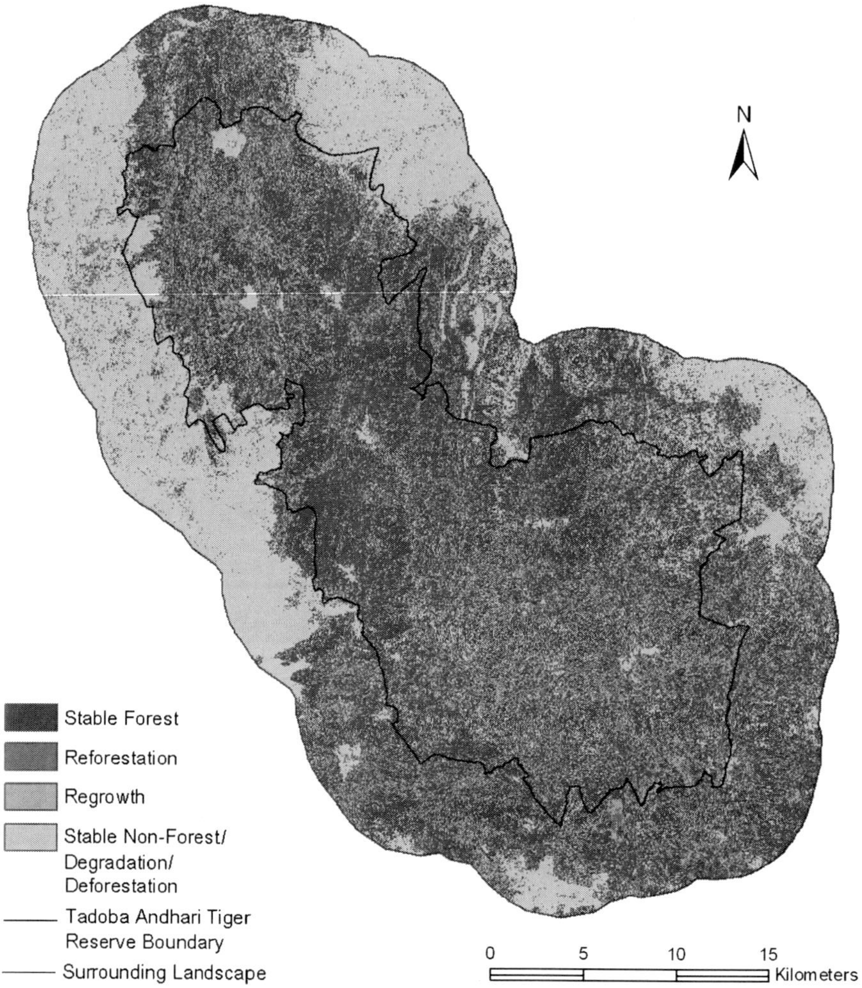


**Fig. 7.5** Map of forest change in Landscape 4 – Tadoba Andhari Tiger Reserve, central Indian plains (*see Color Plates*)

## 7.4 Conclusions

The analyses presented here add to current knowledge of the drivers of reforestation and forest regrowth in tropical or economically developing countries, where discussions of the Environmental Kuznets curve do not apply. Corroborating Rudel and others (2005) and Rudel (this volume), this meta-analysis of reforestation and regrowth using 24 case studies in Bangladesh, India, Nepal, Pakistan and Sri Lanka indicates that reforestation and regrowth in protected and planted forests is driven by the scarcity of forest products, while natural reforestation occurs when marginal agricultural lands are abandoned due to labor scarcities. In addition, regrowth in farm agroforestry areas is increasing due to the difficulty of obtaining timber and fuelwood from forests, both because of their scarcity and the shortage of farm labor.

This analysis contributes new insights to this discussion. While the scarcity of forest products is a critical factor driving the protection of forests across South Asia, a strong conservation ethic also plays a significant role in enabling this protection. Significantly, a large part of this protection and plantation is conducted by local communities, though with some facilitation and support provided by their governments and by NGOs. The forest decentralization programs active across South Asia (in parallel with processes taking place across the globe, Agrawal et al. 2008) have greatly facilitated this process. Resource deprived communities living in conditions of high population density and poverty can organize to protect their environment, even at a significant short term cost to themselves. The reforestation promoted by governments through social forestry, communities through community forestry and Joint Forest Management, and local farmers through agroforestry is more biodiverse than previously acknowledged, being a mix of plantation as well as natural regeneration (Bhat et al. 2001; Maraseni 2008).

These findings are supported by the detailed case studies of four landscapes in Nepal and India. Conducted in a range of biophysical and ecological environments, using standardized methods of investigation at similar spatial and temporal scales, the human drivers of forest regrowth and reforestation have been investigated in detail in these landscapes. Corroborating the meta-analysis, increases in forest area or forest density in these landscapes are largely due to protection and planting, either in government protected national parks and wildlife reserves, or in community protected areas. Some of this protection can be attributed to increasing scarcity of forest products.

In addition, a strong driver for reforestation and regrowth is the increased interest in conservation taken by local communities, national governments, and national and international NGOs. This is due in large part to the awareness of the large scale deforestation that has already taken place, and its negative impacts on biodiversity, soil erosion and environmental change, in addition to impacts on local livelihoods. In at least one case, in the Chitwan landscape, providing incomes through wildlife tourism has helped raise revenues to support forest monitoring by local guards (Nagendra et al. 2004, 2008). Although these efforts have not been distributed as well as they should have, they indicate the potential of positive incentives in encouraging reforestation in already motivated communities. Private players play a relatively small role in reforestation, with only some small private plantations recorded in the Chitwan landscape.

Some regrowth has also taken place through agroforestry, particularly in the Kabrepalanchowk landscape, where leasehold forestry users have been encouraged to plant trees along with agricultural crops in degraded forest lands. In addition, the Mahananda landscape has witnessed fairly widespread, natural regeneration on hill tea estates which are facing a labor crisis. Yet, the major proportion of reforestation and regrowth in these landscapes appears to be taking place in areas protected by national governments and/or local communities.

The potential of local communities to mitigate environmental change has been insufficiently recognized so far. This research indicates that institutions have a powerful capacity to modify otherwise negative socio-economic conditions such as increasing population density, affluence and technology, and if properly incentivized, can act to mitigate global change by promoting increases in forest cover and/or forest density even in densely populated, resource dependent, developing economies.

**Acknowledgments** I thank the Indian Forest Departments of Maharashtra and West Bengal and the Nepal Forest Department, and the local communities living in the study areas for assisting this research. Rucha Ghate, Birendra Karna, Mukunda Karmacharya, Sajid Pareeth and Sugato Dutt provided extensive inputs in the field. Financial support from the Society in Science: Branco Weiss Fellowship, and from a National Science Foundation (USA) grant to CIPEC, is gratefully acknowledged.

## References

Agrawal A, Chhatre A, Hardin R (2008) Changing governance of the world's forests. Science 320:1460–1462

Allen JC, Barnes DF (1985) The causes of deforestation in developing countries. Annals Assoc Am Geogr 75:163–184

Angelsen A, Kaimowitz D (1999) Rethinking the causes of deforestation: lessons from economic models. World Bank Res Obs 14:73–98

Awasthi KD, Sitaula BK, Singh BR, Bajacharaya RM (2002) Land-use change in two Nepalese watersheds: GIS and geomorphometric analyses. Land Degrad Dev 13:495–513

Bhandari B, Grant MR (2007) Land use and population dynamics in the Kalikhola watershed of Nepal. J Rural Community Dev 2:100–109

Bhat DM, Murali KS, Ravindranath NH (2001) Formation and recovery of secondary forests in India: A particular reference to Western Ghats in southern India. J Trop For Sci 13:601–620

Bhattarai M, Hammig M (2001) Institutions and the Environmental Kuznets Curve for deforestation: a crosscountry analysis for Latin America, Africa and Asia. World Dev 29:995–1010

Bhattarai K, Conway D (2008) Evaluating land use dynamics and forest cover change in Nepal's Bara district (1973–2003). Hum Ecol 26:81–95

Brown S, Shrestha B (2000) Market-driven land-use dynamics in the middle mountains of Nepal. J Env Manag 59:217–225

Carter AS, Gilmour DA (1989) Increase in tree cover on private farm land in central Nepal. Mt Res Dev 9:381–391

Carter EJ (1992) Tree cultivation on private land in the middle hills of Nepal: Some lessons from Dolakha District. Mt Res Dev 12:241–255

Das TK, Guha I (2003) Economics of man-animal conflict: A lession for wildlife conservation. ACB Publications and DSA Centre for Regional Economics Studies, Jadavpur University, Kolkata

Dove MR (1993) The coevolution of population and environment: the ecology and ideology of feedback relations in Pakistan. Popul Environ 15:89–111

Ehrlich PR, Ehrlich AH (1990) The population explosion. Simon and Schuster, New York.

Food and Agricultural Organization of the United Nations (2005) Forest Resources Assessment, 2005. Forestry Division, Rome. Accessed at www.fao.org/forestry/fra2005. May, 2007

Foster A, Rosenzweig M (2003) Economic growth and the rise of forests. Q J Econ 118:601–637

Fox J (1993) Forest resources in a Nepali village in 1980 and 1990: The positive influence of population growth. Mt Res Dev 13:89–98

Gautam AP, Shivakoti GP, Webb EL (2004) Forest change, physiography, local economy and institutions in a mountain watershed in Nepal. Environ Manag 33:48–61
Geist HJ, Lambin EF (2001) What drives tropical deforestation? A meta-analysis of proximate and underlying causes of deforestation based on subnational case study evidence. Louvain-la-Neuve (Belgium): LUCC International Project Office, LUCC Report Series no. 4
Geist HJ, Lambin EF (2002) Proximate causes and underlying driving forces of tropical deforestation. Bioscience 52:143–150
Ghate R, Nagendra H (2005) Role of monitoring in institutional performance: forest management in Maharashtra, India. Conserv Soc 3:509–532
Gilmour DA (1988) Not seeing the trees for the forest: a re-appraisal of the deforestation crisis in two hill districts of Nepal. Mt Res Dev 8:343–350
Gilmour DA, Nurse MC (1991) Farmer initiatives in increasing tree cover in central Nepal. Mt Res Dev 11:329–337
Giri C, Zhu Z, Tieszen LL, Singh A, Gillette S, Kelmelis JA (2008) Mangrove forest distributions and dynamics (1975–2000) of the tsunami-affected region of Asia. J Biogeogr 35:519–528
Global Forest Resources Assessment (2005) FAO Forestry Paper 147, United Nations, Rome
Jackson WT, Tamrakar RM, Hunt S, Shepherd KR (1998) Land-use changes in two middle hills districts of Nepal. Mt Res Dev 18:193–212
Jayatissa LP, Guero M-C, Hettiarachchi S, Koedam N (2002) Changes in vegetation cover and socio-economic transitions in a coastal lagoon (Kalametiya, Sri Lanka), as observed by teledetection and ground truthing, can be attributed to an upstream irrigation scheme. Env Dev Sustainabil 4: 167–183
Jensen JR (2000) Remote sensing of the environment: an earth resource perspective. Prentice Hall, Upper Saddle River, NJ
Kauppi PE, Ausubel JH, Fang J, Mather AS, Sedjo RA, Waggoner PE (2006) Returning forests analyzed with the forest identity. Proc Natl Acad Sci U S A 103:17574–17579
Keys E, McConnell WJ (2005) Global change and the intensification of agriculture in the tropics. Global Environ Change Part A 15:320–337
Maraseni TK (2008) Selection of non-timber forest species for community and private plantations in the high and low altitude regions of Makwanpur district, Nepal. Small-Scale For 7:151–161
Mather A (2007) Recent Asian forest transitions in relation to forest transition theory. Int For Rev 9:491–502
Nagendra H, Karmacharya M, Karna B (2004) Evaluating forest management in Nepal: views across space and time. Ecol Soc 10: 24. http://www.ecologyandsociety.org/vol10/iss1/art24/
Nagendra H, Karna B, Karmacharya M (2005) Examining institutional change; leasehold forestry in Nepal. Conserv Soc 3:72–91
Nagendra H, Pareeth S, Ghate R (2006) People within parks: forest villages, land-cover change and landscape fragmentation in the Tadoba Andhari Tiger Reserve, India. Appl Geogr 26:96–112
Nagendra H (2007) Drivers of reforestation in human-dominated forests. Proc Natl Acad Sci U S A 104:15218–15223
Nagendra H, Pareeth S, Sharma B, Schweik CM, Adhikari KR (2008) Forest fragmentation and regrowth in an institutional mosaic of community, government and private ownership in Nepal. Landscape Ecol 23:41–54
Mather AS, Needle CL (2000) The relationship of population and forest trends. Geogr J 166:2–13
Ostrom E, Nagendra H (2006) Insights on linking forests, trees and people from the air, on the ground and in the laboratory. Proc Natl Acad Sci U S A 103:19224–19231
Petit C, Scudder T, Lambin E (2001) Quantifying processes of land-cover change by remote sensing: resettlement and rapid land-cover changes in south-eastern Zambia. Int J Remote Sens 22:3435–3456
Poffenberger M, McGean B, Khare A (1996) Communities sustaining India's forests in the twenty-first century. In: M. Poffenberger M, McGean B (eds) Village voices, forest choices: joint forest management in India. Oxford, Delhi

Poteete A, Ostrom E (2008) Fifteen years of empirical research on collective action in natural resource management: struggling to build large-N databases based on qualitative research. World Dev 36:176–195

Robbins PF, Chhangani AK, Rice J, Trigosa E, Mohnot SM (2007) Enforcement authority and vegetation change in Kumbhalgarh wildlife sanctuary, Rajasthan, India. Environ Manag 40:365–378

Rudel T (2005) Tropical forests: regional paths of destruction and regeneration in the late twentieth century. Columbia University Press, New York

Rudel T, Coomes O, Moran E, Achard F, Angelsen A, Xu J, Lambin E (2005) Forest transitions: towards a global understanding of land use change. Global Environ Chang 15:23–31

Safa MS (2006) The role of NGOs in improving social forestry practice: do they promote livelihood, sustainability and optimal land use in Bangladesh? Small-Scale For Econ Manag Policy 5:207–229

Shankar U (2001) A case of high tree diversity in a sal (Shorea robusta)-dominated lowland for of Eastern Himalaya: floristic composition, regeneration and conservation. Curr Sci 81:776–786

Tucker CJ, Grant DM, Dykstra JD (2004) NASA's global orthorectified Landsat dataset. Photogram Eng Remote Sensing 70:313–322

Turner BL, Meyer WB (1994) Global land-use and land-cover change: An overview. In: Meyer WB, Turner BL (eds) Changes in land use and land cover: a global perspective. Cambridge University Press, Great Britain, pp 1–9

Virgo KJ, Subba KJ (1994) Land-use change between 1978 and 1990 in Dhankuta district, Koshi

Wildlife Circle (1997) Management plan of Mahananda Wildlife Sanctuary, West Bengal. Government of West Bengal, Kolkata

Young OR, Lambin EF, Alcock F, Haberl H, Karlsson S I, McConnell WJ, Myint T, Pahl-Wostl C, Polsky C, Ramakrishnan P, Schroeder H, Scouvart M, Verburg PH (2006) A portfolio approach to analyzing complex human–environment interactions: institutions and land change. Ecol Soc 11: 31. http://www.ecologyandsociety.org/vol11/iss2/art31/

# Chapter 8
# Threats to the Forest Transition in the Midwest United States

**Tom P. Evans, Shanon Donnelly, and Sean Sweeney**

T.P. Evans (✉), S. Donnelly and S. Sweeney
Center for the Study of Institutions, Population and Environmental Change
Indiana University, IN, USA
e-mail: evans@indiana.edu

T.P. Evans
Department of Geography, Indiana University, IN, USA

S. Donnelly
Department of Geography and Planning, University of Akron, OH, USA

H. Nagendra and J. Southworth (eds.), *Reforesting Landscapes: Linking Pattern and Process*, Landscape Series 10,
DOI 10.1007/978-1-4020-9656-3_8, © Springer Science+Business Media B.V. 2010

## 8.1 Introduction

Given the dramatic deforestation occurring in many of the world's developing areas, the prospect for reforestation to offset (now or in the future) some of the negative environmental consequences of deforestation is imminently appealing. There is a growing body of research finding evidence that countries undergo a forest transition, from a phase of rapid deforestation to a period of modest deforestation, and some eventually to reforestation (Mather 1992, Mather and Needle 1998, Rudel et al. 2005; Chapter 3). An examination of the social and environmental dynamics associated with forest transitions can contribute to an understanding of the feedbacks that come into play at different stages of this potential transition. In some places, this dynamic seems to be associated with the creation of non-farm employment that pulls farmers off the land thereby inducing recovery of forests on old and abandoned fields (Mather 1992, Polanyi 2001; Chapter 3). In other places, scarcity of forest products has prompted active reforestation efforts by both government and private owners (Foster and Rosenzweig 2003).

During the 1990s, 38% of the world's countries experienced increases in forest cover, but these reforesting countries differ in the period when the transition occurred and the amount of forest cover remaining when the transition occurred. Some countries started a reforestation phase at 40% of the original forest cover, while others began only when forest cover was nearly 0%. There has been an increasing interest in global processes of reforestation (e.g., Perz and Skole 2003, Nagendra 2007), and with this attention has come a need for elaboration of Forest Transition Theory (Perz 2007; Chapter 4). There are also questions regarding the applicability of Forest Transition Theory across different temporal and spatial scales and the need to distinguish the contributions of short-term vs. long-term dynamics driving trajectories of forest-cover change (Perz 2007).

In many land-change science studies, structured survey instruments are integrated with remote sensing data to explore the attributes of actors that lead to particular land-cover change outcomes (e.g., McCracken et al. 1999, Walsh et al. 2005). However, these social survey data sources are generally unavailable for the periods when many countries transitioned from deforestation to reforestation. For example, in the United States, aerial photography is available as early as the 1930s but data describing household land management practices are sparse for this period and mostly take the form of anecdotal narratives rather than instruments that lend themselves to quantitative analysis. Still, in order to better understand how and why areas transitioned from deforestation to reforestation we need a better understanding of the historical conditions during this period and implications for future land-cover change trajectories.

The overall objective of the research presented here is to describe the trajectories of land-cover change that have occurred in the eastern United States and at a finer spatial scale focusing on the state of Indiana. We summarize the best data available documenting land-cover change to explore the role of actor and landscape heterogeneity in producing aggregate patterns of deforestation and reforestation with an emphasis on key drivers of these historical changes. Then, using a framework of strengths, weaknesses, opportunities, and threats (SWOT), we compare past drivers of reforestation and future potential processes that may affect the trajectories of land-cover change in the near future. We conclude that there are currently several major forces that may modify the recent trend of reforestation that has characterized the past 50–100 years in the region.

## 8.2 Forest Transitions

There has been some progress in articulating the broad-scale social context under which a forest transition occurs in contemporary settings (e.g., Klooster 2003, Perz and Skole 2003). Factors such as population growth, urbanization, income growth, land tenure, and capitalization of agriculture have been linked to deforestation while labor shortage and economic transitions have been linked to forest regrowth (Perz and Skole 2003). While demographic change has commonly been associated with processes of land-cover change (Perz and Skole 2003; Chapter 3), the process of forest regrowth occurred in south-central Indiana in times of both population increase and decrease (Lindsey et al. 1965, Evans et al. 2001). Thus demographic change may certainly be a contributing factor to changes in forest cover, but it cannot entirely explain the trajectories of land-cover change seen in the study area since the forest regrowth process began. Population decline (MacDonald et al. 2000), technological change (Barbier 1997, Mol and Sonnenfeld 2000), and economic sector shifts (Klooster 2003) have been associated with the process of agricultural abandonment and forest regrowth, yet there are many locations where these drivers exist and regrowth has not occurred.

As is the case with many general theoretical approaches, Forest Transition Theory is most often associated with national-level trajectories of land-cover change that oversimplify forest dynamics of pre- and post-settlement forests and mask local-level complexities (Perz 2007). A net increase of area in forest does not necessarily translate to an increase in forest biomass or forest value. Furthermore, forests regrowing on formerly agricultural areas in south-central Indiana have very different species composition, canopy closure, and density than the forests that existed before European-based settlement (Evans et al. 2001). Therefore, a condition of 20% forest cover composed of pre-settlement mature forest should not necessarily be equated with a condition of 20% forest cover composed of post-settlement secondary forest. Another shortcoming of national-scale theories is that they often mask complexities of actors at the local level. While the net result of actions at a local level may result in an increase in forest cover, it is important to understand how different actors may have diverse responses to the same situation. This transition from forested to deforesting or deforested to reforesting occurs at various scales from individual property to counties, states, and nations. Rudel et al. (2005) examined these processes at the national scale using data from the Food and Agriculture Organization, and it is at this level that Forest Transition Theory has primarily been proposed and tested. While national and global datasets offer the benefit of broad spatial extent, they are often hindered by inconsistencies in the quality of data reported and different definitions of what constitutes forest and monitoring inconsistencies between nations. Local-level analysis can provide insight into the efficacy and reliability of national-level data.

Transitions in forest cover are not trivial in their environmental consequences (i.e., carbon sequestration, biodiversity; DeFries et al. 2004, Foley et al. 2005; Chapter 2). During the 1990s, 38% of the world's countries experienced increases in forest cover, but the turnaround has come at very different points in their deforestation trajectory, or at different "trough" points. The difference of when the transition takes place has tremendous implications for the biodiversity of forests that grow back. Rudel et al. (2005) note that the dynamics characterizing northern European transition in the twentieth century are very different from what has been happening in Asia in the past 15 years (Chapters 7, 11). In response to scarcity of forest products and increased flooding, many Asian countries have initiated aggressive reforestation campaigns. In China, this effort was centrally organized (Zhang et al. 2000, Fang and Wang 2001; Chapter 15), while in India it seems that village committees increased the forest cover in a decentralized fashion (Singh 2002, Foster and Rosenzweig 2003; Chapter 7).

Many countries in northern Europe transitioned from net deforestation to reforestation between 1850 and 1980 (Mather et al. 1999), but notably some countries in southern Europe have yet to undergo this transition. Likewise many developing countries, with Brazil mentioned as a prominent example, have not made the transition as yet (Rudel et al. 2005). However, research that aggregates data at the national scale, particularly very large countries such as Brazil or the United States, can miss dynamics of change taking place at subnational scales (Chapter 2). In general, more research has been devoted to the study of diverse macroprocesses leading to *deforestation* than to the study of the dynamics of *reforestation*. Instead of designing careful studies of regions where reforestation has occurred with an effort to sort out the diversity of microlevel processes that cumulate to produce different land-use changes (Turner and Meyer 1991), many scholars and activists have instead simply proposed major policy reforms that they presume will lead to reforestation. Among the most frequent recommendations is for national government interventions to achieve reforestation. Williams (2003), who undertook a massive study of the world history of deforestation, recommends "the need for strong government institutions to implement stated policies and resist elite groups who have traditionally pursued the exploitation of the forest" (p. 498). Many others have called on national governments to take the lead in reversing these dynamics (e.g., Rowe et al. 1992, Deacon 1995, Rice et al. 2001). However, there is evidence that reforestation can occur without large-scale policy prescriptions promoting reforestation, as in the case of some regions of the United States that we explore in the following sections.

## 8.3 Historical Trajectories of Land-Cover Change in the United States

National and state governments have indeed historically been important actors in encouraging reforestation. In the United States, both national and state governments have purchased extensive land to devote to forest regrowth as a result of abandonment from the 1930s to 1950s. Yet in the eastern United States, private landholders own the majority of land covered by forest (in contrast to areas in the western United States) and in Indiana more than 90% of forest cover lies on private lands. This institutional landscape presents significant challenges in understanding the drivers responsible for land-cover change. Areas dominated by private landholdings are managed by a heterogeneous group of actors with diverse preferences and household contexts that result in varying approaches to land management. Public landholdings are often highly fragmented, posing challenges to managers of state and federal forests, yet it is these landholdings that are responsible for the largest contiguous patches of forest in many areas. Net trajectories of land-cover change are the product of the complex decision-making processes driving private and public actors at different management levels.

During the nineteenth century, Indiana and Ohio experienced the most dramatic loss of forest cover in the United States (Figs. 8.1 and 8.2) according to data compiled from the General Land Office (Kellog 1909). Indiana dropped from 85% pre-settlement forest cover to 18% forest cover in 1909, accounting for a 67% decline. Ohio dropped from 90% to 18% during the process of European-based settlement. Other regions of the eastern United States also experienced precipitous declines in forest cover, but none more so than these two states in the Midwest (Fig. 8.3). Macroscale factors certainly explain some aspects of the regional patterns of forest-cover loss. Clearly the westward migration of settlers moving from the East Coast to the Midwest was a demographic trigger in the loss of forest cover.

While Ohio and Indiana experienced the most severe declines in forest cover, many mid-Atlantic and southern states actually had a higher percentage of forest cover prior to European-based settlement (North Carolina, Georgia, Virginia, West Virginia, Pennsylvania, Rhode Island, New Hampshire, Vermont, and Maine). States immediately west of the Mississippi River as well as parts of northern Illinois and Indiana were not heavily forested prior to European-based settlement, as the natural vegetation in these areas is grassland rather than forest. The values shown in Fig. 8.1 are some indication of the reforestation potential in the eastern United States if one only considers biophysical factors and land suitability.

Notably, many states in the lower Mid-Atlantic region did not appear to experience dramatic deforestation according to the pre-settlement and 1909 data (Figs. 8.1–8.3). Soils in the piedmont area of Virginia, North Carolina, South Carolina, and Georgia are of relatively low fertility that may have limited the clearing of forests for agriculture. In addition, the Appalachian Mountains, which run from western North Carolina northeast through Maine (Fig. 8.4), presented topographic constraints as well as highly weathered soils that likely limited the creation of large-scale clearings. Alternatively, these states may have been heavily cleared but already in the process of reforesting by 1909.

Figure 8.5 shows the net changes in state-level forest cover 1909–2005 based on the United States Department of Agriculture Forest Service (USFS) Forest Inventory and Analysis (FIA) and the Kellog (1909) documentation. Despite the tremendous loss of forest cover in the Midwest, states in the Northeast experienced more net forest-cover gain from 1909 to present. In other words, the Northeast was deforested to a higher "trough" point on the forest transition curve, and recovered to

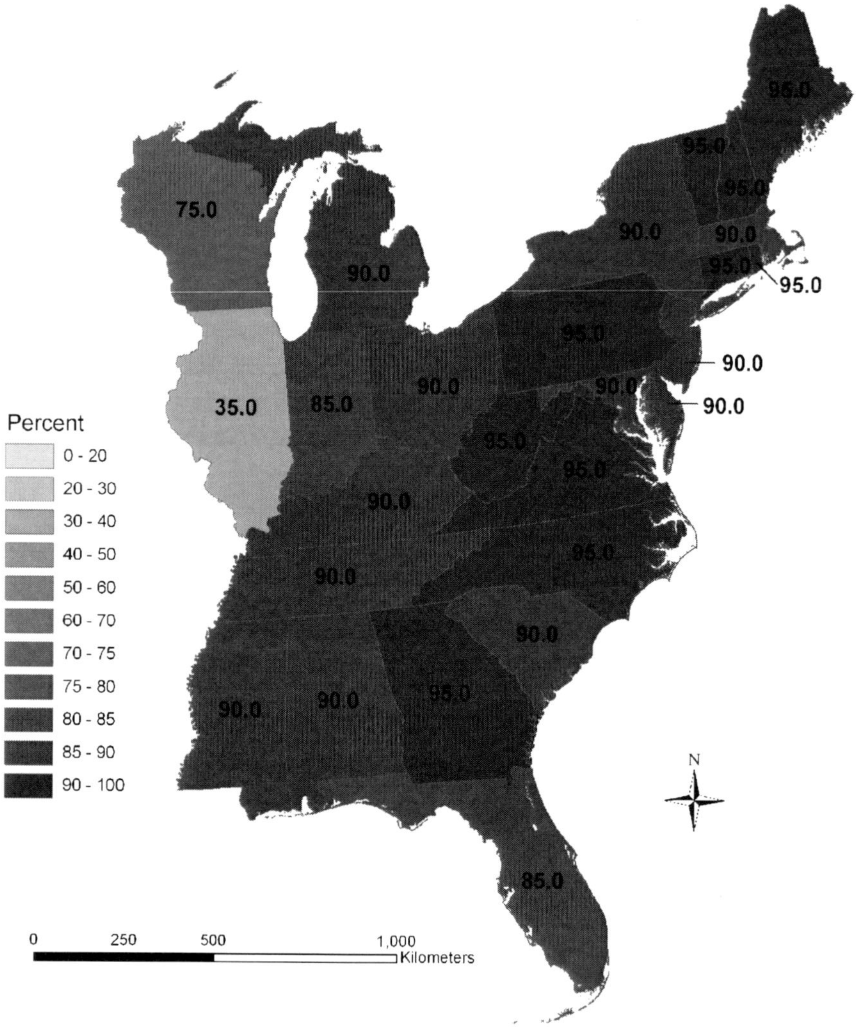


**Fig. 8.1** Pre-settlement forest cover in the eastern United States

a higher reforestation point as well. Notably, three states experienced net deforestation from 1909 to present (Tennessee, North Carolina, and Florida), although given the expected quality of the 1909 data there should be uncertainty as to whether states that show near-stable forest cover are actually in the net deforesting or reforesting categories. Given the paucity of data available for the 1909 time period that could be used for more rigorous statistical analysis and the relatively coarse scale presented here, these data are perhaps little more than interesting storylines. However, there is some value in considering the relative positions of these states at the turn

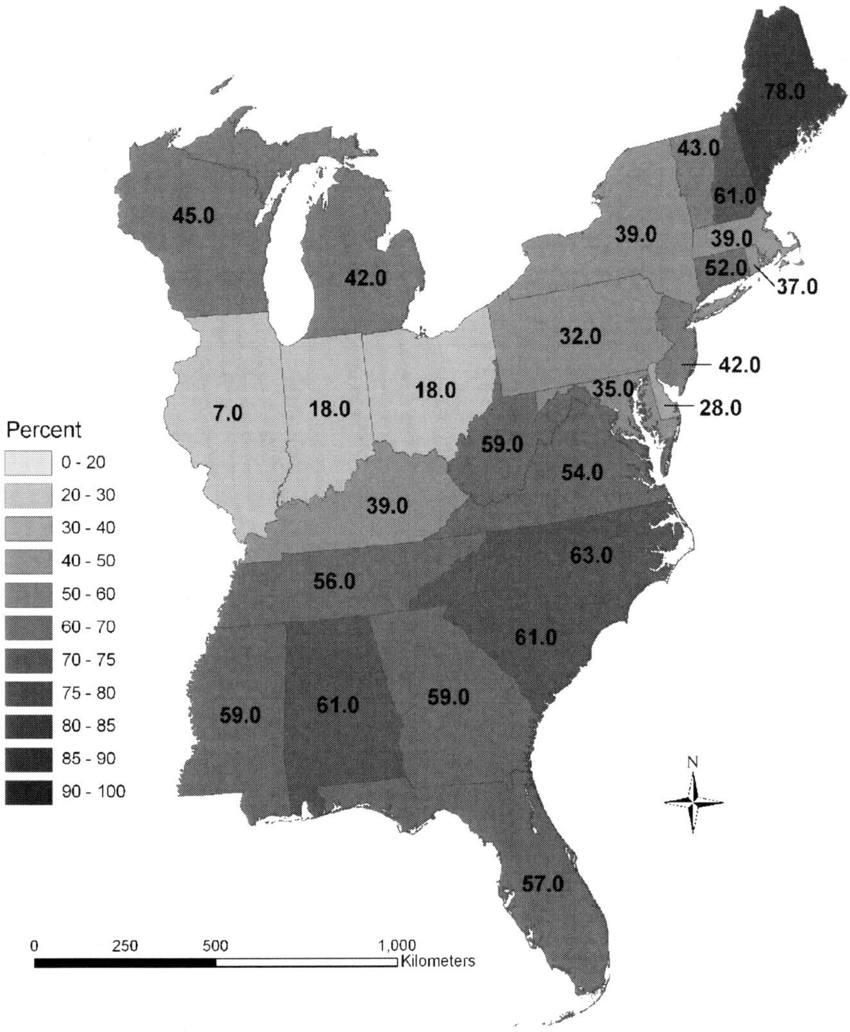


Fig. 8.2 Forest cover in the eastern United States, 1909

of the twentieth century, a time hypothesized to be a transition point from deforestation to reforestation for the United States (Rudel et al. 2005). Clearly there is considerable heterogeneity within the United States and there is some evidence that some regions, the South in particular, that have not yet followed through the transition from deforestation to reforestation. Certainly one explanation for the different trajectories is the scale of analysis. States with high rates of urban development (e.g., Florida) could be offsetting the reforestation that may be occurring in rural areas, suggesting the value of a finer scale of analysis to better understand the drivers of land-cover change.

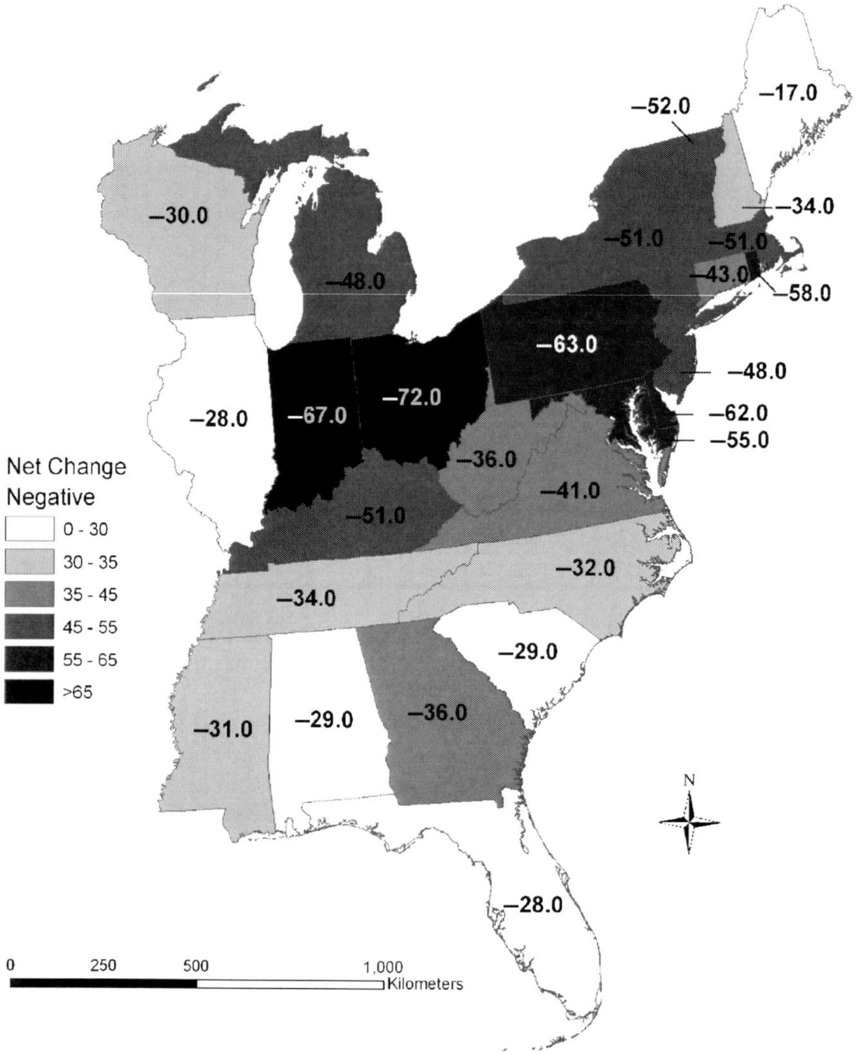


**Fig. 8.3** Net change in forest cover in the eastern United States from pre-settlement to 1909

## 8.3.1 *Trajectories of Land-Cover Change: Indiana, Midwest United States*

Like much of the eastern United States, Indiana was primarily forested prior to the arrival of European settlers in the early 1800s (Kellog 1909, Lindsey et al. 1965). These settlers cleared substantial areas of land for agricultural production (crops and pasture) and for forest products used for construction materials. It is estimated

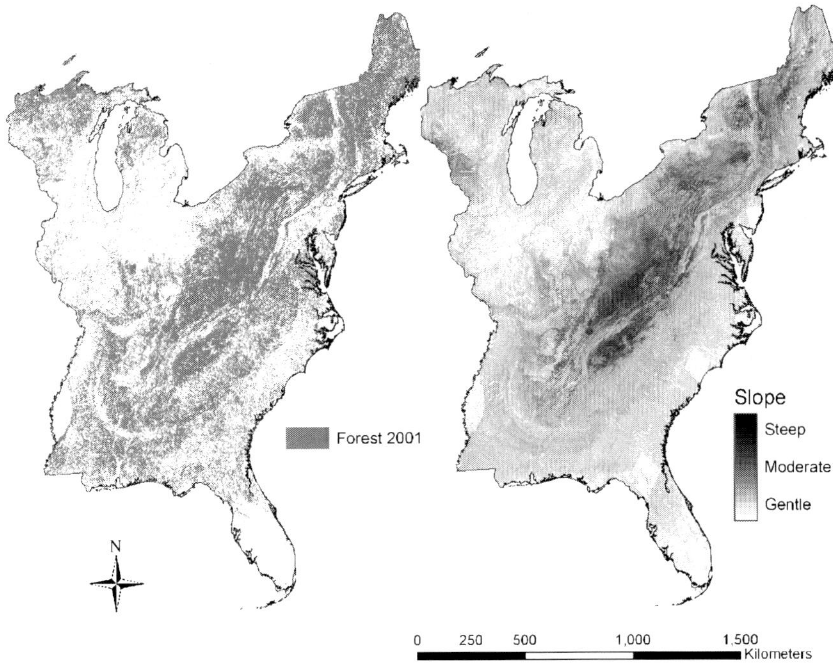


**Fig. 8.4** Forest cover and topography in the eastern United States, 2001. The MRLC classifications of deciduous, evergreen, and mixed forests were recorded as Forest; remaining classifications were recorded as Non-Forest (CGIAR-CSI 2008, MRLC 2008)

that in the early 1800s more than 85% of the state was covered with forest of some type across a wide range of topographic zones with the exception of the northwest corner of the state, which was grassland (Kellog 1909, Lindsey et al. 1965). The process of land clearing continued until the early 1900s, at which time areas marginal for agricultural production were gradually abandoned resulting in a pattern of forest regrowth in areas of low agricultural suitability. The combination of agricultural clearing and timber extraction reduced Indiana's forested land to approximately 560,000 ha, or about 6% of the state by the early 1920s (Nelson 1998). Nelson (1998) estimates that since the 1920s forest cover has increased to roughly 17% (1.6 million hectares). There is some discrepancy in historical estimates with an older report citing forest coverage in the early 1900s of 18% (Kellog 1909), but given that contemporary data are more reliable and show a forest coverage of approximately 17–20% and that there is considerable documentation of reforestation in the twentieth century, we find Nelson's estimates most plausible. Today, Indiana retains only an estimated 0.06% of its old-growth forest from its estimated original forest cover at the time of European-American settlement (Davis 1993). National Agriculture Statistics Service (NASS) data (Fig. 8.6) shows that the majority of forest cover in the state is now relatively young successional forest covering approximately 18–20% of the state and is highly concentrated in the rolling topography of the south-central region.

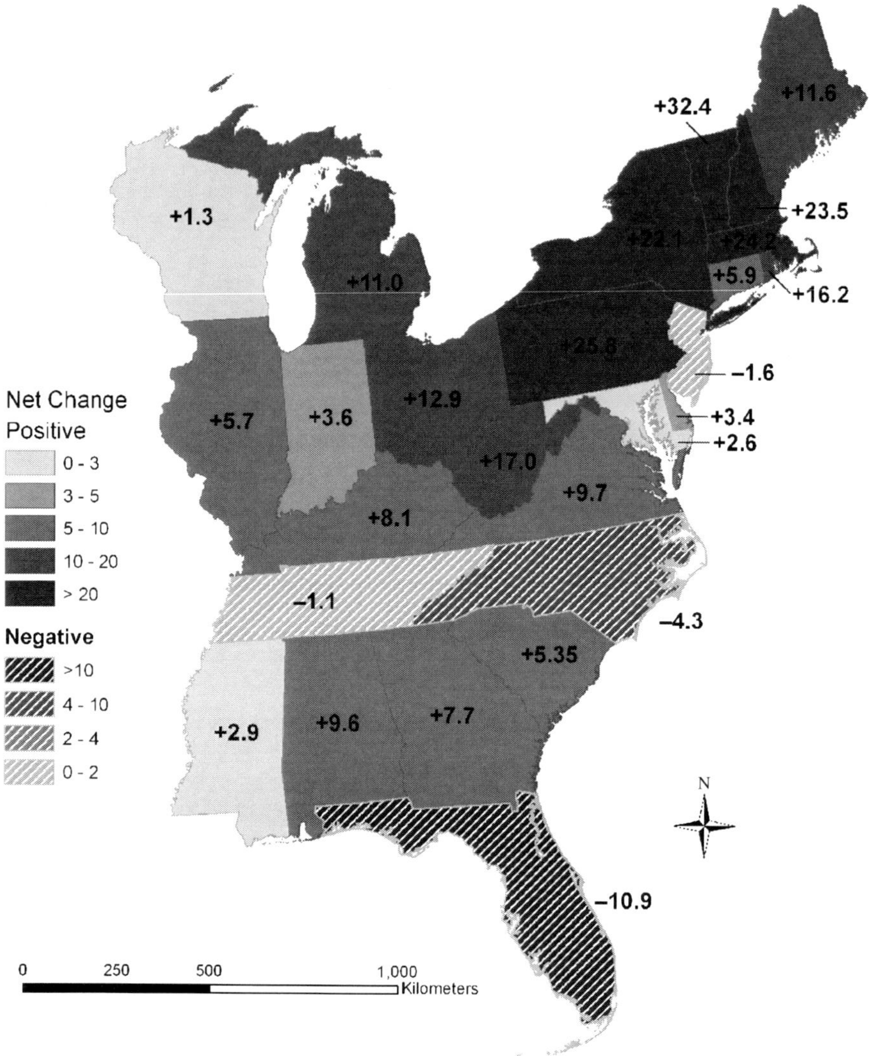


**Fig. 8.5** Net forest change, eastern United States 1909–2005 (USFS 2008)

The NASS data do not provide long-term historical coverage, but the FIA program has collected statistics on land cover in Indiana from 1950 to present. Because the 1950 data exhibit many outliers at the county level, we present data from 1967 to 2005 (Fig. 8.7). FIA data are developed from a set of permanent plots located in each county, although the number of plots varies from county to county depending on the proportion of the county in forest cover. From these plot data the FIA program reports the percent of each county in forest, and data are not disaggregated below the county level due to privacy/confidentiality concerns with the plot locations.

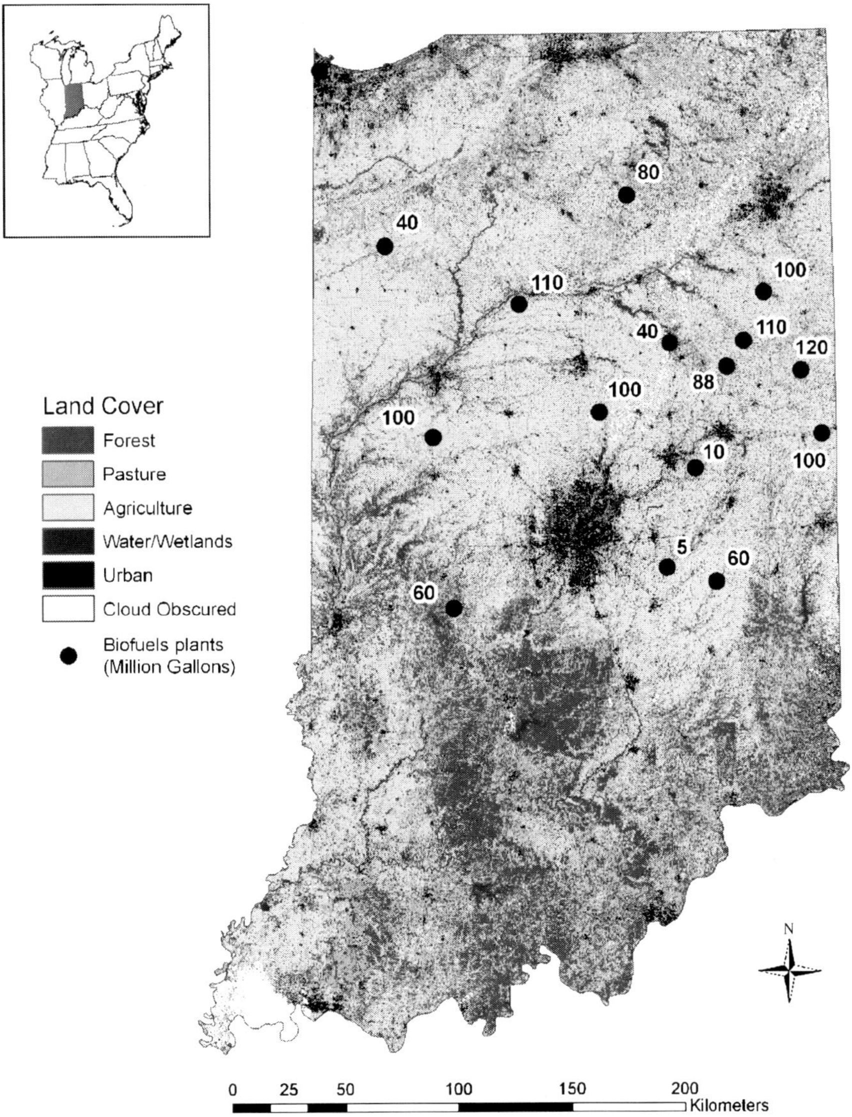


**Fig. 8.6** Land cover in Indiana, 2003, and biofuel production facilities (ISDA 2009, NRCS 2007) (*see Color Plates*)

The accuracy of FIA data has been challenged, but this source provides the best record of county-level data going back to the mid–twentieth century without resorting to the labor-intensive process of interpreting statewide aerial photography. However, in some ways the FIA data are more reliable than land-cover data derived from satellite imagery because the plots are permanently located and revisited each FIA field season.

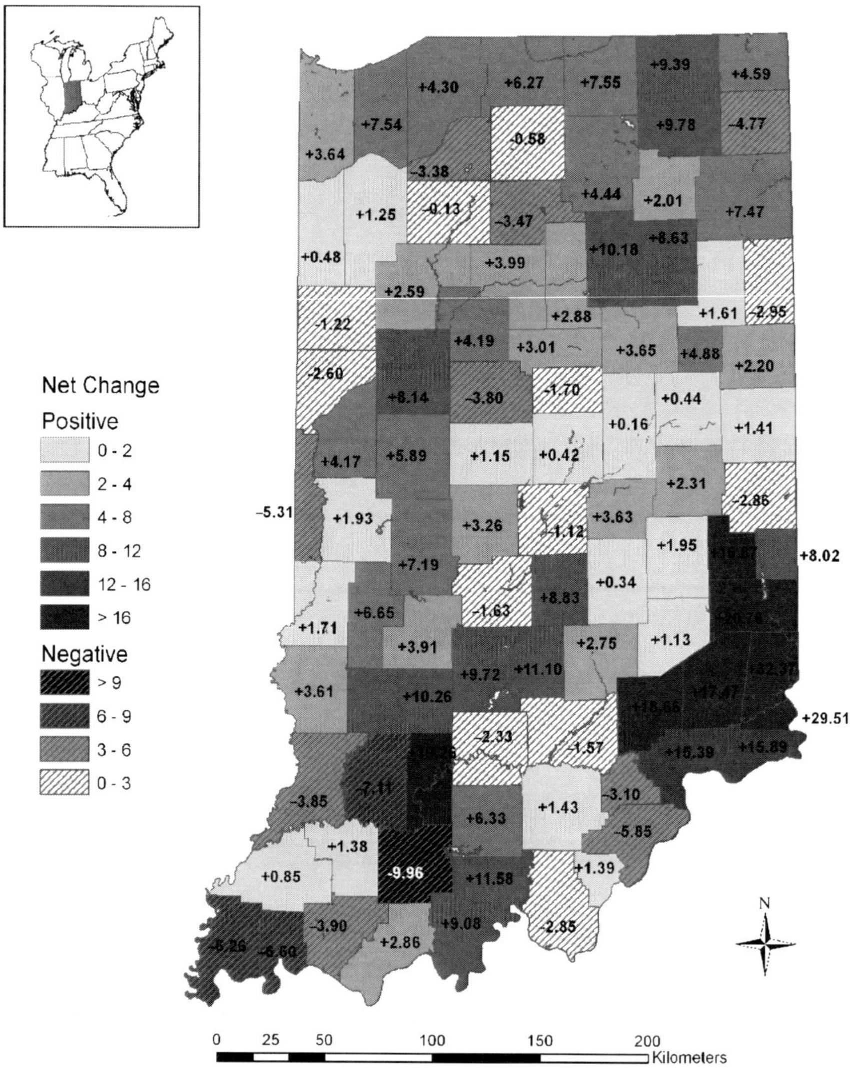


**Fig. 8.7** Net change in forest cover in Indiana, 1967–2005 (Essex 1968, USFS 2008)

Figure 8.7 shows the county-level FIA data for 1967–2005. At this scale it is evident that the 3.6% estimated forest regrowth in Indiana from 1909 to present is a product of a balance between areas of deforestation and reforestation. Areas of reforestation include the south-central and southeast regions and modest but notable increases in forest in selected counties in the northeast. Interestingly, the counties in the southeast part of the state are very close to Cincinnati, Ohio, a major metropolitan area. In fact, the more than 20% increase in forest cover in several of these counties is a result of the creation of the Jefferson Proving Ground, a 22,000 ha World War II munitions installation. Deforestation is evident in the southwest part of the state where mining (especially coal) is a major economic driver. Statewide population grew from approximately 4 to 6 million during this time period, although much of this growth was in urban and suburban areas. Rural population counts in Indiana were relatively stable during this period (increasing from 1.8 million in 1970 to 1.9 million in 1990), which is notable when compared to the positive relationship between population density and deforestation in developing countries (and in the Midwest United States during its frontier phase). Population increase may be associated with deforestation in some areas (Geist and Lambin 2002), but may also be associated with (but not necessarily a driver of) reforestation in other areas. The rural/urban dynamic suggests the important role of local-level conditions, including household decision making, in the process of land-cover change in the region.

#### 8.3.1.1 Timber Production and the Forestry Industry

In the late nineteenth century, Indiana was among the top states in regard to the value of its harvested timber (Streightoff and Streightoff 1916, Parker 1997). By the end of the nineteenth century, old-growth forests that had covered 85–90% of the state prior to European settlement were almost entirely eliminated and had been replaced by farmland (McCracken et al. 1999). Attesting to the state's dynamic forest composition, forests in Indiana were dominated by oaks 50 years ago (CSFES 1953), but the original land surveys prior to widespread European immigration indicated a more balanced mix of oak-hickory and beech-maple forests (Lindsey 1997). Data from USFS statewide inventories conducted in cooperation with the Indiana Division of Forestry are published for 1950, 1967, 1986, 1998, and 2003 (CSFES 1953, Spencer 1969, Spencer et al. 1990, Schmidt et al. 2000, Woodall et al. 2004, Woodall et al. 2005). The surveys varied somewhat in sample design, but they provide compatible estimates of forest area over time. Timberland in Indiana increased by nearly 10% from 1950 to 2003. Over the same period, the volume of growing stock on timberland increased more than 2.5 times. Between 1950 and 2003, the proportion of all timberland in the sawtimber-size class (i.e., stands where the overstory trees are predominantly >11 in. in diameter at breast height) increased from 52% to 73%. These trends are indicative of a forest resource that is maturing and where disturbances to the main canopy are either infrequent or low intensity (e.g., harvest by individual tree selection).

Historically, the Indiana timber industry has experienced dramatic change over the past century. While Indiana produced the most timber of any state in the United States in 1899 (USFS 1990), the timber-production industry underwent a gradual decline in the twentieth century (Clark 1987). Diverse forces had a reinforcing effect on the forestry industry during this time. For example, national markets for wood products changed as the manufacture of automobiles and railroad cars shifted from wood to metal materials. Likewise the industry was characterized by excessive competition and low wages that resulted in the decline of many family-owned sawmills (Clark 1987). But perhaps the most significant factor has been a depletion of high-value hardwood timber, despite the creation of organizations like the Indiana Hardwood Lumbermen's Association (IHLA) to protect the local industry (Clark 1987, USFS 1990). The timber and agricultural industries were dramatically affected by the economic depression of the 1930s. Unlike landowners in the northern part of the state, landowners in south-central Indiana were generally slow to adopt technological changes such as mechanization, which contributed to a general economic decline in the area (Sieber and Munson 1994, Welch et al. 2001).

#### 8.3.1.2 Public Landholdings and Forest Management

Historically, a major force behind the conversion of agricultural land to forest was the acquisition of areas for state and federal forests. Much of this acquisition occurred in the early twentieth century with the development of Hoosier National Forest (HNF) (Welch et al. 2001). During the Great Depression in the 1930s, many farms failed in south-central Indiana where the landscape was less suitable for pasture and crops than in northern Indiana. This situation resulted in a large number of tax delinquencies, and in 1934 the governor of Indiana requested that the USFS purchase lands to create a national forest. A bill was approved in 1935 authorizing the creation of HNF and parcels were purchased from 2,000 landholders in a first phase of development. From 1935 to 1942, the Civilian Conservation Corps initiated a major reforestation effort on hillsides, particularly areas susceptible to soil erosion. However, many of these were coniferous plantings rather than the oak/hickory forests that were more typical species for the region. The current HNF consists of a series of discontiguous USFS-owned properties within a defined purchase boundary within which the USFS may acquire land. The boundaries of HNF have changed over time and land has been exchanged between the federally managed HNF and properties within the Indiana state forest system, including Morgan-Monroe, Yellowwood, Martin, and Ferdinand state forests. Timber harvesting is permitted in HNF and in state forests, although major harvesting activities in HNF management units has slowed due to litigation initiated by non-profit conservation groups in the state (Welch et al. 2001). While the acquisition of land for the creation of HNF is associated with a major conversion of agricultural land to forest, the pace of land acquisition has slowed and there have been proposals to sell HNF property to private landholders, including near urban areas where land values are high and yielding significant economic returns to USFS (Welch et al. 2001).

Six of the current 13 state forests in Indiana were initiated in the 1930s while only one had been established in 1903. The other state forests were started in the 1940s and 1950s, with land acquisition dropping off dramatically by the 1960s. Today HNF has about 80,000 ha, and the Indiana State Forest System encompasses about 60,000 ha. Total federal forestland (HNF, military bases, etc.) is estimated to be 172,000 ha, which accounts for a substantial proportion of the forest in the south-central region of the state. Total state forestland holdings (forests, parks, fish and wildlife areas, nature preserves, etc.) are 130,000 ha (Woodall et al. 2004), and more than 90% of Indiana forests lie on private rather than public landholdings. But while these state and federal forest-management initiatives were no doubt associated with the subsequent increase in forest cover on these public lands, Indiana has also experienced an increase in forest cover on private landholdings.

#### 8.3.1.3 Legislation and Forest Conservation Efforts

The conservation movement in the early twentieth century helped to catalyze activities of non-profit organizations and government agencies concerned with the environment. Indiana's first such organizations focused on sustaining and improving timber production (IHLA and Indiana Fine Hardwoods/American Walnut Association in 1912). The IHLA was then influential in urging the legislature to create the Indiana Board of Forestry in 1901, which evolved to become the current-day Department of Natural Resources.

The Classified Forest Act, passed in 1921, was a particularly significant piece of legislation that targeted the protection of forests on private landholdings. In 1922, State Forester Charles Deam reported to the legislature that by 1920 the area of timberland in Indiana had shrunk "to 1,387,248 acres – an average annual decrease of 92,456 acres. At this rate of clearing, Indiana will be treeless in 15 years. Our area of timber in 1920 was a little over 6% of our area" (IDC 1922:40). Later reports suggest numbers from 1.5 to 2 million acres of forestland (7–10% of the state) (Parker 1997), and these data were instrumental to the motivation to enact legislation that protected forest resources. However, the fragile economic situation of many landowners at the time was an enabling factor and thus it was the combination of these two factors (awareness of declining resource and smallholder poverty) that together resulted in the large conversion of private to public land and then, with time, agriculture to forest.

Another major policy driver affecting land-cover change is the Conservation Reserve Program (CRP), enacted in 1985 (Sullivan et al. 2004). The CRP makes payments to landowners who enroll portions of their landholdings in the program for contract periods of 10 or 15 years. Once enrolled, land cannot be removed from the program without the penalty of having to return any payments received. Different categories of CRP land include protection classes for grassland, shrubland, forest, wetland, and riparian areas. When CRP contracts expire, landowners may either re-enroll their land or remove it from the program without penalty. Thus there are periodic trigger points that lead to pulses of land-use decisions by landowners (Sullivan et al. 2004), a dynamic that is explored in greater detail below.

### *8.3.2 Trajectories of Land-Cover Change in South-Central Indiana*

Land-cover classifications were produced from satellite imagery and aerial photography for analysis of the south-central Indiana region, which is now dominated by forest cover. Using Landsat data, land-cover change analysis was performed on a set of Indiana counties within the boundary of an image footprint covering south-central Indiana (Path 021, Row 033). Forest/Non-Forest supervised classifications were produced for 1984–1985 and 2005–2006 using multiseasonal imagery for each date. Overall, the region experienced a net gain in forest cover of 120,000 ha or 4.37% of the total area during the 20-year period. Deforestation of 121,000 ha (4.39%) was offset by 241,000 ha of reforestation (8.76%).

Much of the reforestation in Indiana occurred before the availability of satellite imagery. In order to develop a spatial representation of land-cover change going further back in time, we acquired historical aerial photography from 1939 to 1997 for Monroe County, Indiana. These photos were visually interpreted for Indian Creek and Van Buren townships, an area of approximately 10 km × 20 km, in roughly 7-year intervals (Fig. 8.8). Historical records suggest that Indiana transitioned from deforestation to reforestation well before the 1939 date, but this is the earliest for which aerial photography is available in this area.

Within the 12 counties for which spatially located CRP data are available, we found that there was significantly more reforestation on CRP-enrolled property than on non-CRP property (Fig. 8.9). A potential criticism of CRP is that it is simply providing a financial benefit to a selected set of landowners for a land-use option that they would select regardless of whether CRP existed or not. Because the aggregate trend from 1985 to present in most of Indiana was one of deforestation, one might conclude that CRP was not a trigger that initiated reforestation. However, the strong signal of reforestation on CRP property evident in Fig. 8.9 demonstrates that while reforestation did occur on non-CRP property (<10%), approximately 40% of CRP property experienced reforestation. Thus it is more reasonable to conclude that the financial benefit afforded by CRP was enough to motivate landowners to convert land from non-forest to forest during the 1985–2006 time period.

Notably, the period of most rapid forest-cover increase occurred between 1930 and 1950, particularly in areas of steep topography (Fig. 8.10), which was a period of decreasing population in Indian Creek Township (Fig. 8.11). It is therefore plausible that population decrease is one factor contributing to agricultural abandonment and forest recovery. Despite this apparent relationship, from 1970 to 2000 there was a dramatic increase in population in Indian Creek during a period when forest cover also increased, albeit slightly. Much of this late–twentieth century population increase is due to low-density residential development on former farms. While there are still few subdivisions (areas of medium-density residential development) in Indian Creek with subacre parcels, there are many large parcels that have been split into 5- to 10-acre parcels. These large residential parcels are not entirely forest but are in general more forested than the former farms from which they were developed. This suggests that population decline can result in forest regrowth but that population increase need not necessarily result in deforestation, at least in a developed economy.

It is perhaps as important to provide plausible explanations for the forest regrowth process during this period of population decline (1940–1950) as during the period of marked population increase (1960–2000). Certainly one explanation for regrowth during the latter period is the relative opportunities provided to landowners of farming,

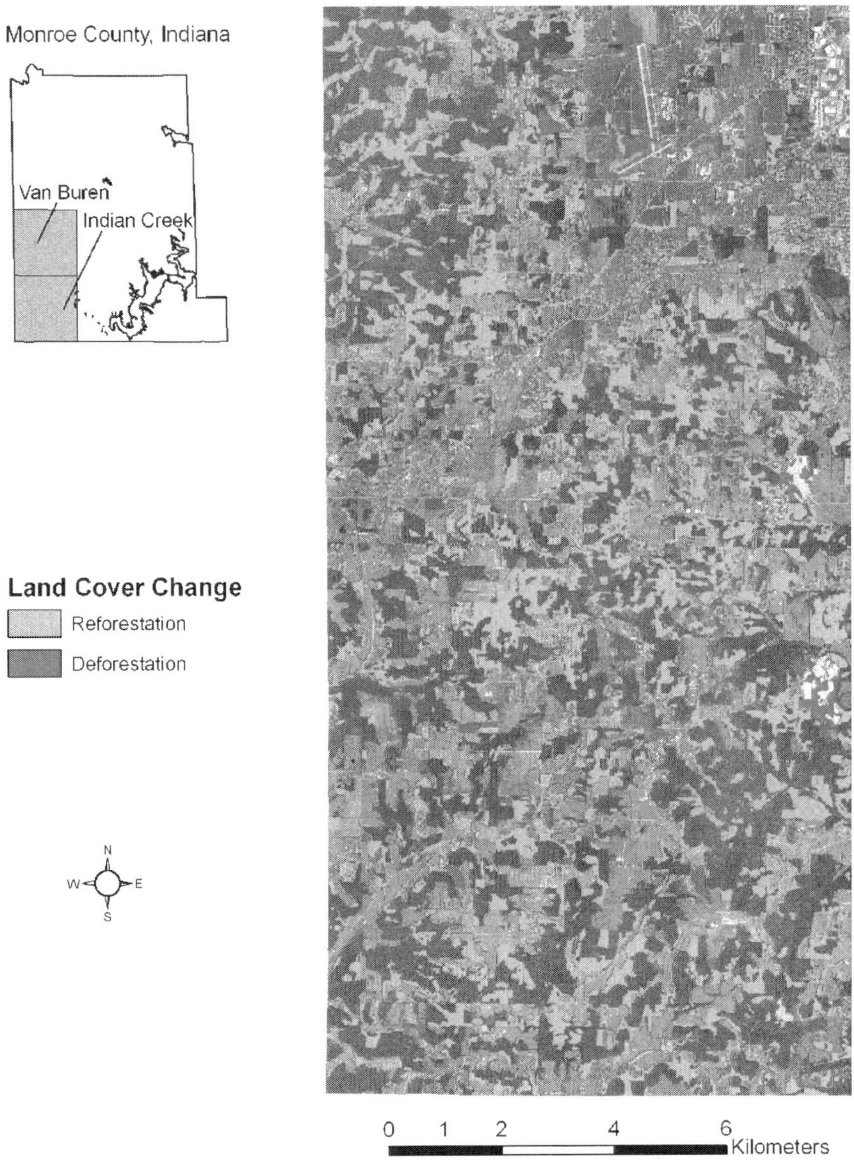


**Fig. 8.8** Land-cover change, 1939–2003, Indian Creek and Van Buren townships, Monroe County, Indiana (*see Color Plates*)

timber harvesting, and wage labor (Fig. 8.12). In particular, wage labor rates increased more rapidly during this period than did prices for aggregate farm products, thus making non-farm employment a more attractive use of labor. Had crop prices increased commensurate with wage labor rates, the slight regrowth observed from 1970 to 1993 may have been one of stable forest cover or even deforestation.

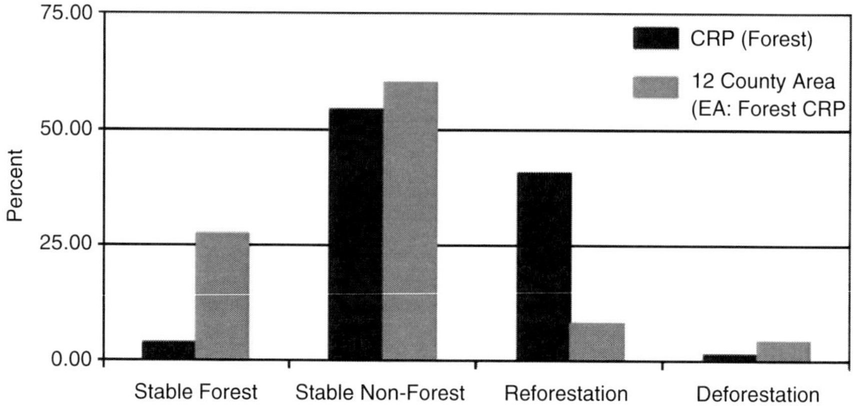


**Fig. 8.9** Reforestation in 12 Indiana counties with land in the Conservation Reserve Program, 1984–2006

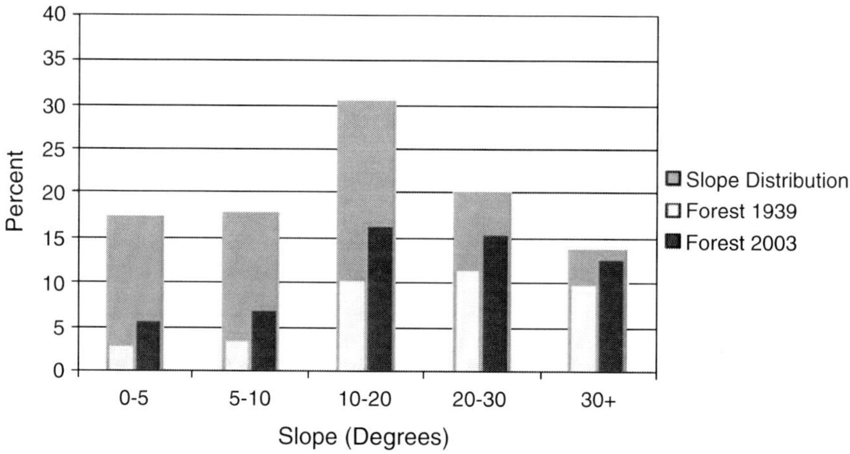


**Fig. 8.10** Land-cover change by slope category, Indian Creek township, 1939–2003

An important aspect of reforestation in south-central Indiana has been the increasing heterogeneity of land-use preferences among smallholders. It is important to recognize the substantial increase in forest cover that has occurred on private landholdings managed by landowners with a diverse range of motivations and incentives. Using agent-based models (ABMs) of land-cover change, prior research found that the heterogeneity of actors was an important factor in reforestation from 1939 to present in south-central Indiana (Evans and Kelley 2004, Evans and Kelley 2008). In this research, an ABM was calibrated to historical land-cover data at the parcel level. The research found vastly different land-use preferences when fitting the model at the household level. In other words, landscape partitions of similar size, topography, accessibility, and agricultural suitability exhibited dramatically

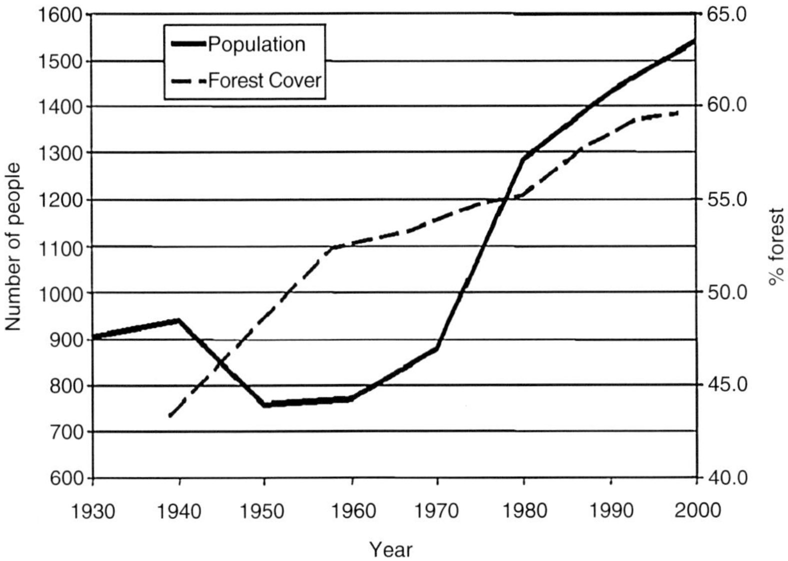


**Fig. 8.11** Population and forest cover in Indian Creek township, 1930–2000

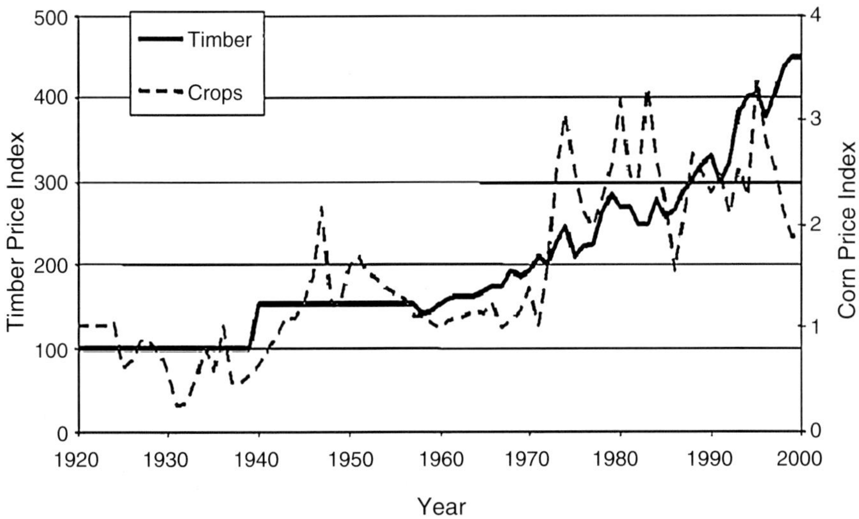


**Fig. 8.12** Aggregated crop and timber prices, 1920–2000 (CBOT 2008)

different trajectories of land-cover change. Contextually, this can be explained by a change in landowner preferences, including the emergence of the aesthetic value of forestland, as well as a diversifying economy where a growing number of landowners received income from off-farm activities. In other words, by the end of the twentieth century we see many landowners whose land-use decision making is motivated by non-pecuniary factors (Evans et al. 2001, Koontz 2001). So ultimately we see a complex landscape in Indiana where diverse social, institutional, economic, and environmental factors have led to an increase in forest cover on both private and public landholdings.

## 8.4 Strengths, Weaknesses, Opportunities and Threats: Future Reforestation Potential in Indiana

The historical pattern of land-cover change in Indiana is the product of diverse actors and diverse actions among actor types. Disentangling the process of reforestation to a single primary driver is not possible, and it is the coupled interaction between social (e.g., demographic change, economic shifts, policy instruments) and biophysical factors (e.g., soils, climate, topography) that are responsible for the pattern of land cover that is now present. We now turn to the question of what land-cover change trajectories are plausible in the near future as a means of addressing what the right-hand side of the forest transition curve may look like for one economically developed region.

To address these potential futures we consider the current context of Indiana within a basic SWOT framework built on strengths, weaknesses, opportunities, and threats. Strengths (internal) and opportunities (external) are "helpful" factors that can contribute to the likelihood of a desired outcome, and weaknesses (internal) and threats (external) are "harmful" obstacles to achieving a desired outcome. SWOT frameworks have been most commonly applied in management contexts. Clearly the process of reforestation is in part a product of intentional planning (as in the case of government-financed plantation projects) but also a by-product of unplanned changes in socioeconomic conditions and context (unemployment rates, commodity prices, cultural preferences). Thus the SWOT framework is in some ways not directly applicable to the process of reforestation because of the unplanned component of land-use and land-cover change. However, we proceed with this framework as a basic approach to identify different factors important to future trajectories of land-cover change in Indiana where "internal" and "external" refer to dynamics within or outside the state.

### *8.4.1 Strengths (Internal) of Future Reforestation Trajectories*

Agricultural abandonment on smallholder-owned land began during the same period as the initiation of publicly managed forests in Indiana. Abandoned areas during this period were generally those most marginal for long-term, sustained agricultural production resulting in a pattern of reforestation that was largely driven by topography. This first phase of reforestation was characterized by a relatively rapid increase in forest cover. The second phase of reforestation in Indiana (roughly 1950 to present) was one driven more by smallholders than public agencies.

As wage labor opportunities developed after World War II, off-farm income became more important than farm income to most households, particularly those near urban areas. This resulted in the conversion of agricultural land to fallow and then successional forest. Much of smallholder-owned forestland is actively managed, including selective timber harvesting resulting in forests with lower basal area than mature forests, but the process still resulted in a net increase in forested land. Indiana has also seen the advent of landowners who are motivated by environmental concerns broadly speaking (Koontz 2001). These landowners are motivated to preserve existing forest cover for recreational activities and to plant trees to create new forest cover. In other words, reforestation in Indiana has resulted from land management decisions by diverse landowners motivated by a wide range of factors.

Over the last 20 years, land trusts have had an increasing influence in Indiana with the development of conservation easements on smallholder parcels and the outright purchase (or donation) of land. Because self-organization by landowners through formal associations is relatively easy and involves low transaction costs, landowners can place a restriction on harvesting timber from their land by simply writing it into their deeds and registering this restriction with county officials (Barton and Silverman 1994, Nelson 2005). Many NGOs, such as The Nature Conservancy, can take a lead to bring attention to the issue of protecting forested land. Land trusts mobilize resources to purchase land from private owners and either protect it themselves or assign it to a national or state government forestry agency. Once NGOs are established and registered, it is then possible for private owners of land to use various forms of conservation easements to assign future development rights to them (Covington 1996, Hoover 2004). The private owners continue to use the land for their own purposes, but they have forgone the possibility of selling land for clearance and development in future years. Land trusts own a relatively small proportion of land, but they are important actors in that properties owned by land trusts generally have high value for conservation/species habitat.

Another private strategy for preserving forested land is common-interest housing developments (e.g., condominiums, cooperatives, intentional communities) where part of the land is jointly owned and managed by the community (Harrison and Richardson 2004). These arrangements have become more prevalent during the past quarter century. In addition, private owners and rural communities with common forests may decide to encourage regrowth or reforest on their land. Studies at regional levels and macrolevels suggest that economic downturns and policies that provide incentives may lead to reforestation in the short term. There is still much to be learned about how individuals and groups reach decisions that result in long-term reforestation.

### 8.4.2 *Weaknesses (Internal) of Future Reforestation Trajectories*

A portion of the loss of forest cover in Indiana is associated with exurban expansion around moderate-sized metropolitan areas. The total population of Indiana is relatively stable compared to other parts of the United States that have experienced dramatic population increases and commensurate impacts on land-cover change such as Atlanta, Georgia, Las Vegas, Nevada, and Phoenix, Arizona. Although there has not been major population growth in Indiana, there has been a redistribution of people from urban cores to proximal suburban areas. Since 1990, Hamilton County, located immediately north of Indiana's state capital of Indianapolis, has regularly ranked among the top 30 counties in the United States in population growth, and from 2000 to 2005 the county ranked number 18. While the population of Hamilton County grew by over 50,000 individuals during this time period, the population of Marion County, which includes the Indianapolis metropolitan area, increased by only 5,000 individuals. Thus, within-state population migration has been responsible for the development of formerly rural areas, primarily for residential development, and this is likely to be a continuing threat to forests if suburban growth continues. A related potential weakness is that Indiana does not have a strong tradition of county-level zoning. Only a small portion of counties empower county-level government to place restrictions on land use, but more counties may develop land-use plans with zoning if this rapid pace of suburban/exurban growth continues. Previous research has shown zoning to be an effective tool for controlling land-use change (Munroe et al. 2005).

### *8.4.3 Opportunities (External) for Future Reforestation*

Many of the legal mechanisms for increasing the amount of land devoted to forest regrowth are themselves derived from common-law precedents reaching back several centuries (Brubaker 1995). While many analysts cannot imagine how private arrangements backed by common-law traditions could facilitate major environmental protection, one of the significant advantages of common law is that it can be tailored to local ecological and social conditions as well as facilitate diverse mechanisms for protecting land without having to rely on central governments and the need for major investments in implementation and sanctioning (Meiners and Yandle 1998). Once an innovative entrepreneurship is shown to be a successful method of increasing forested land, people begin to think of still new ways to increase forest growth without relying entirely on national and state governments (Molnar et al. 2004). North (1986) has examined the relationship between broad legal systems and the relative cost of achieving collective goals in the common-law system of England in contrast to the more centralized Roman-law system of continental Europe. These traditions were brought to the English, Spanish, and Portuguese colonies and continue to impact how human activities and preferences are translated into ecological results. These contrasting legal structures suggest the challenges (and opportunities) that are imposed on various models of forest protection in different national contexts (Elmendorf 2003). Various federal cropland set-aside programs starting in the 1950s periodically encouraged tree planting. The 1990 Farm Bill, with the establishment of the Conservation and Wetland Reserve programs, probably has had the greatest impact recently. The impact of land trusts and NGOs on the protection and restoration of land to forest cover has only recently been a significant factor in Indiana. These groups use the institution of private property to conserve land and encourage reforestation and the preservation of existing forest in Indiana rather than advocate for change. Through fee title (outright ownership) and ownership of the development rights (conservation easements) on real property, they preserve undeveloped land throughout the state and provide landowners the opportunity to preserve their land in perpetuity. The Indiana Chapter of The Nature Conservancy started in 1959, but it wasn't until the late 1990s that the number of land trusts started to increase (eight in 1998, 28 in 2005). The Indiana Heritage Trust Program was created in 1991, and the USFS Forest Legacy Program was initiated in Indiana in 1998. These programs likely accelerated the growth of land trusts and encouraged more forestland conservation in Indiana.

### *8.4.4 Threats (External) to Future Reforestation*

At the "trough" of deforestation in Indiana, around the turn of the twentieth century, state and federal governments were major catalysts in establishing the foundation for reforestation by creating state-managed and federally managed forests. The economic situation of smallholders (tax delinquency) was an enabling factor, but the critical action that resulted in forest increase at this phase was one taken by state and federal actors. The acquisition of land by state and federal actors declined through the twentieth century. While government actors do continue to acquire landholdings, these purchases are primarily completed to reduce the fragmentation of existing landholdings by purchasing key linking parcels rather than to increase the total area under public land management. So while the pattern of public landholdings may change, it is unlikely that the total area under public land management will increase substantially.

A particular threat to forests is recent changes in energy policy in the United States that suggest a shift from fossil fuel–based energy sources to domestic production of biofuels. For many decades the United States has relied primarily on fossil fuels to supply the energy demand for a number of sectors of the economy. However, there is a growing consensus that as a matter of national security, particular attention needs to be directed toward the production of biofuels as a long-term fuel source for vehicles and some industrial processes. The Energy Independence and Security Act of 2007 mandates that the country use 36 billion gallons of biofuels annually by 2022, up from a 2007 level of 6.8 billion gallons (U.S. Congress 2007). Critically, this Act calls for 21 billion of that 36 billion-gallon target to come from advanced biofuels, including cellulosic ethanol. Furthermore, it is estimated that 33% of the corn grown in Indiana in 2008 was used for ethanol production. The transition to biofuel-based production will be affected by the prices for feed- and food-based corn relative to the price for ethanol produced from corn, switchgrass, and woodchips in the context of overall domestic energy prices. Any substantial increase in the use of biofuels as an alternative to fossil fuels will necessitate agricultural intensification and/or agricultural expansion within many regions of the country, regardless of whether the biofuel source is corn, switchgrass, wood products, or some other crop.

Much of the land-use change resulting from an increased reliance on biofuels is projected to occur in the Midwest and southeastern United States (NRC 2007). As one moves west from Ohio and Indiana to Illinois and Iowa, the native vegetation transitions from forest to tall-grass prairie. Much of Iowa, Illinois, and northern Indiana are heavily dominated by corn and soybean production and it is in these areas that changes in energy policy are more likely to result in changes to crop rotation cycles than in changes in land use. Conversely, in the eastern Midwest and the Southeast there is a strong possibility that marginal lands may be placed into biofuel production in the form of switchgrass, wood plantations, or corn. Because there is already a foundation of biofuel production facilities in the Midwest, it is plausible that the production of ethanol from timber-based cellulosic materials may first occur in this area. Forestry research programs that emphasize plantation establishment for either timber production or biomass crops have emerged in the Midwest. In areas that cannot profitably support corn production, forestlands could be cleared for cellulosic feedstock (wood chips) or timber (traditional sawlogs and some wood chips), and then converted to switchgrass or woody biomass plantation production. If cellulosic production technologies improve, there is the possibility that forests would be one resource used to meet the biofuel production targets, which could perhaps serve as an incentive to retain forests or possibly regrow forests. But at this time the production efficiencies for corn are greater than for cellulosic sources and the last 5–6 years have seen a dramatic development of ethanol production facilities in northern Indiana with a growing number in the central region of the state (Fig. 8.6).

The dramatic changes in prices for corn threaten to modify the relevance of government programs designed to protect forest resources. The Conservation Reserve Program described earlier, distributed $1.8 billion in rental payments to U.S. landowners in fiscal year 2008 for the enrollment of properties in the previous year. The payment rate varies from state to state, but in Indiana the average payment was approximately $45/acre. Total national enrollment in CRP totals approximately 6 million hectares, with 51,000 ha enrolled in Indiana (a substantial portion of CRP property is in the West). The U.S. Department of Agriculture reports that 7,570 ha of tree plantings in Indiana alone have been initiated through this program. CRP is a particularly important program in Indiana because of the relatively large proportion of marginal land for agriculture. The state has significant areas that are highly suitable for agricultural production and given recent prices for corn it would require extreme subsidies to provide a sufficient financial incentive to landowners to convert these areas to forest. At the other end of the land suitability spectrum, Indiana does not have major topographic constraints to agriculture in contrast to many states on the East Coast in the Appalachian range. With the exception of the relatively hilly south-central region there are few areas that are unsuitable for crops or pasture. But there are local areas of heterogeneous land suitability that limit the economies of scale possible. These areas present a "low hanging fruit" target for programs like CRP.

While this situation may thus far sound more like an opportunity than a threat, there are a tremendous number of CRP contracts that were due to expire in 2007 and 2008 (Sullivan et al. 2004). Because contracts are 10 or 15 years in length, landowners have few opportunities to take land out of the Conservation Reserve Program without the penalty of having to pay back the revenue they received from those contracts. The confluence of expiring CRP contracts and dramatic increases in price for corn from 2007 to mid-2008 presented a significant threat to forests that have recently regrown on former agricultural lands. Prices for corn have declined since the peaks of mid-2008, but as of mid-2009 corn futures are projected to increase through 2012 (CME Group 2009).

Early in the reforestation phase, when the deforestation rates were also high, government actors were more important to the process. After this initial period, reforestation continued but at a slower rate and reforestation was driven by smallholders (Evans and Kelley 2008). Now we see a new phase where areas highly valued for ecosystem services are under protection (either easements or purchased outright) by land trusts. These phases in the forest transition in Indiana have some implications for understanding potential trajectories of change in developing countries. Today we see many federal governments creating national forests, as well as federally funded programs to initiate reforestation efforts (largely plantations). It is less clear whether smallholder-driven reforestation is occurring in developing countries, but as wage labor opportunities develop it is plausible that this may occur in areas of relatively low agricultural productivity within commuting distance of urban areas. What is less clear is how the balance between deforestation associated with urban growth may outweigh any reforestation driven by these disparate actors.

## 8.5 Conclusion

Integrating numerous sources of land-cover data, it can be seen that Indiana has experienced substantial reforestation during the twentieth century, as has been the case with much of the eastern United States. Reforestation has been concentrated in the south-central part of the state, an area where forests, in contrast to the rest of the state, are the dominant land cover. Much of this reforestation occurred in gaps between existing forest patches resulting in an overall decrease in forest fragmentation in many areas. The forests in south-central Indiana are the product of several diverse actors. The majority of forest in the state lies on private landholdings, but public landholdings are responsible for the largest contiguous areas of forest in the state, which demonstrates the importance of government's role in forest resources. State-level policy has in some cases encouraged regrowth and legal structures have facilitated the acquisition of easements and land titles by non-profit organizations, thereby preserving particularly vulnerable patches of forest.

Despite the trajectories of land-cover change in the past, there are a number of potential threats and weaknesses that may move the state from a trajectory of net reforestation to one of deforestation. Loss of forest to urban growth coupled with a reduction in the rate of rural reforestation suggests that the trend in land cover is already one of deforestation in some regions. Recent institutional forces are potentially critical drivers of land-cover change including recent policy changes related to biofuel production and existing forest preservation. By shifting fuel production to domestic agricultural sources, recently passed energy policy is creating a tighter coupling between fuel supply and land use, which may have unforeseen environmental consequences. The past year has seen tremendous volatility in fuel prices and the boom period of ethanol plant development of early 2008 has slowed somewhat. Certainly some landowners will not modify their land management practices in reaction to these changes. But given the ambitious ethanol production targets implemented in recent legislation, and the fact that approximately one-third of corn production in Indiana was dedicated to ethanol production in 2008, we can expect energy policy to have an increasing impact on land use in the state unless there is a modification to the current national ethanol production targets. Whether these forces ultimately have an impact on state forest cover will depend on the actions of a diverse set of actors including state and federal forest managers, land trusts, and the complex mosaic of private landowners in the state.

## References

Barbier EB (1997) Introduction to the environmental Kuznets curve special issue. Environ Dev Econ 2:369–381

Barton SE, Silverman CJ (1994) Common interest communities: private governments and the public interest. Institute of Government Press, University of California, Berkeley, CA

Brubaker E (1995) Property rights in the defence of nature. Earthscan, Toronto

CBOT (Chicago Board of Trade) (2008) Chicago Board of Trade. http://www.cbot.com (data no longer available at this website but can be purchased from CME Group at http://www.cme-group.com/market-data/datamine-historical-data/datamine.html). Accessed July 21, 2008)

CGIAR-CSI (Consultative Group on International Agricultural Research Consortium for Spatial Information) (2008) SRTM 90m digital elevation data. http://srtm.csi.cgiar.org/, Accessed July 21, 2009

Clark D (1987) Entrepreneurs in hardwood. Indiana Hardwood Lumbermen's Association, Indianapolis, IN

CME Group (2009) Corn futures. http://www.cmegroup.com/trading/commodities/grain-and-oil-seed/corn.html. Accessed July 23, 2009

Covington GM (1996) Conservation easements: a win/win for preservationists and real estate owners. Illinois Bar Rev 84:628

CSFES (Central States Forest Experiment Station) (1953) Forest statistics for Indiana. United States Department of Agriculture Forest Service, Columbus, OH

Davis MB (1993) Old growth in the east: a survey. The Cenozoic Society, Richmond, VT

Deacon RT (1995) Assessing the relationship between government policy and deforestation. J Environ Econ Manag 28:1–18

DeFries RS, Foley JA, Asner GP (2004) Land-use choices: balancing human needs and ecosystem function. Front Ecol Environ 2:249–257

Elmendorf CS (2003) Ideas, incentives, gifts, and governance: toward conservation stewardship of private land, in cultural and psychological perspective. Univ Illinois Law Rev 2003(2):423–505

Essex BL (1968) Forest land in Indiana counties, 1967 (Res Note NC-57). North Central Forest Experiment Station, United States Department of Agriculture Forest Service, St. Paul, MN

Evans TP, Kelley H (2004) Multi-scale analysis of a household level agent-based model of land-cover change. J Environ Manag 72:57–72

Evans TP, Kelley H (2008) Assessing the transition from deforestation to forest regrowth with an agent-based model of land cover change for south-central Indiana (USA). Geoforum 39:819–832

Evans TP, Green G, Carlson L (2001) Multi-scale analysis of landcover composition and land-scape management of public and private lands in Indiana. In: Millington AC, Walsh SJ, Osborne PE (eds) Remote sensing and GIS applications in biogeography and ecology. Kluwer Academic, Norwell, MA

Fang JY, Wang ZM (2001) Forest biomass estimation at regional and global levels, with special reference to China's forest biomass. Ecol Res 16:587–592

Foley JA, DeFries R, Asner GP et al (2005) Global consequences of land use. Science 309:570–574

Foster AD, Rosenzweig MR (2003) Economic growth and the rise of forests. Q J Econ 118:601–637

Geist HJ, Lambin EF (2002) Proximate causes and underlying driving forces of tropical deforestation. Bioscience 52:143–150

Harrison GA, Richardson JJ (2004) Conservation easements in Indiana. Presented at Forest Land Conservation Workshop, Nashville, IN, 10 Nov 2004

Hoover WL (2004) Forest land conservation: an Indiana portfolio. Presented at Forest Land Conservation Workshop, Nashville, IN, 10 Nov 2004

IDC (Indiana Department of Conservation) (1922) Fourth annual report of the Department of Conservation. Indiana Department of Conservation, Indianapolis, IN

ISDA (Indiana State Department of Agriculture) (2009) Biofuels Indiana. http://www.in.gov/isda/biofuels/factsheet.pdf. Accessed July 23, 2009

Kellog RS (1909) The timber supply of the United States (Forest Resource Circular No. 166). U.S. Department of Agriculture, Washington, DC

Klooster D (2003) Forest transitions in Mexico: institutions and forests in a globalized countryside. Prof Geogr 55:227–237

Koontz TM (2001) Money talks–but to whom? Financial versus nonmonetary motivations in land use decisions. Soc Nat Resour 14:51–65

Lindsey AA (1997) Walking in wilderness. In: Jackson MT (ed) The natural heritage of Indiana. Indiana University Press, Bloomington, IN

Lindsey AA, Crankshaw WB, Qadir SA (1965) Soil relations and distribution map of the vegetation of presettlement Indiana. Bot Gaz 126:155

MacDonald D, Crabtree JR, Wiesinger G et al (2000) Agricultural abandonment in mountain areas of Europe: environmental consequences and policy response. J Environ Manag 59:47–69

Mather A, Fairburn J, Needle C (1999) The course and drivers of the forest transition: the case of France. J Rural Stud 15:65–90

Mather AS (1992) The forest transition. Area 24:367–379

Mather AS, Needle CL (1998) The forest transition: a theoretical basis. Area 30:117–124

McCracken SD, Brondizio ES, Nelson D et al (1999) Remote sensing and GIS at farm property level: demography and deforestation in the Brazilian Amazon. Photogramm Eng Rem Sens 65:1311–1320

Meiners RE, Yandle B (1998) Common law environmentalism. Public Choice 94:49–66

Mol APJ, Sonnenfeld DA (2000) Ecological modernisation around the world: an introduction. Environ Polit 9:3–14

Molnar A, Scherr SJ, Khare A (2004) Who conserves the world's forests? Community-driven strategies to protect forests and respect rights. Forest Trends, Washington, DC

MRLC (Multi-Resolution Land Characteristics Consortium) (2008) National land cover database. http://www.mrlc.gov/nlcd.php. Accessed July 21, 2009

Munroe D, Croissant C, York A (2005) Land use policy and landscape fragmentation in an urbanizing region: assessing the impact of zoning. Appl Geogr 25:121–141
Nagendra H (2007) Drivers of reforestation in human-dominated forests. Proc Natl Acad Sci U S A 104(39):15218–15223
NCRS (North Central Research Station, United States Department of Agriculture Forest Service) (2007) Forest inventory and analysis. http://ncrs.fs.fed.us/4801/. Accessed July 23, 2009
Nelson J (1998) Indiana's forests: past, present, and future. The Woodland Steward 7(3). http://www.inwoodlands.org/ (click on "Past WOODLAND STEWARD Articles"). Accessed July 21, 2009
Nelson RH (2005) Private neighborhoods and the transformation of local government. Urban Institute Press, Washington, DC
North DC (1986) The new institutional economics. J Inst Theor Econ 142:230–237
NRC (National Research Council) (2007) Water implications of biofuels production in the United States. National Research Council, Washington, DC
Parker GR (1997) The wave of settlement. In: Jackson MT (ed) The natural heritage of Indiana. Indiana University Press, Bloomington, IN
Perz SG (2007) Grand theory and context specificity in the study of forest dynamics: forest transition theory and other directions. Prof Geogr 59:105–114
Perz SG, Skole DL (2003) Secondary forest expansion in the Brazilian Amazon and the refinement of forest transition theory. Soc Nat Resour 16:277–294
Polanyi K (2001) The great transformation: the political and economic origins of our time. Beacon, Boston, MA
Rice RE, Sugal CA, Ratay SM et al (2001) Sustainable forest management: a review of conventional wisdom (Advances in Applied Biodiversity Science, No. 3). Center for Applied Biodiversity Science Conservation International, Washington, DC
Rowe R, Sharma NP, Browder JO (1992) Deforestation: problems, causes, and concerns. In: Sharma NP (ed) Managing the world's forests: looking for balance between conservation and development. Kendall/Hunt, Dubuque, IA
Rudel TK, Coomes OT, Moran E et al (2005) Forest transitions: towards a global understanding of land use change. Global Environ Change 15:23–31
Schmidt TL, Hansen MH, Solomakos JA (2000) Indiana's forests in 1998 (Resour Bull NC-196). United States Department of Agriculture Forest Service, North Central Research Station, St. Paul, MN
Sieber E, Munson C (1994) Looking at history: Indiana's Hoosier National Forest region, 1600–1950. Indiana University Press, Bloomington, IN
Singh NM (2002) Federations of community forest management groups in Orissa: crafting new institutions to assert local rights. Forests, Trees and People: Newsl 46(September):35–45
Spencer JS (1969) Indiana's timber (Resour Bull NC-7). United States Department of Agriculture Forest Service, North Central Research Station, St. Paul, MN
Spencer JS, Kingsley NP, Mayer RW (1990) Indiana's timber resource, 1986: an analysis (Resour Bull NC-113). United States Department of Agriculture Forest Service, North Central Research Station, St. Paul, MN
Streightoff FD, Streightoff FH (1916) Indiana, a social and economic survey. WK Stewart, Indianapolis, IN
Sullivan P, Hellerstein D, Hansen L et al (2004) The conservation reserve program: economic implications for rural America. United States Department of Agriculture Economic Research Service, Washington, DC
Turner BL, Meyer WB (1991) Land use and land cover in global environmental change: considerations for study. Int Soc Sci J 43:669–677
U.S. Congress (2007) Energy Independence and Security Act of 2007. 110th Congress, 1st session. Government Printing Office, Washington, DC. http://frwebgate.access.gpo.gov/cgi-bin/getdoc.cgi?dbname=110_cong_bills&docid=f:h6enr.txt.pdf. Accessed July 21, 2009
USFS (United States Department of Agriculture Forest Service) (1990) Draft environmental impact statement, land and resource management plan amendment, Hoosier National Forest. United States Department of Agriculture, Washington, DC. Accessed July 21, 2009

USFS (United States Department of Agriculture Forest Service) (2008) Forest inventory data online. http://199.128.173.26/fido/mastf/index.html

Walsh SJ, Rindfuss RR, Prasartkul P et al (2005) Population change and landscape dynamics: the Nang Rong, Thailand, studies. In: Entwisle B, Stern PC (eds) Population, land use, and environment. National Academies, Washington, DC

Welch D, Croissant C, Evans T et al (2001) A social assessment of Hoosier National Forest: manager's report. Center for the Study of Institutions, Population and Environmental Change, Indiana University, Bloomington, IN

Williams M (2003) Deforesting the earth: from prehistory to global crisis. University of Chicago Press, Chicago, IL

Woodall C, Brand G, Vissage J et al (2004) Indiana's forest resources in 2003. United States Department of Agriculture Forest Service, North Central Research Station, St. Paul, MN

Woodall C, Johnson D, Gallion J et al (2005) Indiana's forests 1999–2003. United States Forest Service, North Central Research Station, St. Paul, MN

Zhang P, Shao G, Zhao G et al (2000) China's forest policy for the 21st century. Science 288:2135–2136

# Chapter 9
# Importance of Input Classification to Graph Automata Simulations of Forest Cover Change in the Peruvian Amazon

**Kelley A. Crews and Alexander Moffett**

K.A. Crews (✉)
Department of Geography and the Environment, The University of Texas at Austin, Austin, TX, USA
e-mail: kac@uts.cc.utexas.edu

A. Moffett
Pritzker School of Medicine, The University of Chicago, Chicago, IL, USA

H. Nagendra and J. Southworth (eds.), *Reforesting Landscapes: Linking Pattern and Process*, Landscape Series 10,
DOI 10.1007/978-1-4020-9656-3_9, 

## 9.1 Introduction

Global afforestation/reforestation/deforestation (ARD) rates remain a highly contested sphere (Grainger 2008; Grainger, this volume, Chapter 2), with discrepancies across regions, studies, and time periods. Ostrom and Nagendra (2006) in assessing integrated satellite/field/lab studies point out that incongruities in findings should trigger greater scrutiny, just as coherence among findings should be read to increase trust in those findings. While that point was made in regard to differences in studies across spatial scales, the same can be applied to studies in different localities and across differing temporal scales (in terms of temporal grain or timing and temporal extent or study period). For a period in the scientific literature, it appeared that, particularly in tropical forests, there was a consensus as to the presence of *deforestation* (whether complete or thinning) (e.g., Skole and Tucker 1993; Turner et al. 2004) and the major research efforts were aimed at assessing the rates of loss (e.g., Achard et al. 2002) and understanding the proximate and ultimate driving factors behind those losses (Turner et al. 1995; Laurance 1999; Geist and Lambin 2002; Nagendra et al. 2003).

Near the end of this period the focus followed that of more general landuse/landcover change studies and bifurcated to include a strong emphasis on simulation and scenario modeling (Clarke et al. 1997; Clarke and Gaydos 1998; Verburg et al. 2004; Batty 2005; Andersson et al. 2006; Walsh et al. 2006). Though the challenges in these studies were well understood (Rindfuss et al. 2004; Grainger 2008), issues remained with validation, particularly of historical, remotely assessed components (Brown et al. 2005; Pontius and Malanson 2005). But few until this volume stopped to question whether the underlying assumptions regarding deforestation and nearly ignoring the possibility of reforestation (or afforestation) are justified (see especially Nagendra and Southworth, this volume, Chapter 1). Given the changes in forest policy that appear to be moving towards supporting reforestation efforts (Butler and Laurance 2008; Rudel 2008; Rudel, this volume, Chapter 3), it appears salient to now investigate previous findings, models, and methods.

Considered the leading effort at a global forest inventory, the FAO's Forest Resources Assessment and related reports are used by policymakers, managers, and researchers from around the world (FAO 2001, 2005). In order to complement field- and office-gathered statistics on forest inventory, the FAO utilizes satellite imagery for synoptic assessments. In the 2000 assessment, the FAO improved upon the 1990 report by using a triplicate of images to compare forest changes from 1980 to 1990, 1990 to 2000, and the overall period of 1980 to 2000. In doing so, they reported important issues remaining to be resolved including, notably, that comparable time-series were absent in most countries included in the assessment. Overall figures estimated global forest losses at 9.2 million hectares in the 1980s, with a reduction in losses in the 1990s totaling 8.6 million hectares. Losses in closed forest went from 8.0 million hectares in the 1980s to 7.1 million hectares in the 1990s. But the report goes on to indicate that since the standard error rate was 15%, the end result of this report as to whether there had been a net increase or decrease was not statistically significant. Given the overall inventories of 3.9 billion hectares of forest in 2000 (with net and gross deforestation annual rates of 9.0 and 13.5 million hectares respectively), this lack of confidence perhaps offers some optimism in their finding of an annual increase in forest cover of 4.5 million hectares.

Again, difficulties in assessing, interpreting, and comparing ARD rates is increasingly well documented (Grainger 2008), and much of the problem has been acknowledged by the remote sensing community some time ago, with particular attention paid to difficulties in assessing age groups of forest in fast-growing tropical forests (Foody et al. 1996). The smaller the sample used also can bias findings, particularly depending upon how those samples were collected. The FAO was advised to use 350 samples but due to budgetary constraints was limited to 117 sampling units for the triplicate pan-tropical remote sensing survey (FAO 2001). Ultimately 113 sites were used because of difficulties in finding archived, cloud-free imagery. Compounding difficulties in assembling the time series for each sample unit was also the issue of image seasonality, important even in the oft-considered "aseasonal" tropics (McCleary et al. 2008).

While early assessments of deforestation were aspatial, the vast majority of current deforestation studies rely upon from-to change detection. The underlying premise of a change detection approach is that the start and end year are compatible. In remote sensing terms, the term "anniversary date imagery" is used, meaning that, for example, one should only compare changes across time when seasonal variations have been excluded. For example, in a temperate coniferous forest, it would be improper to compare a summer ("leaf-on") 1975 image with a winter ("leaf-off") 1995 image, since some amount of seeming deforestation would be [incorrectly] detected by comparing asynchronous imagery. In many temperate latitudinal ecosystems where seasons are clearly demarcated and cloud cover is less commonly an issue, obtaining change detection image pairs in the same season can be relatively straightforward. But move toward the equator to the tropics, and the issue becomes cloudier, literally.

## 9.2 Conceptual Framework

Simulation models are increasingly used to forecast land use and land cover change (LULCC) generally and forest change specifically (Rindfuss et al. 2004, Verburg et al. 2004). As with any model, deterministic or stochastic, outputs may vary widely with slight changes in inputs (Wolfram 1984). Such is particularly of concern with spatial simulation models, where input variations can have multiplicative effects when neighbourhood interactions compound the effects of different inputs as translated through transition rules (Brown et al. 2005; Pontius and Malanson 2005). Modeling dynamic systems can thus present difficulties when picking base year or input images. While the Amazon is certainly regarded as a highly dynamic system, in fact much of the narrative of the greater Amazon centers on deforestation (whether complete conversion or thinning) studies, many of which take place in the eastern Amazon where seasonal variability is not as great an issue (Moran 1993; Gerwing 2002). But recent work in the western Amazon has documented strong seasonal pulses not just in precipitation and flooding, but in observable vegetation and forest cover change, potentially associated with green-up events (McCleary et al. 2008). Given the recent climatic oscillations resulting in seasonal drought and even burning of rainforests (Laurance and Williamson 2001; Tapley et al. 2004; Wright 2005), it is even more important to understand how these seasonal shifts may impact deforestation simulation results. This work follows on intra-annual assessments in the northern Peruvian Amazon (McCleary et al. 2008), using graph automata (a generalized form of cellular automata) (Sarkar et al. 2009) to assess the sensitivity of modeling results of forest cover change to randomized discrete intervals of change (increase and decrease) in forest cover.

This work is positioned in the Peruvian Amazon. Like much of the tropical Amazon, the area is riddled with cloud cover year-round, rendering image acquisition difficult without regard to seasonality. And in comparison to her eastern/Brasilian counterparts, the Peruvian Amazon has been reported to be relatively aseasonal, lacking the distinct wet and dry seasons that Brasil rainforests experience. Despite this, vegetation greenups have been preliminarily reported that in fact suggest the need for caution in overlooking seasonality in any of the Amazonian forests long described by narrative of extensive deforestation (McCleary et al. 2008). This work thus seeks to use this section of the northwestern Amazon as a case study to understand how, even in the most presumably aseasonal environments, seasonal vegetation changes can radically impact assessments of change detection, particularly with respect to vegetation and especially with respect to forest cover. Moreover, since a simulation modeling approach is employed, this research also provides information on how the selection of input data can bias results to find more or less change in a particular land cover, such as forest. Lastly, because seasonality is so commonly overlooked in many tropical remote sensing studies (to be fair, the cloud cover issue prevents many researchers from even having the chance to assess seasonality in a given calendar year), this chapter concludes by questioning whether the predominant narrative of tropical deforestation is, in some part, a predictable if previously undescribed bias stemming from ignoring seasonality in tropical forest ecosystems.

## 9.3 Discussion of Methods

### *9.3.1 Cellular Automata*

Cellular automata are discrete dynamical systems that evolve over discrete time steps (Toffoli and Margolus 1987). They have been applied widely in geography to study land use and land cover change (Malanson et al. 2006a; Manson and O'Sullivan 2006; Walsh et al. 2006). Studies have used cellular automata to simulate urban (Clarke et al. 1997; Clarke and Gaydos 1998; Batty 2005; Andersson et al. 2006; Torrens 2006; Xie et al. 2007) and rural development (Malanson et al.

2006b). Cellular automata are composed of sets of spatially distinct cells in a grid system, where the spatially explicit nature facilitates local interactions between the cells. In addition to its location, each cell is assigned one of a finite number of states. At each time step, these states are updated on the basis of both the present state of each cell and the states of the cells that fall within a selected radius. The system as a whole thus evolves on the basis of the evolution of the states of its component cells. Though the individual components are thus quite simple, their interaction allows cellular automata to represent complicated dynamic phenomena (Wolfram 1984) which may then evidence emergent properties of complex systems (Malanson 1999; Manson 2001; Walsh et al. 2006). Formally, a cellular automaton is a graph $G = (V, E)$ in which $V$ consists of a finite set of vertices and $E$ consists of a set of pairs of vertices such that for each $(v_i, v_j) \in E$, both $v_i \in \mathrm{V}$ and $v_j \in V$ and if $(v_i, v_j) \in E$ then $(v_j, v_i) \in E$ (Harary 1969). As applied to the models considered in this chapter, the vertices in $G$ represent cells within a given landscape, while the edges in $G$ represent causal interactions between pairs of cells.

One of the limitations of a cellular automata approach is undue restrictions such as requiring adjacency for cells to impact one another (Gutowitz 1991; Sarkar et al. 2009), despite the acknowledgement in the land use literature that many types of land use changes are in fact triggered by states and events of specific distal areas (Turner et al. 1995). Releasing this restriction adds computational complexity but improves the ability of a model to represent real-world processes.

### 9.3.2 Graph Automata

While cellular automata are sufficiently general to allow for their application to a number of different modeling problems, assumptions implicit within their mathematical framework prove limiting. As noted above, cellular automata models assume both: (i) each cell is subject to the same set of transition rules and (ii) the transition of each cell depends upon the states of all cells that are located within a uniform radius (Toffoli and Margolus 1987; Gutowitz 1991). When applied to the study of LULCC, each of these assumptions can prove to be prohibitive. One way to overcome this limitation is through the use of a more general mathematical structure to represent cellular interactions. Graph automata offer such a structure by allowing for the representation of potential interactions between any set of cells, regardless of their location (proximity) in space (Boccara 2004). Rather than an undirected graph, in which the effects of $v_i$ on $v_j$ are mirrored in those of vj on vi, graph automata are directed graphs in which such symmetry is not required. In addition, the set of rules used to determine the transitions from one state to another need not be identical for each cell. By connecting the cells of cellular automata in a complex network, rather than on a regular geometric grid, graph automata thus offer a more general mathematical framework than that provided by cellular automata (Burks 1970; Gutowitz 1991). Though graph automata were introduced more than 30 years ago (Rosenstiehl et al. 1972, Ng et al. 1974; Milgram 1975; Smith 1976; Wu 1978) they have been used only rarely to model empirical phenomena. It should be noted that while "graph automata" is the name most frequently use to refer to these mathematical structures, they have also been referred to as intelligent graphs (Rosenstiehl et al. 1972), polyautomata (Smith 1976), and web automata (Shah et al. 1973; Milgram 1975). Sarkar et al. (2009) provided the first application of graph automata to the study of LULCC in both the Peruvian Amazon and Andes. To the best of our knowledge, this paper represents the second such application.

## 9.4 Case Study

### *9.4.1 Study Area*

The study area spans the Iquitos, Peru area (centroid 3.74° S, 73.26° W), the largest city in the western Amazon at over 400,000 people, and its upstream associated peri-urban, agricultural, and forested areas. Notable about this area is the lack of overland connection to Lima or Brasil; while roads are used to move goods and people in the area, in general the city is considered to be riverlocked and dependent upon the Amazon (just north/downstream of the confluence of the Marañon and Ucayali Rivers) for transportation. The limited accessibility of the area is likely to affect the dynamics of LULCC as such a relationship has been found elsewhere in the past (Nagendra et al. 2003). The area is comprised of floodplains, ancient floodplains, and slightly upland areas noted for their heterogeneity in soils ranging from white sands to red clays and associated with blackwater and whitewater systems, respectively (Tuomisto et al. 2003; Crews-Meyer 2006).

### *9.4.2 Data*

This study used estimates of landcover derived from a Landsat 7 ETM+ scene (WRS2 Path 006, Row 063) acquired May 31, 2001. The original landcover extraction of this scene as compared to other intra-annual images (March 12 and September 20, 2001) with concomitant accuracy assessments was reported previously (McCleary et al, 2008). The overarching finding of McCleary et al (2008) was that even in supposed "non-seasonal" environments such as the tropical northwestern Amazon (which in contrast to the Brasilian Amazon has no marked dry season), there is fact is a strong seasonal pulse detectable among a variety of landcover classes, including forest cover. Because longer-term change detection studies rely upon a presumption of intra-annual stability, this seasonal/cyclical change in fact may obscure true change estimates when anniversary imagery is unavailable. Sarkar et al. (2009) used a subset of the Iquitos classifications to model future landcover and landuse change under a scenario of increased urban and agricultural expansion in this riverlocked area. This research also used the subset of the

May 31, 2001 Path 006 Row 063 classification centered most closely on Iquitos but including a variety of forest and agricultural areas in order to understand the importance of input imagery into the modeling process. Of the three images used in intra-annual comparison in McCleary et al. 2008, the May 31st image had the least cloud cover that can present problems in spatial simulation modeling. Thus this image was used in analysis here. That is, in cloud-ridden areas such as the Amazon, it is highly unusual to have multiple cloud-free (or low cloud cover) scenes per year; what then are the consequences for a/de/re-forestation research if the scene chosen represents a seasonal shift in greenup that is inappropriate for use as a baseline in longer-term change detection studies?

To this end, the same subset used in Sarkar et al. (2009) was employed here for computational tractability, having an area of 754,961 ha and modeled as a set of 8,612,624 cells. From these, 224,158 cells (roughly 2.6%) were removed from analysis due to cloud cover.

### *9.4.3 Methods*

Graph automata were used to model land cover change in an Amazonian landscape classified using a hybrid unsupervised-supervised approach with Landsat ETM+ imagery (McCleary et al. 2008). While original classification work used a Level-2 classification, here a Level-1 classification was used to ensure comparison with previous work (Sarkar et al. 2009) and to better fit transition rules derived from observations over multiple field seasons and informed by household interviews. An overview of the modeling process is shown in Fig. 9.1. Each vertex in the graph automata represented a 900 $m^2$ cell (30 m pixel). A simple topology was then used to determine the edges of the graph. An edge of weight one was used to connect each pair of physically contiguous cells. Edges of weight two were then used to connect each cell to the set of cells physically contiguous with those to which it was connected with an edge of weight one, and to which it was not yet connected. This process was repeated until edges had been defined for each pair of cells. The resultant assignment of edges to the cells allowed for the representation of the physical distances between cells, with the weight of an edge equal to the distance between the centroids of the cells, as measured in 30 m increments. States were assigned to the cells so as to represent their initial occupation of one of the following states: (i) Forest; (ii) Agriculture; (iii) Water; or (iv) Urban/Bare.

In addition to the original landscape, two additional landscapes were produced by modifying the number of forested cells in the original landscape. In one such landscape, the total number of forested cells was increased by 10% of the initial number of forested cells. Cells that were originally Agriculture or Urban/Bare were randomly selected for conversion, with the probability that each such cell was converted equal to that required for the expected number of forested cells to increase by 10%. Analogous methods were used to create the other landscape, in which the number of forested cells was decreased by 10% from the initial number. The creation of these

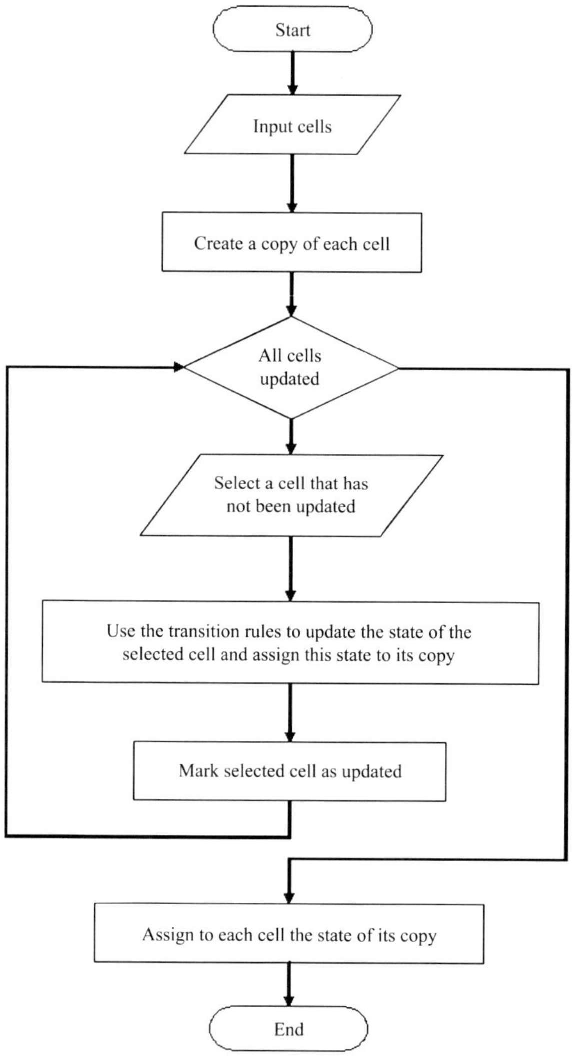


**Fig. 9.1** The landscape dynamics represented in the graph automata model. The flowchart describes the rules used to update the entire set of cells during each time step

landscapes allowed for the investigation of the effects of the initial degree of forest cover on the dynamics of the landscape change. Though increases or decreases in forest cover would not occur completely randomly on the landscape, random assignment was used to avoid introducing unnecessary bias and/or path dependence by assigning states through another means (e.g., intentional placement or simulation).

Following the creation of these modified landscapes, a series of simulations was performed to study the evolution of forest cover in both the original and modified landscapes. In these simulations, the status of each cell in the landscape was updated

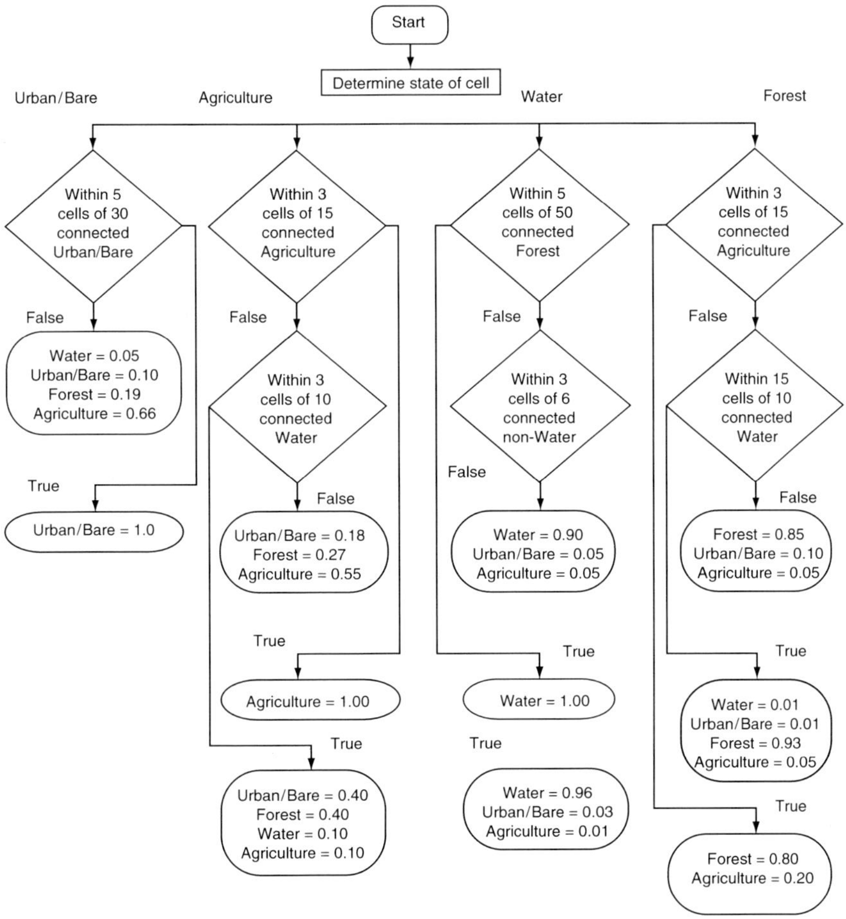


**Fig. 9.2** Transition rules used to update cells. The rules depend upon the initial state of the cell. On the basis of this state, and the states of the adjoining cells, the state of each cell is updated using the probability distributions given in this figure

stochastically, with probability distributions informed by both the current state of the cell and the states of the cells around it. A set of cells was defined as "contiguous" if an edge of weight one connected each cell to at least one other cell in the set. Contiguous sets of cells thus represented spatially clumped sets of cells. The transition rules used to update the states of the cells were based upon the careful analysis of patterns of historical landscape change in the Amazon region (Sarkar et al. 2009). A summary of the transition rules is provided in Fig. 9.2. All transition rules were based upon multiple years of field observation (e.g., tracking how quickly the urban perimeter expands and under what conditions) as well as basic ecological processes for that area's ecosystems and soil types (e.g., regrowth of early successional tree species such as cecropia as varies among soil types). A qualitative description of the rules follows.

Urban and Bare cells had to be grouped together for classification, but in relatively know proportion. Urban/Bare cells close to a cluster of similar cells tend to remain Urban/Bare. Otherwise, they tend to largely become agriculture (primarily the succession of Bare cells, though some Urban conversion has been noted). Due to shifting river flows, Urban cells have seen as taken over by the river, as have exposed Bare cells due to river meanders. Bare cells may also convert to forest roughly 20% of the time, and there have been traces of Urban reclamation into forest parks in the city center.

Forested cells close to small clusters of Agriculture convert to Agriculture at an observed rate of roughly 20% per year, while the rest tend to remain forest (this is without regard to topography or soil type, both mitigating factors in this undulating terrain, as noted by Tuomisto et al (2003)). Forested cells away from Agriculture but close to water may occasionally be converted to Urban (ports), Water (from river meanders), or Agriculture (floodplain farming), though usually (over 90%) stay Forested. Forested cells near neither Agriculture or Water clusters are urbanized (or at least cleared) at 10% per year, and undergo agricultural extensification at roughly 5% per year, with the rest remaining forested.

Agriculture cells on the periphery of Agriculture clusters tend to remain Agriculture. When not connected to other agricultural areas but connected to clusters of Water cells, they tend to convert to Water at 10% per year (river meanders), regrow into Forest at 40% per year (often because early successional growth is misclassified as agriculture), become cleared or converted to Urban occasionally at 40% per year, and stay Agriculture roughly 10% of the time for any given year. Agriculture cells not in proximity to Agriculture or Water cells often are converted to Urban/Bare or Forest (regrowth), and remain Agriculture over half the time.

Water cells connected to even very small clusters of other Water cells predominantly tend to remain Water in the following year, though very rarely (proportionately) may become Urban/Bare (ports, river meanders) or Agriculture (floodplain farming). An exception to this condition occurs if the Water cell is also connected to a large Forest cluster (usually flooded forest), in which case it will remain Water. Otherwise, in general Water clusters remain Water unless very rarely converted to Agriculture or are the result of river meanders that take more than a year to establish vegetation.

The rules were selected so as to represent likely annual changes in land cover. A detailed discussion of the considerations that informed the selection of these specific transition rules is provided in Sarkar et al. (2009). Computational approaches to transition rule selection have also been employed in the study of cellular automata (Li and Yeh 2004).

### *9.4.4 Results*

Maps representing the different landscapes with which the simulations were initiated are provided in Fig. 9.3. In comparing Fig. 9.3a to Fig. 9.3b and Fig. 9.3c, the

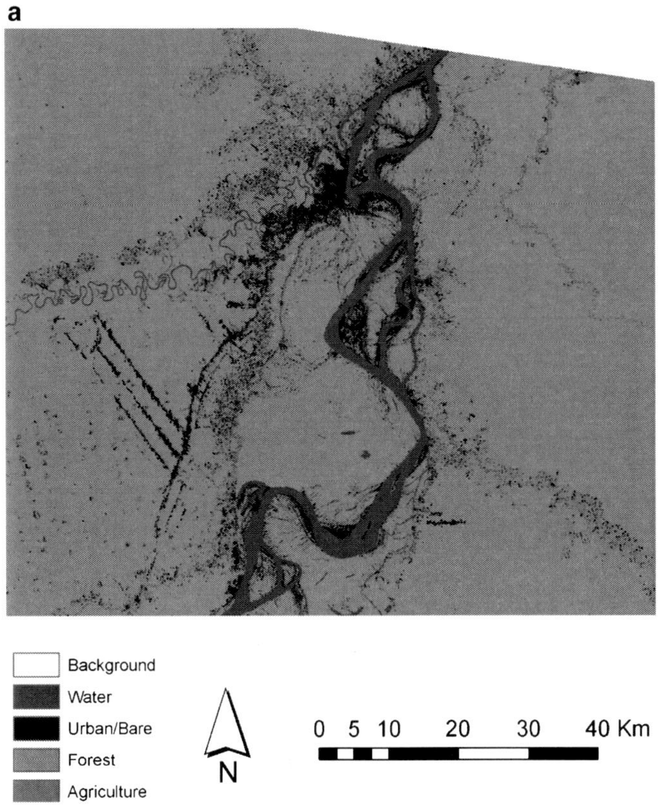


**Fig. 9.3a** Landscapes used to initiate simulation: the unaltered landscape (*see Color Plates*)

general differences in initial forest cover can be observed. However, the modifications depicted in Fig. 9.3b and c appear primarily as background noise and thus cannot be easily identified at the scale at which these figures are presented.

For the original landscape, the simulation was run 1,000 times at annual time steps for 15 years, while 100 simulations were performed for each of the modified landscapes. Transition matrices representing the frequency with which the cells changed from one state to another are provided in Table 9.1. In these matrices, the rows represent the initial states of the cells, while the columns represent the states of the cells after 15 time steps. Entry $(i, j)$ in the matrix thus represents the percentage of cells that initially occupied state $i$ and that subsequently occupied state $j$ following the completion of 15 time steps. A different matrix was produced for each of the initial landscapes, with the comparison of these matrices thus allowing for the analysis of the effects of initial forest cover on the dynamics of forest cover change.

In addition to these transition matrices, the effects of the initial forest cover on the A run was selected randomly from the 100 runs performed using each of the

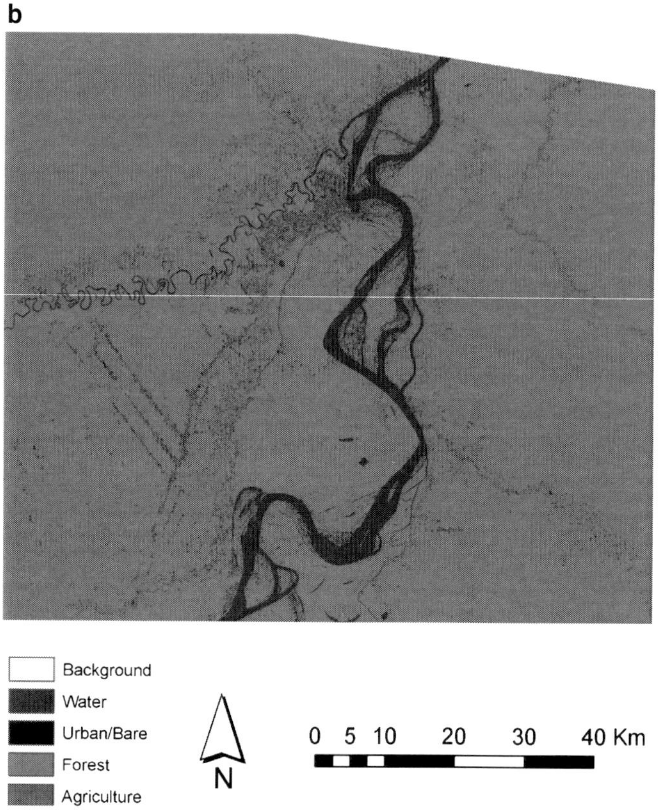


**Fig. 9.3b** Landscapes used to initiate simulation: the landscape altered so as to increase forest cover by 10% (*see Color Plates*)

altered landscapes. Maps depicting the landscape at the end of the fifteenth time step of each randomly selected run are provided in Fig. 9.4, with a corresponding subset for easier visual interpretation shown in Fig. 9.5 (a–e) over multiple conditions.

Figure. 9.4a depicts the simulation results starting with a landscape in which forest cover has been increased from that of the original landscape by 10%. In comparing this landscape with that of the original landscape, the primary differences lie in the formation of several small clumps of Agriculture and Urban/Bare cells. The formation of these clumps of cells follows as a consequence of the transition rules, which specify that connected sets of Agriculture and Urban/Bare cells will feed back upon themselves, thus increasing the number of Agriculture and Urban/Bare cells, as is seen in Fig. 9.4a. In contrast with the Agriculture and Urban/Bare cells, the Water cells are not predicted to be substantially altered over the course of the 15 year period. This result is likewise consistent with the transition rules, which indicate that connected sets of Water cells are likely to remain unchanged during the course of the simulation.

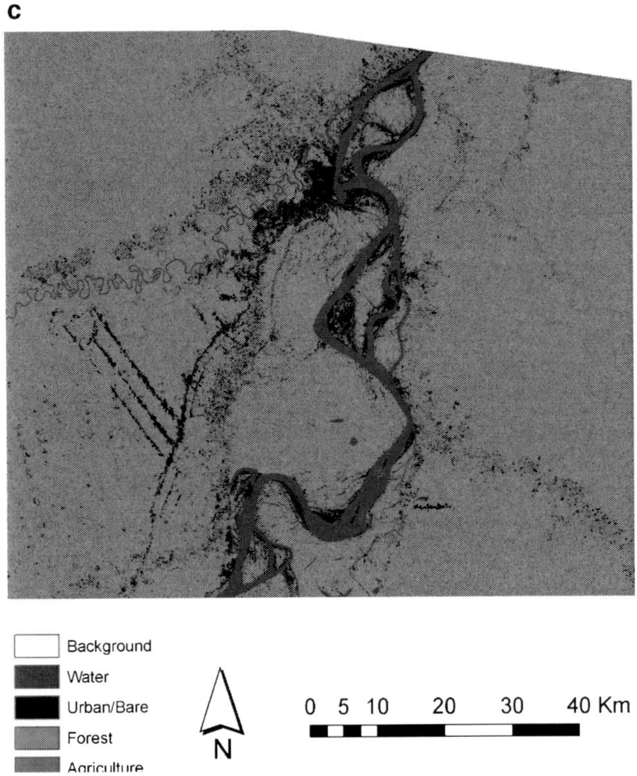


**Fig. 9.3c** Landscapes used to initiate simulations: the landscape altered so as to decrease forest cover by 10% (*see Color Plates*)

Figure. 9.4b depicts the simulation results starting with a landscape in which forest cover has been decreased from that of the original landscape by 10%. As can be seen, the loss of initial forest cover and the replacement of the Forest cells with Agriculture and Urban/Bare cells resulted in a substantial increase in the potential for sets of these connected cells to further transform the surrounding forest, thus generating the large clumps of Agriculture and Urban/Bare cells that dominate Fig. 9.4b. An initial decrease in forest cover can thus lead to substantial further decreases in the number of forested cells. The comparison of Fig. 9.4a with that of Fig. 9.4b thus indicates the potential magnitude associated with changes to the initial conditions of the graph automata. The use of initial conditions that differ in forest cover by only 20% can lead to substantially different simulation results. The importance of such variation to the study of LULCC stems from the propensity of seasonal variation to result in similar alterations of forest cover. For instance, past studies on the seasonal variation of forest cover in the Peruvian Amazon have found the forest cover to diminish by as much as 17.1% between the months of March and September (McCleary 2005). The changes in forest cover considered in

**Table 9.1** Change matrices for all simulations.[a] (a) Transitions from 1,000 simulations starting with the original landscape[b] (adapted from Sarkar et al. 2009). (b) Transitions from 100 simulations starting with a landscape that includes 10% more forest cover than the original landscape. (c)Transitions from 100 simulations starting with a landscape that includes 10% less forest cover than the original landscape

| | | Water (%) | Urban/Bare (%) | Forest (%) | Agriculture (%) |
|---|---|---|---|---|---|
| | | Conditions after 15 years | | | |
| Initial conditions | Water | 3.57 | 1.34 | 0.25 | 0.25 |
| | Urban/Bare | 0.18 | 7.72 | 5.75 | 1.51 |
| | Forest | 1.29 | 11.53 | 51.11 | 1.29 |
| | Agriculture | 0.23 | 0.95 | 3.11 | 7.30 |
| (a) Transitions from 1,000 simulations starting with the original landscape[b] (adapted from Sarkar et al. in press) | | | | | |
| | | Conditions after 15 years | | | |
| Initial conditions | Water | 4.53 | 0.12 | 0.12 | 0.04 |
| | Urban/Bare | 0.37 | 0.21 | 0.85 | 0.16 |
| | Forest | 2.61 | 8.37 | 72.65 | 5.97 |
| | Agriculture | 0.14 | 0.10 | 0.87 | 0.27 |
| (b) Transitions from 100 simulations starting with a landscape that includes ten percent more forest cover than the original landscape | | | | | |
| | | Conditions after 15 years | | | |
| Initial conditions | Water | 3.05 | 1.28 | 0.27 | 0.22 |
| | Urban/Bare | 0.65 | 6.00 | 2.09 | 1.29 |
| | Forest | 1.26 | 11.04 | 45.70 | 15.30 |
| | Agriculture | 0.19 | 0.97 | 2.69 | 5.38 |

[a]Tables sum to 97.38 % as 2.62% of the landscape was removed from analysis due to cloud cover.

[b]Numbers differ from those reported in Sarkar et al. (2009) as they were normalized differently for this manuscript. However, actual cell counts are the same.

this chapter are within the range of the observed changes (using a May image and synthesizing seasonal change, due to cloud cover limitations in the March and September imagery). The same model, using data drawn from different seasons, may thus produce substantially different results.

Table 9.1 indicates that the effects of increasing initial forest cover are nonlinear. Whereas only 45.81% of the initial forest cover was preserved in the original landscape, an increase in the initial forest cover by 10% resulted in the increase in this retention rate to 72.65%. In contrast, the decrease of the initial forest cover by 10% resulted in the retention of 45.70% of the initial forest cover, a retention rate quite similar to that associated with the original landscape. It thus appears that, in the landscape studied in this analysis, decreases in initial forest cover have only a limited effect on the dynamics of subsequent forest cover change, while increases in forest cover result in substantial changes to these dynamics. This type of nonlinear effect is characteristic of complex systems (Manson and O'Sullivan 2006). Further study of the transition rules proposed in this analysis will likely lead to the identification of additional interesting phenomena.

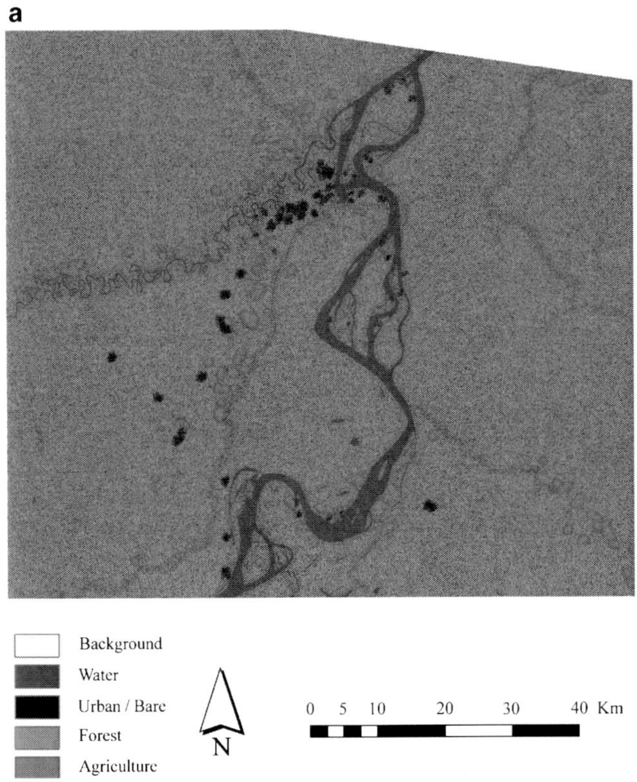


**Fig. 9.4a** Simulation results: a randomly selected result starting with a 10% increase in forest cover (*see Color Plates*)

## 9.5 Discussion

### *9.5.1 Methodological Implications*

The primary benefit of this approach is to improve the potential realism of modeling by releasing the restriction of adjacency for the interactions of cells, but this improvement comes at the cost of increased computational requirements that have in part restricted the scope of this analysis. Using only a subset of the area (754,961 ha) resulted in 8,612,624 cells; combined with a minimal classification scheme (only four classes) and run for only fifteen annual time steps, each run still took 128 min to complete on a Dell PowerEdge 2850 with a 2 GHz Pentium 4 Xeon CPU and 8GB RAM. The 100 simulations for the two input images with varying forest cover thus required more than 2 weeks to complete. Ideally, the work represented here should be repeated using a greater number of input images with smaller incremental change in each image in forest cover (e.g., in 1% increments instead of 10% increments).

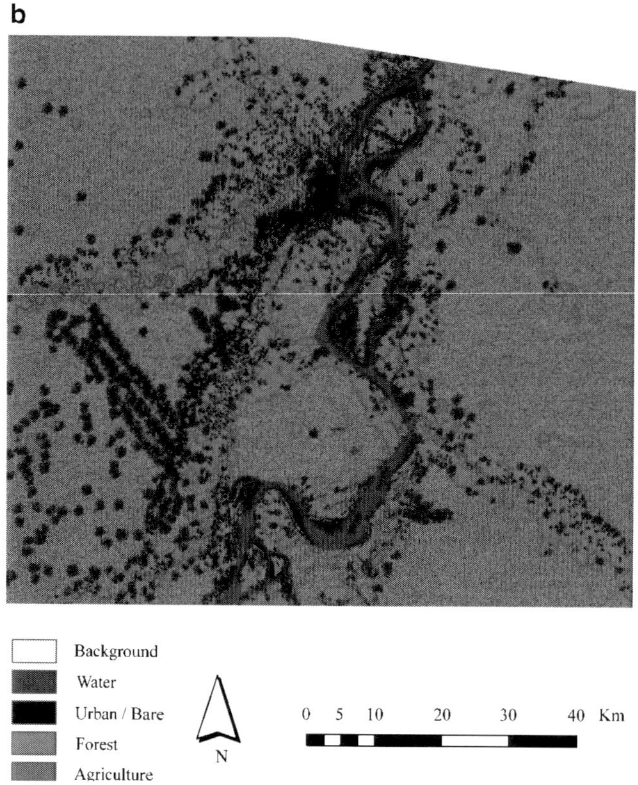


**Fig. 9.4b** Simulation results: a randomly selected result starting with a 10% decrease in forest cover (*see Color Plates*)

To test the robustness of those findings, multiple simulated landscapes at each level of forest cover should be used as well. Running a greater number of simulations obviously would produce an additive effect in processing time. A more realistic implementation of the model should also use a more detailed classification, but increasing the level of classification (to include more classes) and the resulting increase in transition rules would exponentially increase the computational complexity. Further, the generation of transition rules may be limited by increasing to that many classes. For instance, field interviews with local households may not be capable of providing such information if the respondents themselves do not view the landscape with the same "classification scheme" as needed for analysis, highlighting potential pitfalls and biases in original class definition (Robbins 2001).

Further realism could be introduced through the inclusion of the location of a cell as part of the function used to alter the initial landscapes. This step was avoided

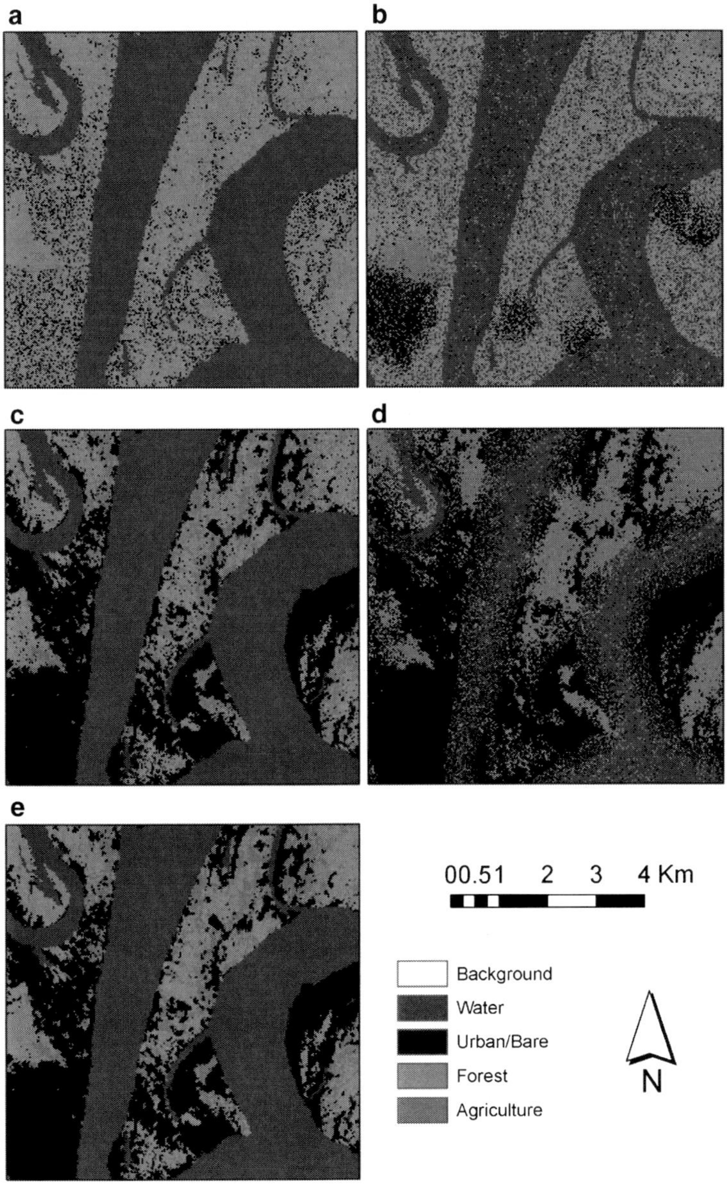


**Fig. 9.5** Simulation results, subset area: the effects of initial forest cover on landscape dynamics when viewed at a finer scale: (**a**) the initial conditions associated with an increase in forest cover; (**b**) a randomly selected landscape produced from the initial conditions in (**a**); (**c**) the initial conditions associated with a decrease in forest cover; (**d**) a randomly selected landscape produced from the initial conditions in (**c**); (**e**) the actual initial conditions, unaltered to reflect an increase or decrease in forest cover (*see Color Plates*)

here so as to avoid bias that could propagate through the simulations, but is feasible and a benefit of a graph automata approach. Also, in this analysis, each forested cell was altered with the same probability: this is ecologically implausible. When changing the forest cover of the original landscape, forested cells selected for alteration were randomly changed to or from Agriculture and Urban/Barren. It is unrealistic to assume either that conversion to these two states should occur with equal probability. The probability with which forested cells are changed to cells of other types is likewise spatially dependent in a manner not captured in this analysis. The representation of spatial heterogeneity within models of land cover change will likely demand the application of different transition rules for different cells. Though the mathematical framework that we present in this paper makes possible the application of these different rules, the ascription of different transition rules to different individual cells will require the empirical analysis of the individual cells. This type of detailed study has not been made for use in this analysis, though the results of such a study could be easily incorporated into our framework. The addition of other "layers" of information (e.g., underlying soils, proximity to logging operations) would improve upon current results but would greatly increase the potential interactions between the cells, thus compounding the problem of computational complexity. In sum, cells are not differentiated to the full extent possible, and the present analysis does not make full use of the attributes of graph automata.

### 9.5.2 *Methodological Implications*

While subsequent analyses of the effects of forest cover on landscape change should examine the effects of altering these conditions to more fully realize the benefits of a graph automata approach, the findings here do provide a proof of concept and support the preliminary conclusion that variation in forest cover observed in one intra-annual time series classification can in fact yield marked differences in simulation results. McCleary et al. (2008) document that by comparing three classifications of images acquired over a span of 6.5 months, significant fluctuation of forest cover in and out of forest classes (particularly with agricultural and other herbaceous classes) occurred in amounts greater than tested here. These results thus show that intra-annual fluctuations are indeed problematic for selection of inputs to simulation models of deforestation and reforestation, and underscores the findings of McCleary et al. (2008) and others that a multitemporal (intra-annual) classification approach is critical for proper assessment of inter-annual changes (Walsh et al. 2001), and perhaps even more so for seeding simulation models that may often be highly sensitive to variations in input cover (amount and distribution) (O'Sullivan et al. 2006). While understanding that simulations are sensitive to changes in inputs (Wolfram 1984), this work provides empirical support of that sensitivity based upon observed (intra-annual) ranges in landscape response in the western Amazon, and suggests that simulation model results of Amazonian forest change be reconsidered in light of the timing of the input imagery.

The larger implications for disentangling past and future assessments of forest changes loom large, especially when revisiting the FAO report finding that, since their standard error was 15% and the net loss of forests was less than 15%, "findings" of deforestation may actually be findings of afforestation or reforestation. Such is not to suggest that all deforestation assessments have been grossly incorrect; rather, there are likely several forces at work which once adequately investigated will explain the discrepancies and provide a more nuanced understanding of ARD processes: (1) seasonal timing of input imagery: even in aseasonal environments, this work has shown that simulation outputs can vary greatly to seemingly small changes in forest cover input; (2) as the FAO reported, non-comparable time-series (e.g., comparing a 1985–1991–1999 triplicate to a 1986–1990–2000 triplicate) complicates interpretation of findings across sites; (3) as per Foody et al. (1996), successional stages can be difficult to assess with a high degree of accuracy unless proper methods are used and adequate field data exist; (4) using the greatest archival extent possible of satellite imagery entails multiple sensor systems and therefore multiple scales of observation, complicated by the multiple scales of ARD footprints that result from very different ARD processes (Foody et al. 1996), and running single-scale studies only may therefore miss important changes in forest cover and quality; (5) many simulation programs treat certain classes (e.g., urban) and certain processes (e.g., deforestation) as terminal – that is, there is no recovery possible in the hard-coded rules despite the fact that there exist many field-based (if anecdotal) mentions of pockets of urban reforestation, for example; and (6) the total range in annual deforestation rates reported for the Amazon alone spans more than the confidence interval. The beauty of simulation modeling is that it provides verifiable/falsifiable proof of the sensitivity of modeling inputs and parameters (e.g., decision rules); verifying the accuracy of the results of multi-temporal analyses remains problematic but critical (Pontius and Malanson 2005). It is hoped that this work and other similar efforts compels researchers to revisit, rerun, and reinterpret their own "deforestation" studies.

**Acknowledgments** Authors are grateful to collaborating team members Sahotra Sarkar for his help in the original formulation of the mathematical framework of the graph automata, Christopher D. Kelley for programming assistance, and Kenneth R. Young for overall field and project analysis and for the chapter photograph. Field assistance and classification work was aided by Amy L. McCleary and Mario Cardozo. This work was supported in part by a National Science Foundation Small Grant for Exploratory Research (SGER) and Doctoral Dissertation Improvement Grant BCS – 0623229 as well as by the GIScience Center at the University of Texas at Austin.

## References

Achard F, Eva HD, Stibig HJ, Mayaux P, Gallego J, Richards T, Malingreau JP (2002) Determination of deforestation rates of the world's humid tropical forests. Science 297:999–1002

Andersson C, Frenken K, Hellervik A (2006) A complex network approach to urban growth. Environ Plan A 38:1941–1964

Batty M (2005) Cities and complexity: understanding cities with cellular automata, agent-based models, and fractals. The MIT Press, Cambridge
Boccara N (2004) Modeling complex systems. Springer, Berlin
Brown DG, Page S, Riolo R, Zellner M, Rand W (2005) Path dependence and the validation of agent-based spatial models of land use. Int J Geogr Inf Sci 19:153–174
Burks AW (1970) Essays on cellular automata. The University of Illinois Press, Urbana, IL
Butler RA, Laurance WF (2008) New strategies for conserving tropical forests. Trends Ecol Evol 23:469–472
Clarke KC, Gaydos J (1998) Loose-coupling a cellular automaton model and GIS: long-term urban growth prediction for San Francisco and Washington/Baltimore. Int J Geogr Inf Sci 12:699–714
Clarke KC, Hoppen S, Gaydos L (1997) A self-modifying cellular automaton model of historical urbanization in the San Francisco Bay area. Environ Plan B 24:247–261
Crews-Meyer KA (2006) Temporal extensions of landscape ecology theory and practice: LULCC examples from the Peruvian Amazon. Prof Geogr 58:421–435
FAO (2001) Proceedings of FAO expert consultation to review the FRA 2000 methodology for regional and global forest change assessment. Forest Resources Assessment Programme, Working Paper 42. (www.fao.org/forestry/fo/fra/index.jsp)
FAO (2005) Global Forest Resources Assessment 2005. Food and Agriculture Organization of the United Nations. Rome, Italy
Foody GM, Palubinskas G, Lucas RM, Curran PJ, Honzak M (1996) Identifying terrestrial carbon sinks: classification of successional stages in regenerating tropical forest from Landsat TM data. Remote Sens Environ 55:205–216
Geist HJ, Lambin EF (2002) Proximate causes and underlying driving forces of tropical deforestation. Bioscience 52:143–150
Gerwing JJ (2002) Degradation of forests through logging and fire in the eastern Brazilian Amazon. For Ecol Manag 157:131–141
Grainger A (2008) Difficulties in tracking the long-term global trend in tropical forest area. Proc Natl Acad Sci 105:818–823
Gutowitz H (1991) Cellular automata: theory and experiment. The MIT Press, Cambridge
Harary F (1969) Graph theory. Addison-Wesley, Reading, MA
Laurance WF (1999) Reflections on the tropical deforestation crisis. Biol Conserv 91:109–117
Laurance WF, Williamson GB (2001) Positive feedbacks among forest fragmentation, drought, and climate change in the Amazon. Conserv Biol 15:1529–1535
Li X, Yeh AG (2004) Data mining of cellular automata's transition rules. Int J Geogr Inf Sci 18:723–744
Malanson GP (1999) Considering complexity. Ann Assoc Am Geogr 89:746–753
Malanson GP, Zeng Y, Walsh SJ (2006a) Complexity at advancing ecotones and frontiers. Environ Plan A 38:619–632
Malanson GP, Zeng Y, Walsh SJ (2006b) Landscape frontiers, geography frontiers: lessons to be learned. Prof Geogr 58:383–396
Manson SM (2001) Simplifying complexity: a review of complexity theory. Geoforum 32:405–414
Manson S, O'Sullivan D (2006) Complexity theory in the study of space and place. Environ Plan A 38:677–692
McCleary AL (2005) Isolating Seasonal Variation in Landuse/Landcover Change Using Multi-Temporal Classification of Landsat ETM Data in the Peruvian Amazon. Masters Thesis. Department of Geography, University of Texas, Austin, TX.
McCleary AL, Crews-Meyer KA, Young KR (2008) Refining forest classifications in the western Amazon using an intra-annual multi-temporal approach. Int J Remote Sens 29:991–1006
Milgram DL (1975) Web automata. Inf Control 29:162–184
Moran EF (1993) Deforestation and land use in the Brazilian Amazon. Hum Ecol 21:1–21
Nagendra H, Southworth J, Tucker C (2003) Accessibility as a determinant of landscape transformation in western Honduras: linking pattern and process. Landscape Ecol 18:141–158

Ng PA, Hsia P, Yeh RT (1974) Sequential Graph Automata. Technical Report 8. Austin: Software Engineering and Systems Laboratory, University of Texas

O'Sullivan D, Manson SM, Messina JP, Crawford TW (2006) Space, place, and complexity science. Environ Plan A 38:611–617

Ostrom E, Nagendra H (2006) Insights on linking forests, trees, and people from the air, on the ground, and in the laboratory. Proc Natl Acad Sci 103:19224–19231

Pontius RG, Malanson J (2005) Comparison of the structure and accuracy of two land change models. Int J Geogr Inf Sci 19:243–265

Rindfuss RR, Walsh SJ, Turner BL, Fox J, Mishra V (2004) Developing a science of land change: challenges and methodological issues. Proc Natl Acad Sci U S A 101:13976–13981

Robbins P (2001) Tracking invasive land covers in India, or why our landscapes have never been modern. Ann Assoc Am Geogr 91:637–659

Rosenstiehl P, Fiskel JR, Holliger A (1972) Intelligent graphs: networks of finite automata capable of solving graph problems. In: Read RC (ed) Graph theory and computing. Academic, New York, pp 219–265

Rudel TK (2008) Forest policy changes in the tropics: an emerging research priority. Global Environ Change 18:253–255

Sarkar S, Crews-Meyer KA, Young KR, Kelley CD, Moffett A (2009) A dynamic graph automata approach to modeling landscape change in the Andes and Amazon. Environ Plan B, 36:300–318

Shah AN, Milgram DL, Rosenfeld A (1973) Parallel Web Automata. Technical Report 231. Computer Science Center, University of Maryland, College Park

Skole DL, Tucker CJ (1993) Tropical deforestation and habitat fragmentation in the Amazon: Satellite data from 1978 to 1988. Science 260:1905–1910

Smith A (1976) Introduction to and survey of polyautomata theory. In: Lindenmayer A, Rozenberg G (eds) Automata, languages, development. North-Holland, Amsterdam

Tapley BD, Bettadpur S, Ries JC, Thompson PF, Watkins MM (2004) GRACE measurements of mass variability in the Earth system. Science 305:503–505

Toffoli T, Margolus N (1987) Cellular automata machines. The MIT Press, Cambridge

Torrens PM (2006) Simulating sprawl. Ann Assoc Am Geogr 96:248–275

Tuomisto H, Ruokolainen K, Aguilar M, Sarmiento A (2003) Floristic patterns along a 43-km long transect in an Amazonian rain forest. J Ecol 91:743–756

Turner BL, Geoghegan J, Foster DR (eds) (2004) Integrated land-change science and tropical deforestation in the Southern Yucatán: final frontiers. Oxford University Press, New York

Turner BL, Skole DL, Anderson S, Fischer G, Fresco L et al (1995) Land use and land-cover change: science/research plan. International Geosphere-Biosphere Program Secretariat, Stocklholm

Verburg PH, Schot PP, Dijst MJ, Veldkamp A (2004) Land use change modeling: current practice and research priorities. GeoJournal 61:309–324

Walsh SJ, Crews-Meyer KA, Crawford TW, Welsh WF, Entwisle B, Rindfuss RR (2001) Patterns of change in land use/land cover and plant biomass: separating intra- and inter-annual signals in monsoon-driven northeast Thailand. In: Millington AC, Walsh SJ, Osborne PE (eds) Remote sensing and GIS applications in biogeography and ecology. Kluwer, Boston, MA, pp 91–108

Walsh SJ, Entwisle B, Rindfuss RR, Page PH (2006) Spatial simulation modeling of land use/land cover scenarios in northeastern Thailand: a cellular automata approach. J Land Use Sci 1:5–28

Wolfram S (1984) Cellular automata as models of complexity. Nature 311:419–424

Wright SJ (2005) Tropical forests in a changing environment. Trends Ecol Evol 20:553–560

Wu AY (1978) Cellular Graph Automata. Ph.D. Dissertation. Department of Computer Science, University of Pittsburgh, Pittsburgh

Xie Y, Batty M, Zhao K (2007) Simulating emergent urban form using agent-based modeling: desakota in the Suzhou-Wuxian region in China. Ann Assoc Am Geogr 97:477–495

# Chapter 10
# Forest Expansion in Northwest Costa Rica: Conjuncture of the Global Market, Land-Use Intensification, and Forest Protection

**Amy E. Daniels**

## 10.1 Introduction

Changes in forest cover have consequences for biodiversity, climate change, and water resources. For many years, the focus of land cover change research in the tropics centered on deforestation processes (e.g., Skole et al. 1994; Lambin 1997). Approaches to studying and quantifying deforestation, however, have sometimes

A.E. Daniels (✉)
School of Natural Resources & Environment, Land Use & Environmental Change Institute, University of Florida, Gainesville, FL, USA
e-mail: adaniels@ufl.edu

H. Nagendra and J. Southworth (eds.), *Reforesting Landscapes: Linking Pattern and Process*, Landscape Series 10,
DOI 10.1007/978-1-4020-9656-3_10, 

belied the complexity of forest cover changes. Forest cover dynamics may be nuanced, multi-directional and exhibit varying degrees of reversibility. Forest degradation may precede outright deforestation (World Bank 1995) or mask long -term, qualitative forest change where net forest cover remains stable (de Jong et al. 2001); secondary regrowth may keep pace with concomitant deforestation (Ramankutty et al. 2007), making "rates" of forest-cover change misleading oversimplifications. Soil degradation that ensues in the wake of intensive non-forest land use may preclude near -term reforestation (Carpenter et al. 2004). Alternatively, forest transitions may occur in which long -term declines in forest cover halt, followed by an enduring expansion of forested area (Mather 1992). The drivers of these respective processes of forest change are often distinct (Grainger 1995), but sometimes overlap or interact (Stern et al. 1992).

Proximate and ultimate drivers of deforestation have been identified through historical-comparative studies and synthetic meta-analyses (Geist and Lambin 2002). Similarly, the concept of forest transitions has emerged to explain how and why forest expansion through secondary regrowth occurs (Mather 1992). The concept holds that economic development drives off-farm migration and urbanization, contributing to spontaneous forest regeneration on agricultural lands due to land abandonment and a lack of labor (Rudel et al. 2005; Chapter 3). An alternative, though not mutually exclusive, pathway to the forest transition posits that scarcity of forests and forest-derived products, along with changing social norms promoting "improved" land management, may motivate reforestation through tree planting (Rudel et al. 2005) and/or the establishment of forest reserves (Perz 2007).

Studies from developed, industrial countries yield evidence to support these mechanisms of forest transitions (e.g., Foster 1992; Staaland et al. 1998; Andre 1998; Mather et al. 1999). Questions abound, however, regarding how relevant developed world experiences and observations are in developing countries (Koop and Tole 1999). Perz (2007) highlights the need to incorporate other contextual biophysical and institutional explanations of forest expansion if the forest transition concept is to prove meaningful in the developing-world forest management and conservation dialogue. National studies with aggregated data are unlikely to address important questions about regional developing country forest change. Sub-national case studies are critical to understanding the relationship between development and forest cover change (Klooster 2003).

In the region of Mexico and Central America, forest cover dynamics are complex with evidence of forest recovery in some places (Chapter 5). Costa Rica is a good case in point since it has undergone net forest expansion in recent years. The country experienced centuries of forest loss, most pronounced from 1950 to the mid-1980s, and had one of the highest deforestation rates in the world at 3.9% per year during this period (Leonard 1986). The inflection point in Costa Rica's forest area curve occurred in the late 1980s (Kleinn et al. 2002), representing an opportunity to study the process of reversing the country's net forest loss to a trend of forest expansion Over the last several decades, Costa Rica built a world renowned protected area network covering over a quarter of the national territory. The protected area network and other environmental protection policies undoubtedly play a vital role in abating deforestation and facilitating forest regrowth. Yet forest expansion does not appear to be limited to protected areas (Daniels 2004; Arroyo-Mora et al. 2005), suggesting that other important processes merit further analysis.

In this study I focused on identifying dominant land cover conversion sequences that explained net forest expansion in the Tempisque Basin of northwest Costa Rica. I analyzed the physical landscape setting, along with the socioeconomic and policy context for these conversions to explore the factors that drove forest expansion in the landscape from 1975 to 2000. This time period was chosen in light of the differences in data availability for important factors related to forest cover dynamics like forestry incentives and a more recent program that pays for forest ecosystem services (see Section 10.3 for details). I quantified the contribution of forest in protected and non-protected areas to net land cover changes and to the landscape's dominant trajectories. This allowed me to partly disentangle the effect of protected area establishment from other processes driving forest expansion in order to develop a conceptual model of forest recovery.

By contrasting this model with explanatory mechanisms posited by the forest transition concept I examined how well the latter accounts for observed forest expansion in the Tempisque Basin. Many of the significant studies on forest transitions fail to explicitly consider the biophysical landscape setting and pattern of forest expansion, or how these factors interact with the land tenure and policy arena (e.g., Mather 1992; Mather and Needle 1998; Rudel et al. 2005). Hence, my approach in this case study contributes to the development of a nuanced forest transition concept that is more robust in the developing world context.

## 10.2 Tempisque Basin: Geographical and Historical Setting

Costa Rica's Tempisque Basin lies in the northwestern province of Guanacaste, comprising 10% of the national territory (5,404 km$^2$, Fig. 10.1). Mean temperature is 27.5°C and an annual mean precipitation of 1,817 mm falls between May and November (Mateo-Varga 2001). The Tempisque River runs roughly north to south through the center of the basin, increasing in volume in the southern watershed with the confluence of the Bebedero River and several other important tributaries. Thirteen Holdridge life zones (Holdridge 1967) are found in the basin, along with myriad habitats like tropical dry forest, moist pre-montane forest and vast seasonal wetlands. The Cordillera Guanacaste defines the eastern border of the watershed, reaching over 2,000 m in elevation.

With more productive soils and more readily cleared vegetation than the rainforest-covered regions of the country, the Tempisque Basin is the only part of lowland Costa Rica to be continuously inhabited since the beginning of the colonial era (Peters 2001). Significant forest clearing had taken place in Guanacaste by the seventeenth century (Boucher et al. 1983). The Tempisque Basin is thus hardly a contemporary agricultural frontier. Since European arrival cattle ranching was the primary cause of deforestation by generating demand for grazing land; timber was

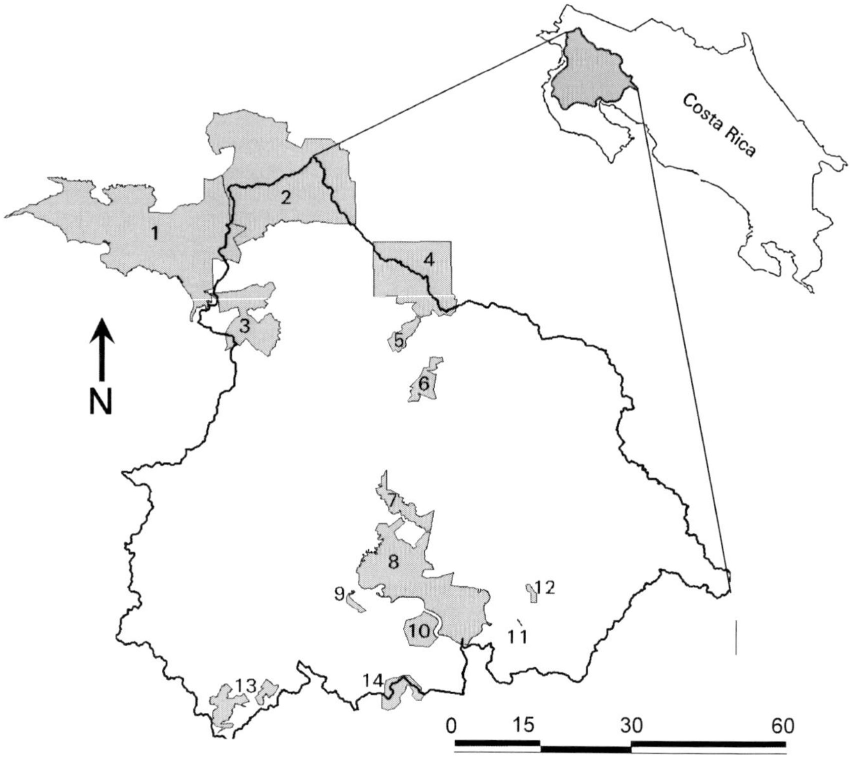


**Fig. 10.1** Map of Tempisque river basin in northwest Costa Rica (5414 km$^2$). Shaded polygons represent protected areas within, or that partly intersect with, the basin. For this analysis, portions of protected areas falling outside the basin were excluded. Protected areas are labeled as follows: (**1**) Santa Rosa National Park (38,656 ha, 1966+), (**2**) Guanacaste National Park (34,651 ha, 1991), (**3**) Horizontes Experimental Forestry Station (7330 ha), (**4**) Rincon de la Vieja National Park (14,161 ha, 1974), (**5**) La Virgen State Farm (1,923 ha), (**6**) Las Delicias State Farm (1,378 ha), (**7**) Lomas de Barbudal Biological Reserve (2,645 ha, 1986), (**8**) Palo Verde National Park (18,410 ha, 1977+), (**9**) Mata Redonda Lagoon (372 ha, 1994), (**10**) Corral de Piedra Wetland (2,484 ha, 1994), (**11**) Madrigal Lagoon (12 ha, 1994), (**12**) Taboga Forest Reserve (303 ha, 1978), (**13**) Diria Nacional Forest (13,402 ha combined, 1991), (**14**) Barra Honda Nacional Park (2,297 ha, 1974)

a secondary motive. Land clearing was reinforced by settlement policies that made partial forest clearing integral to obtaining formalized land rights.

Certain physical constraints and socio-economic characteristics of the Tempisque Basin played an integral role in calcifying the extensive cattle ranching land use system that dominated the region for centuries. Because of the long, harsh dry season, transhumant grazing was adopted where cattle were herded upslope to cloud-bathed, evergreen pastures during the dry season. Alternatively, herds were moved down slope to graze in the green floodplain of the Tempisque River. This extensive pattern of low-intensity resource management was evidenced by the basin's land tenure patterns: 13 ranchers held 11,000 ha or more; one rancher alone controlled nearly 134,000 ha (Edelman 1985) or about a quarter of the basin. Minimal investment was made in land and thus production per unit area remained low for these *latifundios*. Beef yields, for example, averaged about 168 kg/ha, relative to a possible yield of 273 kg/ha (Edelman 1985).

Agricultural census data shows that ranch size and output efficiency were negatively correlated in northwest Costa Rica, suggesting that land concentration was as much speculative as it was production oriented (Taylor 1980). This begs the question of why and how *latifundios*, a key element of a seemingly irrational land use system that is intimately tied to forest-cover patterns, persisted for so many centuries without intensifying production. Edelman (1992) details the logic of Guanacaste's *latifundios*, highlighting the importance of institutional rent (see de Janvry 1981) including subsidized production and highly-favorable credit conditions.

Favorable credit was facilitated by the powerful cattlemen lobby's connections to the nationalized banking system. With extensive landholdings as collateral, ranching operations were as much about concentrating credit, often invested in enterprises other than ranching, and about harvesting subsidies as it was about cattle production (Taylor 1980). Since land was passed down through families, sometimes dating as far back as initial land grants, the cost of land was often not a consideration in land-use decision making (Edelman 1981).

The pattern of land cover and tenure resulting from the persistence of this extensive land use system in the Tempisque Basin (up until 1970s) superficially resembled that of a "hollow frontier". The hollow frontier model attempts to explain the persistence of deforested land despite population declines. It posits that settlers clear land for agriculture, degrade soils through poor land use management, and then abandon crop cultivation plots to move further into the forest (James 1959; Casetti and Gauthier 1977). Land concentration occurs in the wake of this out-migration to the frontier, or in recent years for some places, to urban centers (Browder and Godfrey 1997). Despite depopulation, forests fail to regenerate because the land use adopted by the large landholders, generally cattle ranching, requires little labor to stay in production.

While similar to a hollow frontier in appearance and with regard to the pattern of land tenure, a key distinction for the Tempisque Basin is that land concentration largely *caused* out-migration rather than resulting from it. Low-intensity cattle ranching in the area required about six worker days per hectare per year (Taylor 1980), encouraging the highest negative migration rates in the country (–22%, Universidad de Costa Rica 1976).

## 10.3 Forces of Landscape Change

Several forces aligned to set the context for landscape change in the Tempisque Basin. Completion of the Pan-American Highway in the 1950s facilitated land-based connectivity to national markets (Peters 2001). In 1953 the cattlemen's lobby effectively pushed for legislation permitting the export of live cattle (Edelman 1985), later followed by beef once the first packing house was established. Sugar cane cultivation was promoted in the Tempisque lowlands after implementation of the 1959 US-Cuban trade embargo. By the late 1970s, Costa Rica was the fourth largest supplier of beef to the U.S. (Peters 2001).

The average price for beef on the international market was \$2.37/kg during the 1970s; but in the early 1980s the price of beef on the international market fell precipitously (1985–1999 mean price was \$1.36/kg, Arroyo-Mora et al. 2005). At the same time, interconnected with the global recession, Costa Rica experienced an acute economic crisis while having one of the highest *per capita* debt loads in the world (Harrison 1991). Interest on Costa Rica's external debt rose 15% in a short time and the nation defaulted on international loans (Lara et al. 1995). In the structural adjustments that ensued at the behest of international lending institutions, Costa Rica reduced public spending, including subsidies for cattle production.

In the mid-1970s, construction began for Costa Rica's largest hydroelectric plant just outside the eastern border of the basin, which now generates a quarter of the nation's electricity (capacity 372 MW) (DeWitt 1977). After electricity generation, water began being channeled to the lowlands of the Tempisque Basin in the late 1980s via some 234 km of canals in a project known by its Spanish acronym, PRAT (*Proyecto Riego Arenal Tempisque*). The government redistributed land outside of the eastern floodplain of the Tempisque River in 5 ha plots for flooded rice cultivation using water from PRAT. Otherwise, irrigation water is distributed to existing agro-industry operations, which was already extracting water from the Tempisque River and the underground aquifer.

In addition to providing much-needed electricity for economic and social development, PRAT greatly affected land use potential in parts of the basin. Year-round cultivation became possible with steady irrigation, increasing yields and profitability. This encouraged land use intensification at a time when cattle ranching was hardly viable. Irrigation raised yields from 3 to 8 MT/ha for rice and from 8 to 14 MT/ha for sugar cane (Amador et al. 2003). Furthermore, crop diversity expanded to include cantaloupe, watermelon, sod, and occasionally vegetables like onion and bell pepper.

By the 1970s, the establishment of protected areas also acted as a force of landscape change by protecting many remaining forested areas and facilitating forest regeneration in others. It was the Forestry Law of 1969 (no. 4465) that created a legal mechanism for establishing protected areas through executive decrees. Since then, the park system has grown in area and morphed through various administrative structures (Boza 1993; Campbell 2002), having many implications for national development. Parks are often the centerpiece for tourism development at the local and regional levels, which has generally helped garner public support for conservation. The shift of Costa Rica's historically agrarian economy toward a service economy related to tourism was highlighted in 1994 when the tourism industry surpassed all other sectors of the economy in earning foreign currency (Brockett and Gottfried 2002).

In the Tempisque Basin, the first national parks were established in the early 1970s, save a part of Santa Rosa first protected in the mid-1960s. By 2000, 14 protected areas that at least partly intersect the basin had been established across multiple categories including national forests and an experimental reforestation reserve (Fig. 10.1, Table 10.1). These areas comprise 47,340 ha or 9.5% of the basin. While Costa Rica's protected area system does include several inhabited indigenous reserves, none fall within the Tempisque Basin. Some area incorporated into parks in the basin, like parts of Guanacaste National Park and Palo Verde National Park, had long been used as pasture. These areas were left to naturally regenerate or were managed under experimental and assisted forest regeneration regimes (Janzen 2002).

A final, critical influence relevant to land use and forest cover entailed a series of revisions to the forest regulatory framework. From 1969 to 1994, an evolving series of tax-credit-based incentives was developed to encourage reforestation and forest protection on private property. These incentives were not universal, but rather implemented by landholder application upon management plan approval. Results of tax credits were not always as intended, particularly in the early years, since natural forest could be cleared and replaced by plantations to receive the tax benefit. In 1996, a new forestry law (no. 7575) was passed transforming these tax credits into direct payments for compliance with one of several modalities of forest-based land use (see Daniels et al. 2009 for details). That is, landholders began being compensated for the sale of environmental services (i.e., payments for environmental services, PES). The new forest law also disallowed changes in forest land use; extant forest cannot be cut down even on private property (article 19) without a permit.

Implementation and monitoring of both the incentives and prohibitions provided for in Law 7575 is continuously evolving. Measuring the impact of the incentives on forest cover, particularly prior to the PES scheme created in the new forestry law, has been difficult. Records of properties that received tax credits were not spatial and program administration has changed multiple times through the years making it difficult to find reliable documentation. Data for the first phase of PES (1997–2000) consist of GPS points taken within ~ 1 km from the actual contracted forest. Furthermore, forest contracted for PES may be divided into multiple, distinct patches. Only in 2006 did PES rules begin to require a map of the actual forest contracted for ecosystem service provision.

**Table 10.1** Protected area in the Tempisque basin over the three dates of land cover analysis

| Year | Protected Area (ha) intersecting watershed | Protected Area (ha) contained in watershed | Percent of Watershed[a] |
|---|---|---|---|
| 1975 | 55,114 | 6,208 | 1.2 |
| 1987 | 76,472 | 28,176 | 5.6 |
| 2000 | 127,561 | 47,340 | 9.5 |

[a] After eliminating pixels that were not represented across all three image dates either due to cloud cover in one or more dates or WRS1 scene truncation. The latter eliminated most of Santa Rosa and Guanacaste National Parks.

In this analysis, "non-protected landscape" refers to all land outside of public protected areas. This designation obviously under represents forest protection measures by not explicitly including forests conserved due to tax incentives, PES and forest protection afforded by article 19 of the 1996 forestry law. These latter forms of forest protection are dealt with implicitly to the extent that non -spatial data would allow. Also, the appropriate selection of image dates facilitates understanding of how the discussed factors may have impacted forest cover.

## 10.4 Methods

This analysis used the classification of Landsat satellite imagery to quantify land cover changes that occurred from 1975 to 2000 in the Tempisque Basin and to determine net area gains or losses of major land cover classes. Single-date land cover classifications were compiled into a trajectory layer where the value of each pixel indicated the specific land cover conversion sequence that occurred in that specific location across the three dates of analysis. The biophysical and economic landscape setting was calculated for dominant trajectories to discern geographic patterns associated with particular conversion sequences. These generalized patterns were used to construct a generalized conceptual model of landscape change.

### *10.4.1 Land-Cover Classification and Change Detection*

Four Landsat images were obtained to assess land cover dynamics over time in the Tempisque Basin. All images were acquired from the early dry season since cloud cover greatly obscures land surface features during the wet season. The first two images, 1974 (path 17, Row 53) and 1975 (Path 16, Row 53), were collected by the multispectral sensor (MSS) aboard the earliest Landsat satellite. These images were stitched together as a mosaic to create a baseline land cover representation for the study region. Since 85% of the resulting mosaic was comprised of the 1975 scene, this mosaic is referred to as the 1975 image. The middle time point was a 1987 image from the Thematic Mapper (TM) sensor; the final date was a 2000 image collected by the Enhanced Thematic Mapper plus (ETM+) sensor.

Image dates were chosen to facilitate an understanding of the drivers of landscape change. The 1975 image represents the baseline landscape largely prior to protected area establishment or forestry tax credits, and when extensive cattle ranching was still thriving. The 1987 image coincides with the inflection of the national forest transition curve for Costa Rica, the beginning of PRAT-based irrigation, and after the crash of the international beef market. The 2000 image serves as a follow-up date, by which time all protected areas in the basin had been established, but prior to effects from the systematic application of the new forestry law (Law 7575). Data availability for PES after 2000 lends itself to a spatially explicit analysis since maps of properties were collected (though not the actual forest patches within properties). For this reason, the present study stops at year 2000.

Geometric rectification of the 2000 image used GPS-based ground control points and root mean square error (RMSE) was below 0.5 or 15 m. Other dates were then co-registered to the 2000 image (RMSE $\leq$0.5). To ensure that data variance across the series corresponded to changes on the ground, rather than differences at the sensor level, all images were corrected for atmospheric haze and sensor bias (Jensen 1996). Corrections were also made for non-anniversary image dates. Solar energy strikes the earth at distinct angles that vary as the planet moves through its annual revolution. The differences in reflected energy due to date alone were removed. These resulting calibrated data were unitless in the form of reflectance (albedo), the ratio of energy reflected by a surface to the energy received. To ensure uniform spatial extent across dates, pixels corresponding to clouds occurring in any image date were eliminated from the other dates. The 1975 scene boundary was used in subsetting images to the watershed extent because its footprint truncated the northern extent of the watershed (by 110 km$^2$). These adjustments resulted in an extent of 5,153 km$^2$ (95.35% of actual basin).

Training samples for the 2000 classification were collected during fieldwork in 2001 (323 reference points). Historic photo mosaics corresponding to the dates of the satellite imagery were used to generate training samples for the 1975 and 1987 image (393 and 368 reference points, respectively). Over half of the land cover training data were set aside for accuracy assessment. During fieldwork, 29 semi-structured land use history interviews were conducted throughout the watershed to construct agricultural calendars, verify image interpretation and better understand land-use systems and changes therein. Land cover classes employed were crop (c), wetland (w), grassland (g) and forest (f) (Table 10.2). Note that there are no natural dry grasslands and grassland cover thus generally corresponds to cattle ranching land use.

**Table 10.2** Description of land cover classes (abbreviations in parentheses)

| Class | Description |
|---|---|
| Crop (c) | Dominant crops include rice, sugar cane and melon; includes all phenological stages of each crop (including bare soil associated with pre- and post-harvest crop cover on agricultural plots |
| Wetlands (w) | RAMSAR definition of wetland was employed[a] with the exception that flooded rice fields were considered crops; includes open water, mangroves, flooded forests, fresh marshes with emergent and floating vegetation, fresh meadows of grasses and sedges; includes all phenological stages of each sub-classes |
| Grasslands (g) | Includes strict pasture of grasses only, pasture with occasional trees, recently fallowed pasture scrub (< 3 years) still dominated by grasses, grassy areas along roads, and finally, regions frequently burned that cannot support appreciable woody growth |
| Forest (f) | Limestone forests, deciduous lowland forest, evergreen lowland forest, and pre-montane moist forest; rainy season canopy closure $\geq$35% |

[a] Areas of marsh, fen, peatland or water, whether natural or artificial, permanent or temporary, with water that is static or flowing, fresh, brackish or salt, including areas of marine water the depth of which at low tide does not exceed 6 m.

Land cover was independently determined for each image using a rule-based classification (Daniels 2006). This approach was developed because statistical clustering of spectral data alone failed to accurately identify the land cover classes of interest. The rule-based classification technique incorporated domain knowledge, spatial relationships and two distinct clustering algorithms. Overall classification accuracy exceeded 90% for each date and class-wise accuracy was ≥85% for each date. For details see Daniels (2006). Area for each land cover class was compared across the three-date image series to determine net-area changes across the study period. Also for each date, land cover inside and outside of protected areas was calculated for comparison.

### 10.4.2 Trajectory Analysis

Land cover composition was assessed for each of the three dates to determine trends of net-area change (e.g., net increase in forest cover). These net-area trends then guided the spatially explicit analysis of land cover conversion sequences (i.e., what conversion processes accounted for the observed net changes in land cover).

Land cover across all three dates was compiled into a single raster layer of land cover trajectories where each possible sequence of land cover conversion was given a unique value. For example, a single pixel followed over the three original land cover maps may have transitioned from grassland (g) to forest (f) from 1975 to 1987, and then remained forest through 2000. This trajectory was given a unique integer value and labeled "g-f-f" to indicate the sequence. For this analysis, a total of $4^3 = 64$ unique trajectories were possible. The area of each trajectory was calculated for protected and non-protected parts of the landscape by summing the pixels in each of the two categories.

### 10.4.3 Dominant Explanatory Trajectories

Because of the relatively large number of possible trajectories (i.e., 64 classes) in a three-image time series, criteria were developed to determine the dominant trajectories in the landscape. If a particular land cover experienced a net-area gain from 1975 to 2000, the goal was to understand which trajectories accounted for the increase. For instance, for a net crop area increase over the study period, what was the predominant land cover converted *to* cropland that explains the net increase in crop area? More generally, for an increase in land cover *x*, the areas of all trajectories ending in *x* for year 2000 (explanatory trajectories) were compared. Similarly, if land cover *y* experienced a net area decrease over the study period, all possible explanatory trajectories starting with that land cover in 1975 were compared to understand what conversion processes explain the decrease in land cover *y* by 2000.

**Table 10.3** Of the possible 64 trajectories (left), only the dominant explanatory trajectories were retained for analysis of each trend of net land cover change. Explanatory trajectories account for the majority of net area change (gain or loss)

| | Explanatory Trajectories[1] | | |
|---|---|---|---|
| All possible trajectories | Forest Gain | Cropland Gain | Wetland Loss |
| 1. g-g-g | g-f-f | g-c-c | w-c-c |
| 2. g-g-f | g-g-f | g-g-c | w-w-c |
| 3. g-g-c | f-f-f | c-c-c | w-w-w |
| n. etc. | | | |
| 64. w-g-c | | | |

Dominant explanatory trajectories were identified as those accounting for the majority of net-area change for a particular land cover (Table 10.3).

### 10.4.4 *Landscape Setting for Trajectories*

Five raster layers were generated to provide spatial data for key physical and socioeconomic landscape gradients related to land cover trajectories based on previous research (see Daniels and Cumming 2008). The resolution of each of these layers matched the pixel size for land cover data (30×30 m). Layers included a digital elevation model (DEM), slope, distance to nearest road (d.rd), distance to large forest (d.forest), and distance to nearest population center (d.popctr).

The DEM (1 m vertical resolution) was obtained from the Organization for Tropical Studies and slope was computed from it. Public roads were obtained from the *Instituto Geografico Nacional* (IGN) of Costa Rica. GPS coordinates were recorded for the six main population centers, accounting for over 95% of the population in the watershed. Large forests were defined as those on the first image with an area ≥50 ha. These patches were queried and isolated from the 1975 land cover map via an eight neighbor rule for patch definition. ArcView Spatial Analyst calculated distance surfaces. For dominant trajectories means ($\mu$) and standard deviations ($\sigma$) for each of the landscape setting variables were calculated.

## 10.5 Results

### 10.5.1 *Land Cover Area: Net Trends*

Net changes in land cover areas revealed the following trends in the basin: an increase in crop area, a decrease in wetland and grassland, and an increase in forest (Fig. 10.2). Crop area increased 47,829 ha from 1975 to 2000, almost exclusively

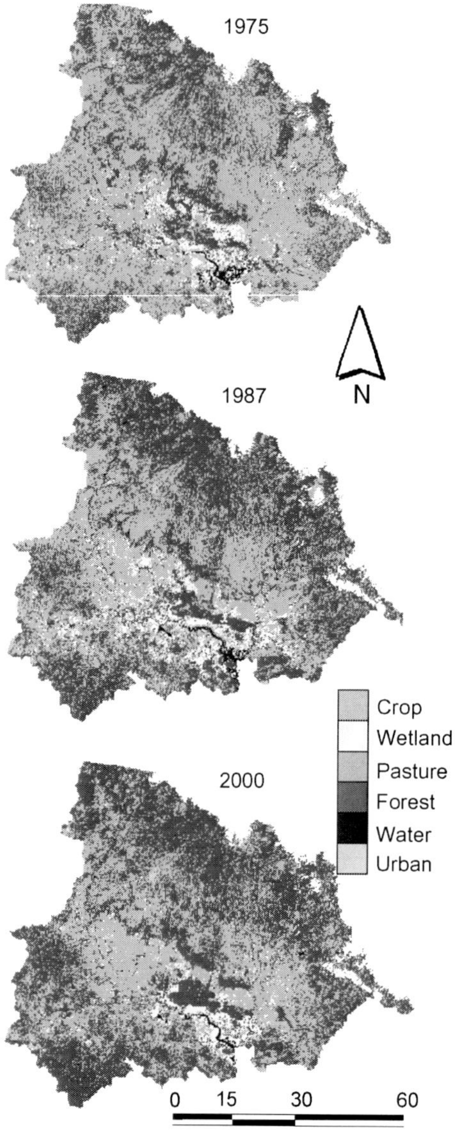


**Fig. 10.2** Land cover maps (plus water) for 1975, 1987, and 2000. Note water and urban classes were not employed in trajectory analysis (*see Color Plates*)

in the non-protected landscape (Fig. 10.3) indicating that no threat of incursion on parks existed. Wetland area decreased by 7,019 ha. Nearly half of the wetlands remaining by 2000 fell within protected areas. By 2000, grassland had decreased to nearly half its 1975 extent, a loss of 141,102 ha. The proportion of grassland falling in protected areas increased with each date, however, since land acquired

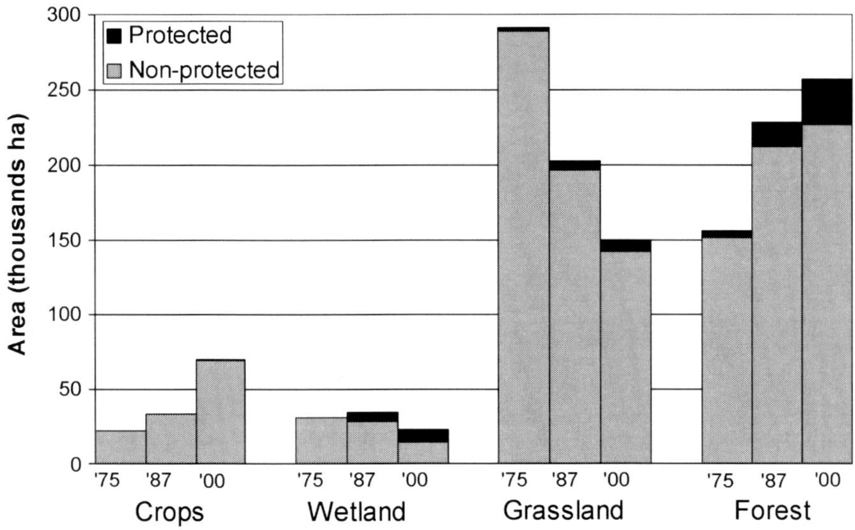


**Fig. 10.3** Trends of net area land-cover change for 1975, 1987, and 2000 in the Tempisque river watershed. The *black region of each bar* indicates the contribution of each land cover class within protected areas

for parks within the basin often required restoration from non-forest uses like cattle ranching.

For the 101,404 ha increase in forest, an increasing proportion fell within protected areas. While only 11 ha or 0.1% (Fig. 10.4) of modern PES was invested in reforestation in the basin, this modality was the focus of all past iterations of forest incentives, save only Forest Conservation Tax Vouchers or CPB which was the direct precursor of PES (year 1995 only). Unfortunately, reliable data are not available for estimating how much of this net increase in forest may be attributable to early reforestation incentives (Fig. 10.4). Of the 256,708 ha of forest by 2000, nearly 12% (26,466 ha) fell within protected areas.

## *10.5.2 Explanatory Trajectories*

The dominant trajectories accounting for the net area increase in cropland were grassland conversions (g-c-c with 14,290 ha and g-g-c with 16,116 ha, Fig. 10.5). Only about one-fifth of total 2,000 crop area had been cropland throughout (c-c-c), indicating a substitution of crop cultivation for former pastures with grassland cover. Net loss of wetlands was largely accounted for by conversion to cropland, with nearly 4000 ha converted by 1987 (w-c-c) and half again as many by 2000 (2,335 ha for w-w-c). Wetland conversion represents a minor fraction of the total area converted to crop cover in the Tempisque Basin over the study period.

| Incentive[1] | Active Period | Basin (ha) | National (ha) |
|---|---|---|---|
| IDR | 1969 - 1988 | unknown | 35,597 |
| CAF | 1986 - 1995 | unknown | 38,086 |
| CAFA | 1988 - 1995 | unknown | 33,818 |
| CAFMA | 1993 - 1995 | unknown | 22,120 |
| CPB | 1995 | 17,000[2] | 22,199 |
| PES | 1997-2000 | 7,853[3] | 256,520 |

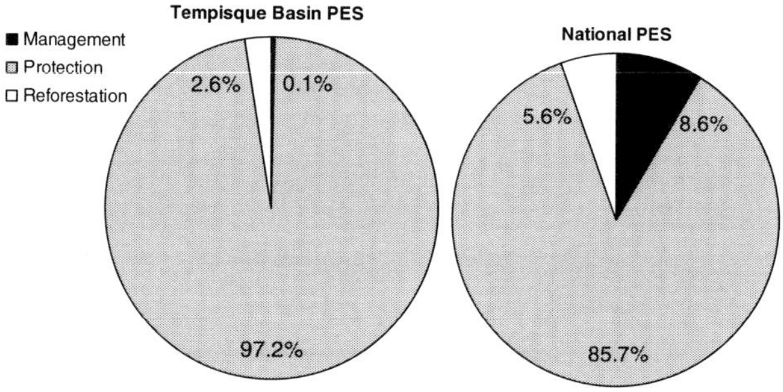


[1]See Daniels et al. 2009 for details on the evolution of incentives. Abbreviations are acronyms for the following titles that have been translated to English: Income Tax Deduction, Certified Forestry Tax Voucher, Advanced Certified Forestry Tax Voucher, Advanced Certified Forest Management Tax Voucher, Forest Conservation Tax Voucher, and Payments for Environmental Services.

[2]Estimated by Ing. Francisco Ramirez, Area de Conservacion de Guanacaste (SINAC). No other data were available.

[3]These data are from 1999 and 2000 either because no projects were located in the basin from 1997-1998 or because of poor data management when PES field administration changed from SINAC to FONAFIFO.

**Fig. 10.4** Area protected through various forms of forestry incentives for the Tempisque basin and at the national level. The *pie charts* illustrate the break-down of 1997–2000 PES area across the three forestry modalities

Yet the proportion of wetlands lost through agricultural conversion (6,415 ha) is much greater, 21% of their original extent.

The importance of protected area establishment for wetland conservation (w-w-w) is evidenced by the fact that over half of conserved wetlands fell within protected areas (Fig. 10.5) (see Daniels and Cumming 2008). Reforestation trajectories were dominant in explaining both net loss of grasslands and net increase in forest. A majority of grassland reforestation (77,224 ha) occurred from 1975 to 1987 (g-f-f) and over half again as much grassland (45,981 ha) reforested by 2000 (Fig. 10.5). Protected areas played a more significant role in reforestation trajectories than in explaining other trends of land-cover change. Eight percent and 7.5% of trajectory area for g-f-f and g-g-f, respectively, occurred within protected areas, sizeable considering that parks comprise less than 10% of the basin area. Forestry incentives were likely important too, though their contribution cannot be estimated directly (Fig. 10.4). Figure 10.6 illustrates that of the total 2,000 forest area within protected areas, 16,102 ha (53%) of it was already forest in 1975 and remained forest throughout. This f-f-f trajectory in protected areas explained more of the 2,000 forest area than

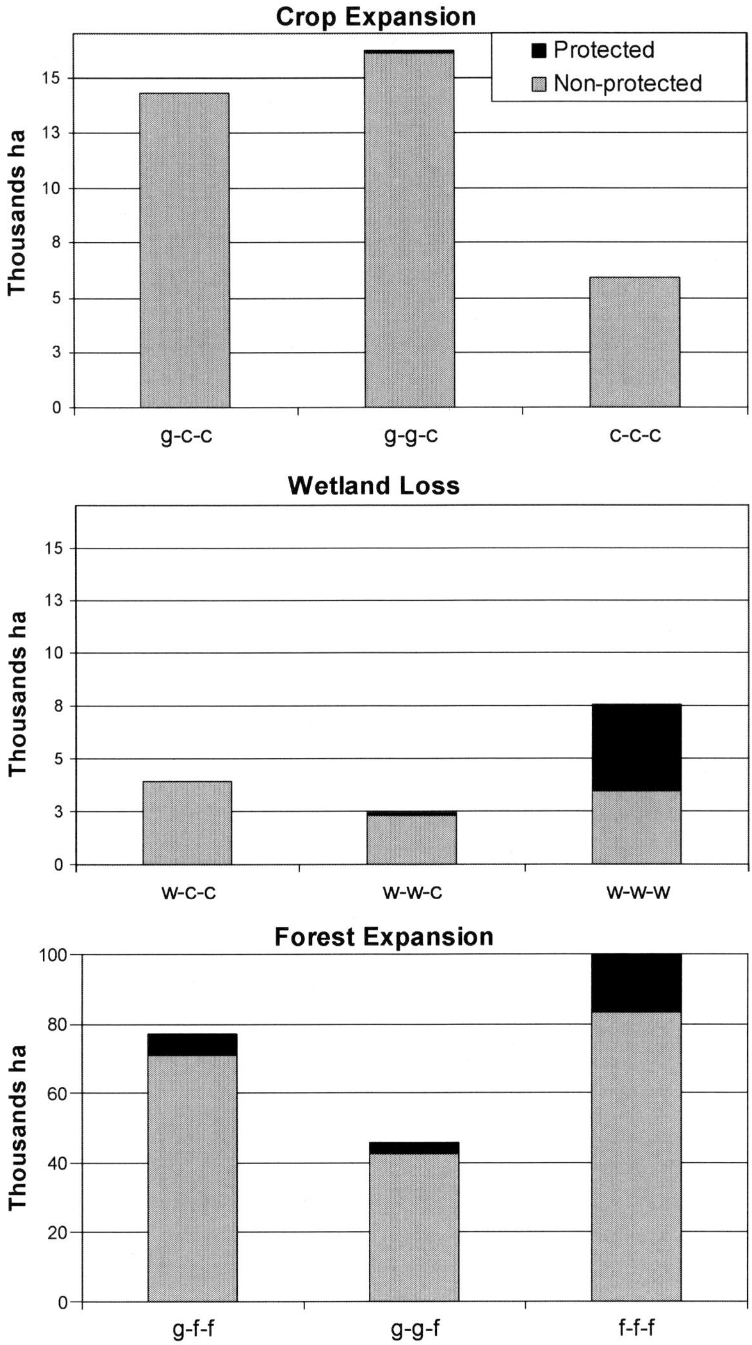


**Fig. 10.5** Dominant land cover trajectories explaining the observed net area changes in land cover. The *black region of each bar* indicates the contribution of respective trajectories within protected areas. The three letters in each trajectory name represent land cover for 1975–1987–2000, respectively, using these abbreviations: c = crop, g = grassland, f = forest, and w = wetland

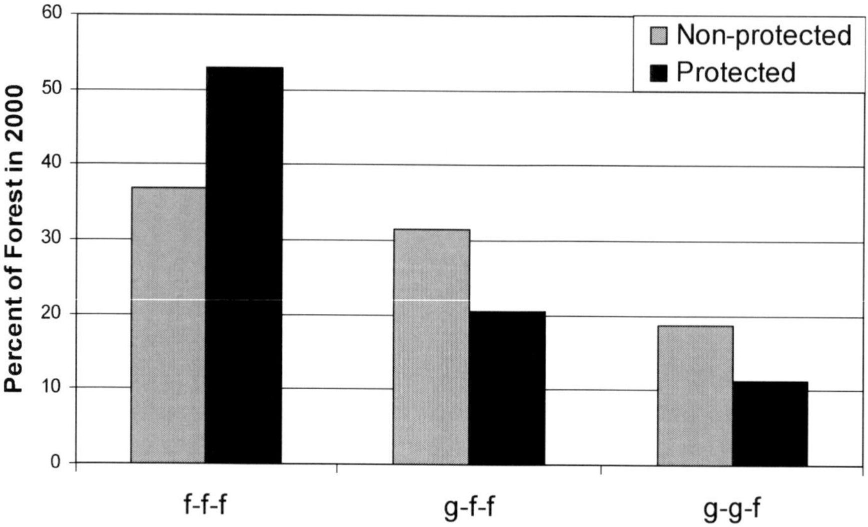


**Fig. 10.6** Graph of dominant trajectories explaining the expansion of forest observed by year 2000 for protected and non-protected areas of the landscape. The bars indicate the percent of non-protected (*gray*) or protected (*black*) forest in year 2000 explained by each trajectory

grassland reforestation by 1987 (g-f-f; 6,211 ha or 20%) or by 2000 (g-g-f; 3,451 ha or 11%) combined. The forest conservation trajectory (f-f-f) explained 37% (83,384 ha) of non-protected 2,000 forest area, compared with 50% for combined g-f-f and g-g-f trajectories (31% or 71,013 ha and 19% or 42,531 ha, respectively).

### *10.5.3 Protected and Non-protected Landscape Comparisons*

Results from comparing physical and socioeconomic landscape variables for protected and non-protected areas of the Tempisque Basin revealed that areas established as parks are different from the surrounding landscape (Table 10.4). Park land on average had a higher elevation (232 vs. 191 m) and occurred on slightly steeper slopes relative to land outside of parks (12 vs. 10%). Protected areas also occurred at greater distances from population centers (16.1 vs. 10.2 km) and roads (3.2 vs. 1.5 km). Protected areas pixels were, on average, substantially shorter distances from large forest patches (0.4 vs. 1.1 km).

### *10.5.4 Landscape Setting for Dominant Trajectories*

The means ($\mu$) and standard deviations ($\sigma$) of landscape variables are listed for each landscape trajectory (Table 10.5) indicating spatial patterns with regard to the physical or socioeconomic landscape setting. On average, trajectories accounting

**Table 10.4** Population means (μ) of key variables representing physical and socioeconomic landscape setting in the Tempisque basin, stratified by protected and non-protected status. D. popctr, D.rds and D.forest signify distance to nearest population center, nearest road and large forest patch (> 50 ha in 1975), respectively

| | N | Elevation (m) | Slope (%) | D.popctr (m) | D.rd (m) | D.forest (m) |
|---|---|---|---|---|---|---|
| Non-Protected | 3,813,062 | 191 | 10 | 10,197 | 1,527 | 1,108 |
| | | (205) | (16) | (6,121) | (1,379) | (1,409) |
| Protected | 394,436 | 232 | 12 | 16,129 | 3,198 | 415 |
| | | (291) | (18) | (5,988) | (2,248) | (790) |

Standard deviations (σ) in parentheses.

**Table 10.5** Population means (μ) of variables representing physical and socioeconomic landscape setting in the Tempisque Basin, stratified by unique land cover trajectories. Tempisque basin for D.popctr, D.rds and D.forest signify distance to nearest population center, nearest road and large forest patch (> 50 ha in 1975), respectively. The letters in each trajectory name represent land cover for 1975–1987–2000, respectively, with the following abbreviations: c=crop, g=grassland, f=forest, and w=wetland

| | N | Elevation (m) | Slope (%) | D.popctr (m) | D.rd (m) | D.forest (m) |
|---|---|---|---|---|---|---|
| g-f-f | 858,042 | 239 | 14 | 11,119 | 1,527 | 1,010 |
| | | (202) | (18) | (6,661) | (1,451) | (1,297) |
| g-g-f | 510,910 | 175 | 12 | 9,961 | 1,416 | 1,261 |
| | | (182) | (17) | (6,315) | (1,475) | (1,419) |
| w-c-c | 43,977 | 13 | 1 | 11,450 | 2,555 | 1,430 |
| | | (12) | (2) | (4,350) | (1,547) | (,1236) |
| w-w-c | 27,295 | 13 | 0 | 9,434 | 1,810 | 1,537 |
| | | (9) | (2) | (5,133) | (1,480) | (1,320) |
| g-c-c | 158,955 | 20 | 1 | 7,891 | 1,397 | 2,050 |
| | | (16) | (2) | (3,700) | (1,033) | (1,669) |
| g-g-c | 180,829 | 46 | 2 | 8,420 | 1,153 | 1,871 |
| | | (73) | (5) | (4,838) | (1,027) | (1,567) |
| f-f-f | 1,105,410 | 281 | 14 | 12,035 | 2,000 | 398 |
| | | (252) | (18) | (6,356) | (1,609) | (994) |
| g-g-g | 934,159 | 145 | 7 | 9,755 | 1,600 | 1,357 |
| | | (154) | (13) | (6,096) | (1,689) | (1,403) |
| c-c-c | 65,815 | 18 | 1 | 7,698 | 1,326 | 2,250 |
| | | (15) | (2) | (3,435) | (1,063) | (1,573) |
| w-w-w | 84,159 | 9 | 0 | 17,310 | 2.946 | 1,375 |
| | | (6) | (2) | (6,897) | (1,596) | (1,066) |

Standard deviations (σ) in parentheses.distance to infrastructure (D.rd 2 km and D.popctr 12 km) only by the wetland conservation trajectory.

for wetland loss (w-c-c and w-w-c) occurred closer to population centers and roads than conserved wetlands (w-w-w). Conversion to cropland also occurred on slightly higher, more-sloped land relative to wetland conservation (w-w-w). The latter trajectory occurred at the lowest elevation (μ=9 m) and slope (μ=0%), and in sites that were most isolated relative to infrastructure like roads (μ=2.9 km) and towns (μ=17.3 km).

Explanatory trajectories for cropland gains were those of grassland conversion, where conversion to cropland in the first time step (g-c-c) occurred at lower elevation, gentler slopes and farther from forest and infrastructure, relative to conversion that occurred in the second time step (g-g-c). The dominant grassland loss trajectories were g-f-f and g-g-f, though clearly conversion to cropland played a significant role. Reforestation in the first time step (g-f-f), relative to the second time, occurred at higher elevations (239 vs. 175 m), steeper slopes (14 vs. 12%), farther from population centers (11.1 vs. 10.0 km), but effectively the same distance from roads (1.5 vs. 1.4 km).

On average the g-f-f sites occurred closer to existing large forest patches (1.0 km) compared with grassland reforestation in the second time step (1.3 km). Forest conservation (f-f-f) occurred at highest elevations (281 m) and steep slopes (14%). This trajectory was also closer to extant forest (398 m) than any other trajectory, in part by definition. And finally, f-f-f trajectories were surpassed in Sites that were grassland throughout (g-g-g) occurred at elevations (145 m) and slopes (7%) intermediate to grasslands that reforested and grasslands converted to cropland. Similarly, g-g-g sites were intermediate, relative to conversion to cropland and reforestation trajectories, with regard to distance from large forest patches (1.4 km). On average, sites characterized by the g-g-g trajectory were more distant from population centers (9.8 km) and roads (1.6 km) than grasslands converted to cropland.

## 10.6 Discussion

Gains in cropland were greatest from 1987 to 2000, after the implementation of PRAT which underscores the importance of regional irrigation infrastructure in changing the land use potential for this seasonally arid landscape. Logically, wetland loss was also greatest in the second time step since wetlands were predominantly converted to crops. Greater gains in forest area were seen in the first time step corresponding to forestry tax credits, the establishment of several protected areas, and decreased profitability of cattle ranching resulting from a significant decline in beef prices and reduced production subsidies.

Grassland loss was greatest in the first time step, effectively reciprocal to gains in forest area. Since this time step included the crash of beef prices, this indicates a tradeoff between cattle ranching and forest-related land uses. The decline of grasslands in the second time step was also substantial, however, owing to continued reforestation and intensification facilitated by PRAT. Grasslands in the lowland areas of the watershed were converted to crop cultivation. Also in the second time step, incremental additions of forest on private land were facilitated through forestry incentives and PES. For example, about 204 ha of reforestation were afforded during this period through the reforestation modality of PES (Fig. 10.4).

Results suggest that conversion processes had unique domains within the broader landscape. Wetland conservation (w-w-w) occurred on the lowest-lying, flattest and most isolated regions of the basin, whereas those wetlands first converted to cropland (w-c-c) were on higher, gently sloping land. These patterns reflect the importance of both drainage and accessibility for crop cultivation. In especially boggy sites, crop yields may be diminished by seasonal flooding. Investment required for wetland drainage may also deter conversion in such locations.

Grassland areas undergoing intensification through conversion to cropland were located on lower and flatter land. In contrast, grasslands on higher, sloped lands—marginal areas for intensive cultivation or pasture use—generally became reforested. These findings concur with results from an analysis of forest cover change for a broader region of northwest Costa Rica (Arroyo-Mora et al. 2005), and with those described in Kozak (Chapter 11) for the Polish Carpathians. Further, in agreement with patterns observed in developed-world forest transitions, both agriculturally-marginal and isolated lands preferentially experienced reforestation (Mather and Needle 1998). Grassland areas where forest regenerated were farther from population centers and roads, but closer to large patches of extant forest, than was grassland that converted to cropland.

A substantial portion of the net increase in forest area observed by year 2000 occurred due to reforestation and forest conservation on privately held land. This provides a quasi-control for the influence of protected areas on forest expansion in the landscape, lending support to evolving forestry incentives and the decreased profitability of cattle-ranching as drivers of forest recovery. The latter set a favorable context for the transformation of the dominant land use system, reinforced by PRAT's implementation. Low land rent from cattle ranching also contributed to spontaneous forest regrowth by facilitating the retirement of marginal areas of the landscape. Coupled with the forestry tax credits and later PES, relatively unprofitable cattle ranching enhanced the economic attractiveness of forest-related land use. Since no systematic spatial records are available for landholders that participated in the forestry-incentives program, their historic contribution to forest recovery was not explicitly quantified (but see Sanchez-Azofeifa et al. 2007 which used GPS points taken in the general vicinity of PES projects from 1997 to 2000, as opposed to polygons needed for this spatially explicit analysis).

Protected areas were established preferentially in locations where forest had been conserved; on average, parks had higher elevations, steeper slopes and were more isolated. This is not unique among analyses of protected area site selection patterns (e.g., Rouget et al. 2003; Southworth et al. 2004). Precisely these characteristics contribute to the persistence of forest cover by making such areas less competitive for economically driven and deforesting land uses. These results, in addition to the forest persistence trajectory (f-f-f) on marginal lands in the non-protected landscape, suggest that forest in some now-protected locations would have been conserved during the study period due to the landscape's physical constraints alone.

In terms of forest expansion within protected areas, reforestation of grasslands nearly equaled extant forest that became park land. Yet trends in the non-protected landscape suggest that some of these upland grass areas may have reforested even without protected status. This exemplifies Grainger's point (Chapter 2) where the withdrawal of agricultural institutions by itself—in this case institutional rent associated with cattle ranching—may facilitate forest regrowth by default. Further, the rise of forestry institutions and the modified role of agricultural ones contributed to a policy context where forests were more viable economically than they had been (conditions also noted to facilitate forest investment in Chapter 14).

Thus, the history of this landscape underscores that forest cover expands and shrinks, in part, with economic trends that influence the dominant land use system(s). Establishing protected areas in perpetuity ensures that at least some minimum forest remains regardless of the economic context, at least in Costa Rica where protected area boundaries are generally respected. While there is forest cover in the non-protected landscape that appears relatively unthreatened either due to geographic setting or short-term (5 year) protection through PES, deforesting or forest degrading land uses for these such areas is conceivable in the future. Further, most of the basin's non-protected forest is in varying successional stages. Protected area establishment may not have been determinant of forest expansion in the basin, but is likely a critical element for the long-term recovery of old-growth tropical dry forest—one of the most endangered ecosystems in the world.

The unique landscape niches associated with particular trajectories, along with the broader drivers of landscape change, can be synthesized in a conceptual model explaining observed forest expansion in the Tempisque Basin (Fig. 10.7). Grasslands in sufficiently flat, accessible areas intensified from pasture use to crop cultivation; on more marginal, isolated grasslands, or those falling within protected areas, reforestation occurred. Forces behind forest recovery in the basin are multiple and interrelated as with other case studies in this volume (e.g., the combination of economic and political drivers of Vietnam's transition, see Chapter 14).

The causal structure that had long ensured the maintenance of cleared land got rearranged (see Chapter 4) in the Tempisque Basin. The conjuncture of protected area establishment, declining beef prices, diminished subsidies for beef production, implementation of regional irrigation and the advent of forest incentives manifested in significant forest recovery. The decreased profitability of cattle ranching appears to have set a favorable context, whereby its synergy with land-use intensification and forest protection, acted at a critical moment in the history of the landscape to favor forest expansion. Had nothing changed regarding the economics of cattle ranching, the impact of forest incentives and irrigation on forest cover in the Tempisque Basin would have probably been diminished.

Models attempting to explain forest cover dynamics may fail to address mechanisms of non-forest land cover change concomitant with changes in forest area. Understanding the relationship between different conversion processes in landscapes where forest expansion has been observed will contribute to a more nuanced understanding of forest recovery. In the case of the Tempisque Basin, considering the landscape as a forest/non-forest dichotomy would have failed to reveal the clear trend of land use intensification, a critical component of landscape change. Intensification, while beneficial for forest recovery, has been detrimental to wetlands

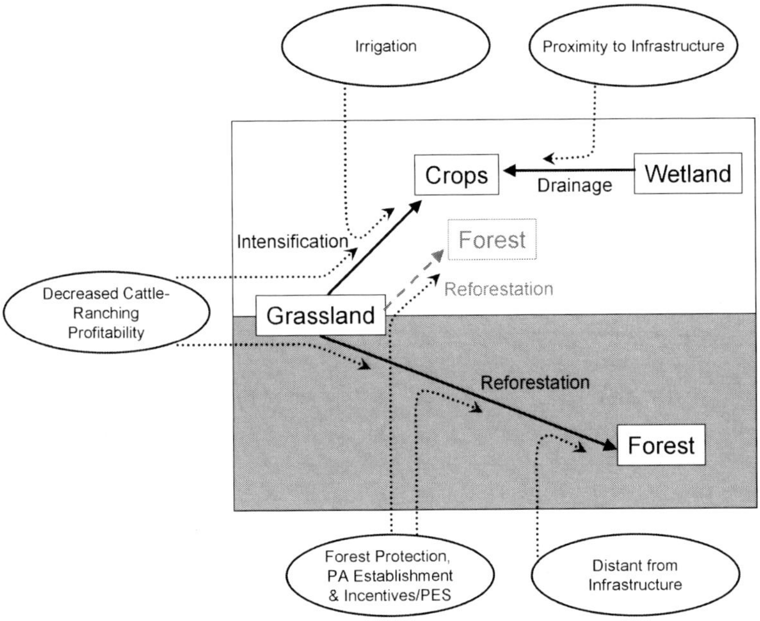


**Fig. 10.7** Conceptual model of dominant trends accounting for forest expansion in the Tempisque basin. The landscape is represented by the large rectangle, divided into uplands (*gray*) and lowlands (*white*). The four land cover classes analyzed in this study are represented by small rectangles with arrows depicting the dominant trajectories (e.g., grassland conversion to crops). *Ovals* around the landscape depict the forces driving observed processes of landscape change. The dominant reforestation trend occurred in upland areas (*black*). Lowland reforestation was a minor factor in accounting for forest expansion (gray)

given that roughly two-thirds of non-protected wetlands were converted to cropland. The ways in which land cover patterns are affected by shifts in the dominant land use system (e.g., from extensive cattle ranching to intensive use) must be anticipated to ensure policy coherence. For instance, a system of wetland conservation or restoration on private land—analogous to tax incentives or, more recently, the PES scheme for forests—may mitigate negative tradeoffs from the indirect competition between reforestation and wetland conservation, processes that interact despite occurring in distinct niches of the landscape.

The proposed model of forest expansion (Fig. 10.7) begs several important questions about the longevity of the observed forest recovery in the Tempisque Basin. The unique landscape niches of different trajectories highlight the need for spatially conscious management decisions regarding how shifts in land use are facilitated or constrained by the physical landscape. Significant portions of the non-protected landscape, occurring on sites marginal for cultivation or pasture use, appear to comprise "conditional forests," land that is forested depending upon the economic and policy context. If forest protection (article 19 of Law 7575) is not fully implemented, if PES initiatives are eliminated, and/or if extensive (as opposed to intensive) cattle ranching is profitable again to the extent that it had been, will forests on marginal areas persist as implied in the enduring forest recovery concept of a "forest transition"?

Forest protection mechanisms and incentives like PES are clearly critical to facilitating secondary forest regeneration and conservation. In Costa Rica, these factors were locked in at an optimal time when land-use economics were being reorganized. Recent trends (i.e., post-2000) certainly support this notion in that there are now more head of cattle in Guanacaste than in the best years of extensive ranching model of the past—yet on far less pasture. The Ministry of Agriculture (MAG) still subsidizes cattle production, but of the intensive sort. Synthesizing production patterns and forest recovery patterns suggests that Costa Rica has crafted relatively effective institutions for addressing historic deforestation threats. But are these institutions equipped to conserve the observed forest recovery for the long-term (see Daniels et al. 2009)?

Demographic factors have often figured prominently in explaining forest cover trends (Meyer and Turner 1992). Indeed, about half of the variance in the extent of deforestation can be explained by population in the long run, though the relationship is far from simple or static (Mather et al. 1999). Similarly, one of the major forest transition pathways is thought to be the demographic trend of out migration in rural landscapes where forest regeneration occurs in the wake of a diminished labor pool and associated land abandonment (Rudel et al. 2005, Chapter 6). In northwest Costa Rica, however, Harrison (1991) found that forest cover was not correlated with population trends from 1950 to 1984, the most intense period of deforestation (Harrison 1991). Given the labor-saving, extensive land tenure and land use system in place during the centuries of deforestation, the lack of correlation is not surprising and noted to characterize much of Latin America (Sloan 2007).

The relationship between forest cover and population in recent years for the Tempisque Basin has not been examined. Yet from 1984 to 2000, corresponding roughly to the second time step in this analysis (1987 to 2000), the population growth rate for Guanacaste was nearly 2% (compared with less than 1% from 1973 to 1984, roughly the first time step in this study). Trends of net emigration diminished four-fold in the period from 1995 to 2000 relative to trends in the 1970s. Net negative migration even reversed in some cantons to net positive immigration. With half again as many people in 2000 as at the beginning of the study (represented by 1973 census data) and a declining or reversing trend for out-migration in the province, forest recovery does not appear to be driven by depopulation. This suggests that population patterns were independent of deforestation or reforestation of the Temspique Basin. A carefully designed analysis of demographic trends with disaggregated data and corresponding land cover maps is needed to test this.

Other broad classes of explanations for forest transitions discussed in the literature find greater support in the Tempisque Basin. Grainger (1995) describes several mechanisms that resulted in forest transitions over the course of economic development. Agricultural intensification leads to abandonment of marginal lands which, in turn, reforest. He attributes this process to industrialization/urbanization and decreased competitiveness of small scale agriculture.

While intensification was certainly an important component of forest expansion in the Tempisque Basin, its drivers were distinct from Grainger's generalized scenario. This study illustrates the importance of context-specific factors like PRAT, in explaining forest transitions.

Grainger also describes a mechanism whereby improved land management and a shift in attitudes about forest that occurs through development. This certainly characterizes the case of the Tempisque Basin, and Costa Rica more generally, over the last several decades (also see Chapter 7). The shift in forest valuation was not driven only by nature-based tourism, but was facilitated by exogenous forces like the global recession, crash of the beef market and reduced public spending. The latter circumstances brought about a new economic context and a reorganization of national development goals. Hence, according to the experience of the Tempisque Basin, a robust forest transition theory cannot neglect to incorporate both context specificity and non-local influences on land use and land cover.

Agriculture—extensive or intensive—is hardly the major threat to forests anymore in the Tempisque Basin. Results illustrate that low-intensity land uses account for an ever-diminishing area in the Tempisque Basin. Perhaps the greatest uncertainty regarding the future of forest cover and the longevity of the observed forest expansion is landscape gentrification and associated real estate development. The forestry law's ban on forest land -use change proves costly to enforce and arguably altogether unrealistic. Property values in the basin have increased steeply from real and speculative land investment but PES does not exceed the opportunity cost of land development which externalizes ecosystem degradation. About 24,853 ha (Fig. 10.4) of forest were protected through tax vouchers (17,000 ha) and direct payments (7,853 ha), representing 10% of the 2000 forest area in the basin. Yet once these ecosystem service contracts expire, the fate of these forested lands is uncertain.

A critical but largely absent dimension in forest transition theory is the role of regional and international timber trade in facilitating forest recovery at local and regional scales. At the national level, Costa Rican timber imports have increased dramatically in recent years (Barrantes and Salazar 2007), underscoring that forest recovery does not imply a net decrease in timber demand. Analogous to the forest transition path of many developed nations, forest conservation and regeneration in Costa Rica could be simply afforded by displaced deforestation and/or expansion and intensification of timber plantations elsewhere.

Ultimately, context-specific case studies must be linked to broader global forest cover trends and the supply/demand of forest goods in order to appropriately contextualize regional or national-level forest recovery, its driving forces, and its degree of permanence. These are all critical points for post-Kyoto REDD initiatives (see also Chapter 2). As of now, the issues of land development pressure, the timber trade and deforestation displacement are not explicitly addressed in the forest transition literature for developing countries.

The phenomenon of forest recovery in the Tempisque Basin is instructive. It casts local forest expansion, not as an inevitable byproduct of economic development as suggested by forest transition theory, but as a process driven by the conjuncture of key global, national and local factors affecting land use patterns like the international beef market; national forest protection policies and irrigation development; and local land use intensification respectively. Evidence from this research suggests that patterns of forest cover like those characterized by the "hollow frontier" and "forest transition" concepts are not necessarily competing models in explaining forest-cover trends after the passing of an agricultural frontier. Rather, these models may describe distinct phases of a longer-term process of forest change in different economic and development contexts. The challenge, insofar as the provision of forest goods and services is concerned, lies in understanding the levers that maintain a favorable context for the long-term conservation of this forest recovery in a dynamic, developing economy.

**Acknowledgments** This research was funded through a Fulbright Fellowship and a NASA pre-doctoral fellowship. Many thanks are owed to the Organization for Tropical Studies, Jose Cubero of FONAFIFO, and Francisco Ramiriz and Orlando Matarrita of SINAC/MINAE for their assistance and support. Helpful comments and suggestions by reviewers are especially appreciated.

## References

Amador JA, Chacon RE, Laporte S (2003) Climate and climate variability in the Arenal river basin of Costa Rica. In: Diaz HF, Morehouse BJ (eds) Climate and water: transboundary challenges in the Americas. Kluwer, Boston, MA

Carpenter FL, Nichols JD, Sandi E (2004) Early growth of native and exotic trees planted on degraded tropical pasture. Forest Ecol Manag 196(2–3):367–378

Andre MF (1998) Depopulation, land-use change and transformation in the French Massif Central. Ambio 27:351–353

Arroyo-Mora JP, Sanchez-Azofeifa GA, Rivard B et al (2005) Dynamics in landscape structure and composition for the Chorotega region, Costa Rica from 1960 to 2000. Agric Ecosyst Environ 106:27–39

Barrantes A, Salazar G (2007) Usos y aportes de la Madera en Costa Rica. Oficina Nacional Forestal. San Jose, Costa Rica

Boucher DH, Hansen M, Risch S et al (1983) Agriculture. In: Janzen DH (ed) Costa Rican natural history. University of Chicago Press, Chicago, Il

Boza MA (1993) Conservation in action – past, present, and future of the national park system of Costa-Rica. Conserv Biol 7:239–247

Bray DB (2009) this volume Forest cover dynamics and forest transitions in Mexico and Central America: towards a great restoration? In: Nagendra H, Southworth J (eds) Reforesting landscapes: linking pattern and process. Springer, New York

Brockett CD, Gottfried RR (2002) State policies and the preservation of forest cover: Lessons from contrasting public-policy regimes in Costa Rica. Lat Am Res Rev 37:7–40

Browder JO, Godfrey BJ (1997) Rainforest cities: urbanization, development, and globalization of the Brazilian Amazon. Columbia University Press, New York

Campbell LM (2002) Conservation narratives in Costa Rica: conflict and co-existence. Dev Change 33:29–56

Casetti E, Gauthier HL (1977) Formalization and test of hollow frontier hypothesis. Econ Geogr 53:70–78

Daniels AE (2004) Protected area management in a watershed context: a case study of Palo Verde National Park, Costa Rica. Thesis, University of Florida, Gainesville, FL

Daniels AE (2006) Incorporating domain knowledge and spatial relationships into land cover classifications: a rule-based approach. Int J Remote Sens 27:2949–2975

Daniels AE, Cumming GS (2008) Conversion or conservation? Understanding wetland change in northwest Costa Rica. Ecol Appl 18:49–63

Daniels AE, Esposito V, Bagstad K et al (accepted) A decade of PES: building on Costa Rica's model and applying lessons learned. Ecol Econ, special issue

De Janvry A (1981) The agrarian question and reformism in Latin America. The Johns Hopkins University Press, Baltimore, MD

de Jong W, Freitas L, Baluarte J et al (2001) Secondary forest dynamics in the Amazon floodplain in Peru. Forest Ecol Manag 150:135–146

DeWitt RP (1977) The Inter-American Development Bank and political influence, with special reference to Costa Rica. Praeger, New York

Edelman M (1981) Apuntes sobre la consolidacion de las haciendas en Guanacaste. Avance de Investigación No. 44. Instituto de Investigaciones Sociales, Facultad de Ciencias Sociales. Universidad de Costa Rica, San Jose, Costa Rica

Edelman M (1985) Extensive land-use and the logic of the Latifundio – a case-study in Guanacaste province, Costa-Rica. Hum Ecol 13:153–185

Edelman M (1992) The logic of the latifundio: the large estates of northwestern Costa Rica since the late nineteenth century. Stanford University Press, Stanford, CA

Foster DR (1992) Land-use history (1730–1990) and vegetation dynamics in Central New-England, USA. J Ecol 80:753–772

Geist HJ, Lambin EF (2002) Proximate causes and underlying driving forces of tropical deforestation. Bioscience 52:143–150

Grainger A (1995) The forest transition: an alternative approach. Area 27:242–251

Grainger A (2009) The bigger picture – tropical forest change in context, concept and practice. In: Nagendra H, Southworth J (eds) Reforesting landscapes: linking pattern and process. Springer, New York

Harrison S (1991) Population growth, land use and deforestation in Costa Rica, 1950–1984. Interciencia 16(2): 83–93

Holdridge LR (1967) Life zone ecology. Tropical Science Center, San Jose, Costa Rica

James PE (1959) Latin America. Odyssey Press, New York

Janzen DH (2002) Tropical dry forest: area de conservacion de Guanacaste, northwestern Costa Rica. In: Perrow MR, Davy AJ (eds) Handbook of ecological restoration, vol 2, restoration in practice. Cambridge University Press, Cambridge, UK

Jensen JR (1996) Introductory digital image processing: a remote sensing perspective. Prentice Hall, Upper Saddle River, NJ

Kleinn C, Corrales L, Morales D (2002) Forest area in Costa Rica: a comparative study of tropical forest cover estimates over time. Environ Monit Assess 73:17–40

Klooster D (2003) Forest transitions in Mexico: institutions and forests in a globalized countryside. Prof Geogr 55:227–237

Koop G, Tole L (1999) Is there an environmental Kuznets curve for deforestation? J Dev Econ 58:231–244

Kozak J (2009) Forest cover changes and their drivers in the Polish Carpathian mountains since 1800. In: Nagendra H, Southworth J (eds) Reforesting landscapes: linking pattern and process. Springer, New York

Lambin EF (1997) Modelling and monitoring land-cover change processes in tropical regions. Prog Phys Geogr 21:375–393

Lara S, Barry T, Simonson P (1995) Inside Costa Rica – The essential guide to its politics, economy, society and environment. Resource Center Press, Albuquerque, NM, USA

Leonard HJ (1986) Recursos naturals y desarrollo economico en America Central: un perfil ambiental regional. Seria tecnica, informe tecnico/CATIE, No. 127

Mateo-Varga J (2001) Caracteristicas generales de la cuenca del Rio Tempisque. In: Jiménez JA, Gonzalez E (eds) La Cuenca del Rio Tempisque, Perspectivas para un Manejo Integrado. Organización para Estudios Tropicales, San Jose, Costa Rica, pp 32–72

Mather AS (1992) The forest transition. Area 24:367–379
Mather AS, Needle CL (1998) The forest transition: a theoretical basis. Area 30:117–124
Mather AS, Fairbairn J, Needle CL (1999) The course and drivers of the forest transition: the case of France. J Rural Stud 15:65–90
Meyer WB, Turner BL (1992) Human-population growth and global land-use cover change. Annu Rev Ecol Syst 23:39–61
Meyfroidt P, Lambin EF (2009) Forest transitions in Vietnam and Bhutan: causes and environmental impacts. In: Nagendra H, Southworth J (eds) Reforesting landscapes: linking pattern and process. Springer, New York
Nagendra H (2009) Reforestation in the human dominated landscapes of South Asia. In: Nagendra H, Southworth J (eds) Reforesting landscapes: linking pattern and process. Springer, New York
Perz SG (2007) Grand theory and context-specificity in the study of forest dynamics: Forest transition theory and other directions. Prof Geogr 59:105–114
Perz SG, Almeyda AM (2009) A tri-partite framework of forest dynamics: hierarchy, panarchy and heterarchy in the study of secondary growth. In: Nagendra H, Southworth J (eds) Reforesting landscapes: linking pattern and process. Springer, New York
Peters G (2001) La cuenca del Tempisque: una perspective historica. In Jiménez JA, Gonzalez E (eds) La Cuenca del Rio Tempisque, Perspectivas para un Manejo Integrado. Organización para Estudios Tropicales, San Jose, Costa Rica, pp 1–21
Ramankutty N, Gibbs HK, Achard F (2007) Challenges to estimating carbon emissions from tropical deforestation. Global Change Biol 13:51–66
Rouget M, Richardson DM, Cowling RM (2003) The current configuration of protected areas in the Cape Floristic Region, South Africa – reservation bias and representation of biodiversity patterns and processes. Biol Conserv 112:129–145
Rudel TK, Coomes OT, Moran E (2005) Forest transitions: towards a global understanding of land use change. Global Environ Change 15:23–31
Rudel TK (2009) this volume Three paths to forest expansion: a comparative historical analysis. In: Nagendra H, Southworth J (eds) Reforesting landscapes: linking pattern and process. Springer, New York
Sanchez-Azofeifa GA, Pfaff A, Robalino JA et al (2007) Costa Rica's payments for environmental services program: intention, implementation and impact. Conserv Biol 21:1165–1173
Sloan S (2007) Fewer people may not mean more forest for Latin American forest frontiers. Biotropica 39(4):443–446
Skole DL, Chomentowski WH, Salas WA et al (1994) Physical and human dimensions of deforestation in Amazonia. Bioscience 44:314–322
Southworth J, Nagendra H, Carlson LA et al (2004) Assessing the impact of Celaque National Park on forest fragmentation in western Honduras. Appl Geogr 24:303–322
Staaland H, Holand O, Nellemann C et al (1998) Time scale for forest regrowth: abandoned grazing and agricultural areas in southern Norway. Ambio 27:456–460
Stern PC, Young OR, Druckman D (1992) Global environmental change: understanding the human dimensions. National Academy Press, Washington, DC
Taff GN, Muller D, Kuemmerle T, et al. (2009) this volume Reforestation in Central and Eastern Europe after the breakdown of socialism. In: Nagendra H, Southworth J (eds) Reforesting landscapes: linking pattern and process. Springer, New York
Taylor JE (1980) Peripheral capitalism and rural–urban migration: a study of population movements in Costa Rica. Lat Am Perspect 7(2/3):75–90
Universidad de Costa Rica (1976) La poblacion de Costa Rica. Instituto de Investigaciones Sociales, Editorial Universidad de Costa Rica, San Jose, Costa Rica, pp 90–91
World Bank (1995) Mexico resource conservation and forest sector review. Natural Resources & Rural Poverty Operations Division Country Department II Latin America & the Caribbean Regional Office, Washington, DC

# Chapter 11
# Forest Cover Changes and Their Drivers in the Polish Carpathian Mountains Since 1800

**Jacek Kozak**

## 11.1 Introduction

In many developed countries, forest area has been increasing slowly since the nineteenth century, an increase that has been a clear reversal of forest cover trends in the previous period of dominant deforestation, referred to as the 'forest transition'

J. Kozak
Institute of Geography and Spatial Management, Jagiellonian University, Kraków, Poland
e-mail: jkozak@gis.geo.uj.edu.pl

H. Nagendra and J. Southworth (eds.), *Reforesting Landscapes: Linking Pattern and Process*, Landscape Series 10,
DOI 10.1007/978-1-4020-9656-3_11, 

(Mather 1992; Mather and Needle 1998). Agricultural land was abandoned or purposefully converted to forests, reflecting the economic and social development of societies, changing population pressures, and national policies attempting to overcome the scarcity of forest products (Mather 1992; Grainger 1995; Mather et al. 1999; Rudel et al. 2005; Chapters 2 and 3).

Mountains pose a range of difficulties for agricultural activities. Shifting from agriculture to other activities and migrations to lowlands therefore was and has continued to be a common practice in mountain areas, leading to land abandonment, forest expansion and afforestation in these regions, especially in Europe (Piussi 2000; MacDonald et al. 2000). Lichtenberger (1978) concludes that one of the triggers leading to the transformation of mountain areas and subsequent land abandonment was the liberation of peasantry in European countries during the eighteenth and nineteenth centuries. Another factor stimulating the conversion of agricultural land to forests was the changing economic status of forests. New visions of forests were emerging with forests being perceived as valuable and strategic resources to be protected from the destructive influence of agricultural expansion. Moreover, scientists started to promote cause-effect relationships between deforestation and environmental hazards, such as flooding and soil erosion. Such connections inspired measures to increase forest cover in mountain catchments (Whited 2000; Mather and Fairbairn 2000). In the near future, the establishment of protected areas in mountain regions might sustain a slow recovery of forests in some areas (Maggi 2005). Further, possible strategies to mitigate carbon emissions may stimulate interest in allocating land for further forest cover increases and intensifying afforestation actions (Rudel et al. 2005; Canadell and Raupach 2008).

This paper attempts to investigate forest cover changes, and in particular forest cover increases, occurring in the Polish Carpathian Mountains in Central Europe since 1800 and to identify their drivers. Future scenarios for forest cover change are also discussed. The quantification of forest cover changes in the region was based primarily on case studies using geographically referenced data (maps, satellite imagery, aerial photos). Where possible, available official statistics were used to confirm the results; however the complex political history of the region makes statistical comparisons difficult due to changes in political and administrative boundaries. Forest cover change rate values were calculated according to the formula given in Puyravaud (2003), accepted by the Food and Agriculture Organization in global forest assessments.

## 11.2 The Polish Carpathians

The Carpathians form a prominent mountain arc in Europe, encompassing the Hungarian lowlands and forming a huge curve that stretches from the Danube close to Vienna to the Danube again between Romania and Serbia. The area of the region exceeds 200,000 km$^2$, including lower lying areas of foothills and uplands (Kondracki 1989). The Polish Carpathians cover the northernmost section of the mountains

(Fig. 11.1), stretching for more than 300 km from Cieszyn in the west to Przemyśl in the east, with an area of approximately 20,000 km² (Warszyńska 1995).

A major part of the Polish Carpathians belongs to the Outer Carpathians and is composed of relatively low foothill zones with elevations up to 500–600 m (Pogórze Karpackie), and higher and steeper middle mountains (Beskidy Zachodnie and Bieszczady) with elevations predominantly up to 1,000–1,400 m, culminating in the Babia Góra Mountain (1,725 m). The Tatra mountains (Tatry), the highest part of the entire Carpathian arc, have a different character, with elevations exceeding 2,500 m in Slovakia (Gerlach, 2,655 m) and reaching 2,499 m in Poland (Balon et al. 1995). Before human influence, most of the Polish Carpathians were covered with forests extending to the timberline at 1,400–1,500 m (Towpasz and Zemanek 1995).

Since medieval times, the northern slopes of the Carpathians have been under the control of the kingdom of Poland. Although the presence of humans in valley floors, basins, and the foothill zone was marked thousands of years before (Valde-Nowak 1988; Pietrzak 2002), the regular and intensive colonisation that induced large-scale deforestation of the Carpathian area accelerated in the thirteenth and fourteenth centuries (Polskie Towarzystwo Leśne 1965; Podraza 1981; Dobrowolska 1985). High mountain areas in the northern Carpathians were also affected, and the upper timberline was depressed by grazing activities (Sokołowski 1928; Środoń 1948). According to the estimates of Podraza (1981), by the end of the eighteenth century, forests made up 23% of the Polish Carpathians in the foothill zone and 34% in the mountain zone.

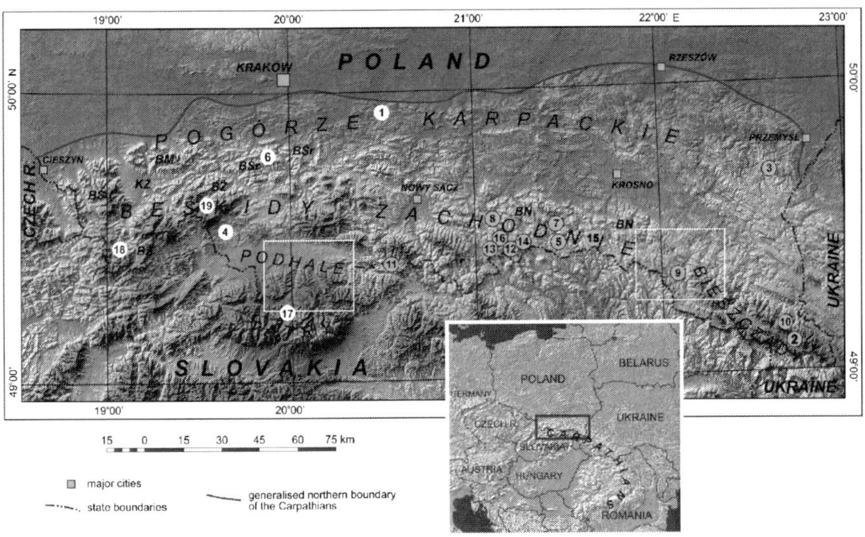


**Fig. 11.1** The Polish Carpathians. Numbers in circles indicate locations from Tables 11.1 through 11.3, unless the region is not labelled on the map. Grey circles denote areas depopulated in the 1940s. White rectangles show areas presented at Fig. 11.2. Abbreviations: BŚ – Beskid Śląski; BŻ – Beskid Żywiecki; BŚr – Beskid Średni; BM – Beskid Mały; BN – Beskid Niski; KŻ – Kotlina Żywiecka

## 11.3 From the Early Nineteenth Century to the 1930s: Towards Stabilisation of Forest Cover

In the late eighteenth century, southern Poland, including most of the present-day Polish Carpathians, became a part of the Austrian Empire, forming the province of Galicia. In the nineteenth century, societal changes increased the mobility of rural populations (in particular, with the liberation of peasantry in 1848) and, together with technological advances (e.g., construction of railways, industrialization), started to heavily impact mountain forests. At the end of World War I (1918), Poland won its independence, taking control of the northern Carpathians along with the newly established Czechoslovakia.

### *11.3.1 Drivers*

New Austrian land regulations with critical implications for forestry and forest management went into force in the Polish Carpathians as early as the end of the eighteenth century, reducing the impact of agriculture on forests (Żabko-Potopowicz 1956). Along with the liberation of peasantry in 1848, a sequence of regulations terminating the traditional rights of use of forest areas by rural communities (e.g., for grazing) stimulated afforestation of some mountain pastures by the second half of the nineteenth century (Jostowa 1972). The regulations also imposed significant restrictions on deforestation in the form of the Forest Act of 1852 (Polskie Towarzystwo Leśne 1965). As a result of this legislation, forested land became isolated within the overall land property as a separate domain and forestry started to compete in economic terms with agriculture for land resources (Polskie Towarzystwo Leśne 1965; Broda 1985).

In the late nineteenth century drivers of forest cover change in the Carpathians were similar to those operating in western Europe: increasing mobility of rural populations after the liberation of peasantry and expanding industry that attracted workers to cities and stimulated migration from rural areas, both of which led to land abandonment. In addition, there was significant migration abroad, mostly to North America (Górz 2002; Turnock 2003). The internal and foreign migrations weakened the earlier demand for land, referred to as "land hunger" (Gawlikowski and Zabierowski 1979; Górka 1995).

The population losses caused by World War I may have induced further land abandonment and forest expansion in some areas. According to Gawryszewski (2005), the population of southern Poland declined from almost 8 million in 1910 to 7.5 million in 1919, reaching the 1910 level in 1925, and only then steadily increasing to 9.2 million in 1939. Similar patterns were replicated in mountain communities. For example, Kozak et al. (2007a) described a decline in population of two communes in the Beskid Żywiecki mountains in the Polish Carpathians between the late nineteenth century and the 1920s, followed by a slow increase in the 1930s. Soja (2001b) described a similar pattern in the Lemko Land, an ethnic region in the southeastern part of the Polish Carpathians. The increased population in rural areas in the Polish Carpathians in the 1920s and 1930s and the brevity of the period of time before the disaster of World War II probably prevented significant recovery of forests in the first half of the twentieth century (WW II).

## 11.3.2 Case Study Evidence and Change Estimates

On the basis of statistical and forest data, Broda (1985) concluded that forest area in Galicia decreased from 28.2% in 1815 to 25.4% in 1913. Similar decreases in forest area in the nineteenth century were also suggested by Żabko-Potopowicz (1956) and Wiącko (1956), who however pointed to uncertainties in the estimates of forest area in Galicia. Further, for the period between World War I and World War II, Broda (2000) noted a small decrease in forest area in southern Poland (from 25.4% in 1923 to 23.3% in 1937). Frequently, the period until World War II has been described as a period of continuous devastation of Carpathian forests due to the lack of proper management, and several studies have claimed that deforestation for agricultural purposes occurred in the region throughout the nineteenth century and the first half of the twentieth century (Polskie Towarzystwo Leśne 1965).

Evidence of forest cover decrease can be found in several regions of the Polish Carpathians (Table 11.1). In most cases the decrease is small, and can quite likely be attributed to data inaccuracies. Several authors confirm that deforestation for agricultural purposes still played an important role in the Polish Carpathians in the late nineteenth and early twentieth centuries, as it did in the Podhale region (Górz 1994) and in the commune of Pcim, approximately 40 km south of Kraków, where available statistics show that forest area was slowly decreasing between 1880 and 1900 (German and Sadowski 2005).

Stronger evidence for the changes of forest cover in the Polish Carpathians can be derived from the analysis of cartographic data and available statistics, which indicate either a slow increase in forest cover or a stabilisation of forest cover beginning in the nineteenth or early twentieth century (Table 11.1). In areas settled well before the seventeenth and eighteenth centuries, including wider basins and low foothills, deforestation had already ceased by the eighteenth century (Pietrzak 2002). Forest cover increase or stabilization is also evident in visual comparisons of various cartographic data. For example, in the village of Zawoja in the Beskidy Mountains, the upper boundary of settlements and agricultural land was already well established as early as the beginning of the nineteenth century, and very little difference was found between cadastral maps compiled in 1844 and maps of that area in the 1920s (Mrazkówna and Kubijowicz 1925). Similarly, Kaim (2007) compared cadastral information compiled in the mid-nineteenth century for the village of Szczawnica (with a scale of 1:2,880) with a detailed topographical map created in 1937 (with a scale of 1:20,000), and found that new forest patches appeared in this area between these two times.

**Table 11.1** Annual forest cover change rates in the Polish Carpathians in the nineteenth and early twentieth centuries

| Region/location | Map Ids | Area (ha) | Period | Annual forest cover change rate (%) | Source of forest cover data/ references |
|---|---|---|---|---|---|
| Bochnia – Brzesko foothill area | 1 | 8,220 | 1787–1820 | –0.38 | Pietrzak (2002) |
| | | | 1820–1868 | 0.38 | |
| | | | 1868–1897 | –0.20 | |
| Nowy Sącz region | | – | 1900–1931 | –0.20 | Gawlikowski and Zabierowski (1979) |
| Village Ustrzyki Górne | 2 | 3,300 | 1785–1852 | –0.15 | Augustyn (2000) |
| | | | 1852–1900 | 0.01 | |
| Beskid Niski | | 200,000 | 1869–1900 | –0.06 | Soja (2001a) |
| | | | 1900–1931 | 0.40 | |
| Villages in the Bieszczady Wysokie Mts. | | 6,170 | 1852–1903 | –0.03 | Wolski (2007) |
| Pogórze Przemyskie | 3 | 62,000 | 1824–1936 | –0.02 | Janicki (2004) |
| Beskid Średni | | 29,780 | 1861/2–1912/4 | 0.32 | Ostafin (2008) |
| | | | 1912/4–1930s | 1.39 | |
| Orawa | 4 | 18,500 | 1820s–1930s | 0.34 | Kozak (2003) |
| Upper Wisłoka catchment | 5 | 34,000 | 1900–1931 | 0.46 | Lach and Wyżga (2002) |
| Village Trzebunia | 6 | 2,400 | 1845–1930 | 0.55 | Ostafin (2003) |

In conclusion, throughout the nineteenth and early twentieth centuries, forest area in the Polish Carpathians showed signs of stabilisation. Stabilisation can be concluded from the apparently contradictory results of numerous studies, which present either a continuous decrease of forest area in the Polish Carpathians or a slow increase after 1800. Deforestation and forest expansion occurred simultaneously, in general balancing each other at a regional scale. A reversal in the sign of the forest cover change rate in several locations however documents a clear transition phase between periods of prevailing decrease and periods of prevailing increase in forest area.

## 11.4 The Communist Period, 1945–1989

World War II brought about substantial losses in population and induced major political changes in the region. Poland and Czechoslovakia lost claim to much of the northern Carpathians, which were captured by the Soviet Union. Shifts in political boundaries resulted in re-settlement actions, which led to depopulation of the eastern part of the Polish Carpathians. Poland, like all countries in the region, was forced to adapt to the communist system, features of which included a centrally controlled economy, nationalisation and collectivisation.

### *11.4.1 Drivers*

Re-settlement actions were carried out in the region in the 1940s. First, at the end of and soon after World War II, Polish citizens were forced to migrate westwards to Poland, while Ukrainians from eastern Poland migrated eastwards to the Soviet Union. Then, at the end of 1940s, the Rusyn ethnic groups living in the northern Carpathians were forced to migrate into northern and western Poland (Snyder 1999; Soja 2001b). The outcome of these re-settlement actions was a large-scale and substantial depopulation of the area stretching eastwards of Nowy Sącz along the Polish boundary, an area that consists mostly of the Beskid Niski and Bieszczady mountain ranges (Fig. 11.1). For example, in the so-called Lemko Land, the population declined from 108,000 in 1931 to 31,000 in 1950 (Soja 2001b); however, the magnitude of the depopulation and the further re-development of population was significantly varied spatially. In the village of Hańczowa in the Beskid Niski mountains, the population fell from 800 to under 100 after the re-settlement, then grew slowly to 300 in the 1950s and 550 in 2003 (Reiser 2006). In the villages of the upper part of the Wilsznia catchment in the Beskid Niski mountains, the population fell from 863 in 1931 to 101 in 1950, and then declined to 66 in 1998 (Maciejowski 2001). In the Bieszczady Mountains, many villages were completely abandoned (Wolski 2007).

In the post-war period between 1950 and 1988, the population in the Polish Carpathians increased in most communes, including rural areas, in spite of an outflow of population from rural areas to urban centres located either in the Carpathians or, more frequently, in the industrial districts in the foreland (Długosz and Soja 1995). Related declines in population in the Polish Carpathians were noted mostly in the first 15–20 years of the post-war period, along with a rapid industrialization forced by the communist regime. For example in the commune of Zawoja, the population fell from more than 6,000 in 1946 to 5,400 in 1960 and then started to increase (Nanowski 1974), with the growth trend continuing today (Bański 2005). Equally important from the viewpoint of land use and land cover changes in the region were changes in the employment structure, which led to a slow decline in the number of people active in agriculture. For example, between the 1930s and the 1980s, the percentage of the population employed in agriculture in the Podhale region in the Polish Carpathians dropped by 24%, while the total population increased by 70% (Krakowska 1994; Górz 1994).

As in most of post-war Poland, agriculture in the Carpathian area was made collective to only a small degree. In 1970, in roughly the middle of the communist period, the share of private farms in the provinces of southern Poland comprising the Polish Carpathians exceeded 90% (Główny Urząd Statystyczny 1971). While most of the forests in Poland were nationalised, a large proportion of forests in the Polish Carpathians remained either private or owned by groups of farmers, especially in the province of Kraków, where private forests constituted approximately 50% of the total forest area in 1970 (Główny Urząd Statystyczny 1971). Only in areas where re-settlement actions took place were former private farms nationalised to a large degree.

The high share of private land distinguished the Polish Carpathians from regions in neighbouring countries, where in general land was claimed by the state after World War II (Chapter 6). Although the agricultural land in Poland was not nationalised, the attitude of the communist regime towards private agriculture was driven by ideology throughout this period, and farmers were subjected to a number of restrictive regulations. At the same time, a view was promoted that the mountains were too heavily deforested and devastated due to a long period of inappropriate forest management (Polskie Towarzystwo Leśne 1965). Therefore, an optimization of land use and land cover mosaic in mountain areas was recommended based on carefully planned afforestation (e.g., Adamczyk et al. 1980). This strategy was referred to as the so-called 'lowering of the forest-agricultural boundary' (Galarowski and Kostuch 1965) and was in line with trends in Polish forestry and the demand to increase the forest proportion at the national scale. This increase did in fact occur in Poland; forest cover grew from 21% in 1945 to 28% in 1987 (Smykała 1990).

### *11.4.2 Case Study Evidence and Change Estimates*

The pattern of forest cover change in the post-war period differed significantly between regions that were and were not depopulated in the 1940s (Kozak et al. 2007a).

For depopulated areas in the Polish Carpathians there is considerable evidence for a substantial increase in forest area, both in the records of foresters (e.g., Kulig 1956; Opieliński 1989) and in a number of studies carried out with the use of maps, satellite images or aerial photographs (Fig. 11.2). Conversion of agricultural land to forests was not only restricted to afforestation; a high proportion of new forests developed as a consequence of spontaneous forest succession on abandoned land. Depending on the area and period studied, forest cover change rates varied from 0.38% to 2.15% (Table 11.2). In most cases, forest cover change rates exceeded 1%, and were relatively consistent across the entire depopulated region.

Outside depopulated areas, the annual forest cover change rates were significantly lower than those in the depopulated areas (Fig. 11.2) and varied from 0.08% to 0.77% (Table 11.3).

Forest expansion in the Polish Carpathians has been a spatially selective process, especially when it has resulted from a natural succession on abandoned lands. In the second half of the twentieth century forest expansion (or afforestation) typically occurred in areas with higher elevation, on steeper slopes and close to

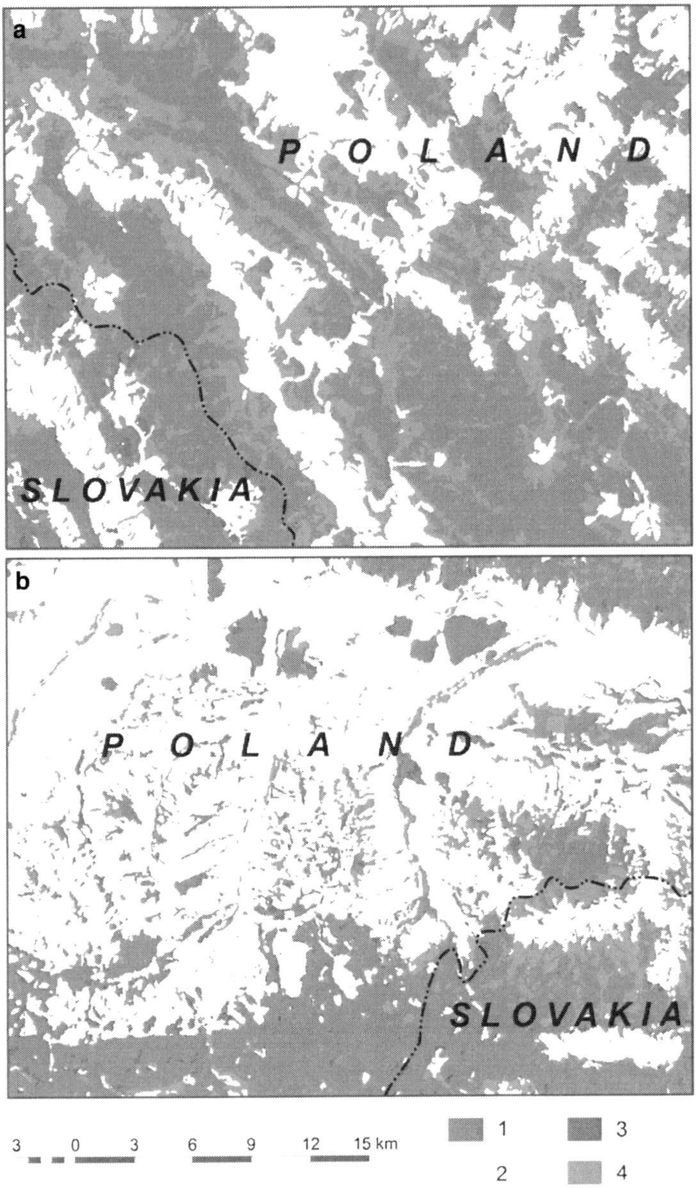


**Fig. 11.2** Forest cover changes in the depopulated areas (Beskid Niski and Bieszczady Mountains), 1936–1992 (**a**) and outside the depopulated areas (Podhale Region), 1931–2000 (**b**), after Kozak et al. (2007a). 1 – forests, no change in the analysed period; 2 – open land, no change in the analysed period; 3 – deforestation; 4 – forest expansion (*see Color Plates*)

forest edges (Kozak et al. 2004; Kozak 2005; Ostapowicz 2007). In the Western Beskidy Mountains, net forest change between 1933 and 1995 increased steadily with elevation up to 20% except elevations between 900 and 1,100 m, where

**Table 11.2** Annual forest cover change rates in the Polish Carpathians after World War II in depopulated areas

| Region/location | Map ids | Area (ha) | Period | Annual forest cover change rate (%) | Source of forest cover data/ references |
|---|---|---|---|---|---|
| Villages in the Bieszczady Mts. | | – | 1935–1998 | 0.38–1.87[a] | Angelstam et al. (2003) |
| Pogórze Przemyskie | 3 | 62,000 | 1936–1990 | 0.74 | Janicki (2004) |
| Magurski NP | 7 | 20,000 | 1935–1999 | 0.77 | Kardaś (2000) |
| Commune Uście Gorlickie | 8 | 28,800[b] | 1935–1995(9) | 0.77 | Woś (2005) |
| Commune Komańcza | 9 | 45,500[b] | 1935–1995(9) | 0.79 | Woś (2005) |
| Beskid Niski – E part | | 45,400 | 1933–2000 | 0.85 | Kozak et al. (2007a) |
| Village Caryńskie | 10 | 240 | 1852–1996 | 0.98 | Wolski (2001) |
| Beskid Niski – W part, Bieszczady Mts. | | 82,600 | 1936–1992 | 1.09 | Kozak et al. (2007a) |
| Małe Pieniny | 11 | 2,100 | 1936–2004 | 1.26 | Kaim (2007) |
| Villages in the Bieszczady Wysokie Mts. | | 6,170 | 1960–1980 | 1.26 | Wolski (2007) |
| Village Hańczowa | 12 | 2,100 | 1933–2003 | 1.36 | Reiser (2006) |
| Upper Wisłoka catchment | 5 | 34,000 | 1938–1995 | 1.42 | Lach and Wyżga (2002) |
| Catchment of Biała Dunajcowa | 13 | 5,000 | 1933–1975 | 1.47 | Warcholik (2005) |
| Catchment of Regetówka | 14 | – | 1933–1975 | 1.58 | Warcholik (2005) |
| Catchment of Wilsznia | 15 | 2,300 | 1938–1998 | 2.12 | Maciejowski (2001) |
| Catchment of Ropa | 16 | 2,000 | 1933–1975 | 2.15 | Warcholik (2005) |

[a] 'Forests' in 1998 include also areas of forest regeneration.
[b] Główny Urząd Statystyczny 2007.

forests were dominant already in the 1930s; above 1,000 m more than 80% of open areas existing in the 1930s were forested in 1995 (Fig. 11.3; Kozak 2005). In particular, the increase in forest cover modified the course of the upper timberline and the forest-agriculture boundary. Throughout the twentieth century, advances of the upper timberline were observed in the Tatra mountains (Wężyk and Guzik

**Table 11.3** Annual forest cover change rates in the Polish Carpathians after World War II outside depopulated areas

| Region/location | Map ids | Area (ha) | Period | Annual forest cover change rate (%) | Source of forest cover data/ references |
|---|---|---|---|---|---|
| Tatrzański NP | 17 | 198,000 | 1977–1999 | 0.08 | Paterek and Olędzki (2005) |
| Beskid Śląski, Kotlina Żywiecka, Beskid Żywiecki[a] | | 189,900 | 1933–1995 | 0.13 | Kozak (2005) |
| Orawa | 4 | 18,500 | 1931–2001 | 0.15 | Kozak (2003) |
| Beskid Średni | | 29,780 | 1930s–2003 | 0.21 | Ostafin (2008) |
| Beskid Śląski, Kotlina Żywiecka | | 91,600 | 1931–1986 | 0.22 | Kozak et al. (2007a) |
| Podhale – E part | | 82,600 | 1931–2000 | 0.29 | Kozak et al. (2007a) |
| Commune Rajcza | 18 | 13,100[b] | 1935–1995(9) | 0.37 | Woś (2005) |
| Commune Zawoja | 19 | 12,900[b] | 1935–1995(9) | 0.39 | Woś (2005) |
| Village Trzebunia | 6 | 2,400 | 1934–2002 | 0.39 | Ostafin (2003) |
| Bochnia – Brzesko foothill area | 1 | 8,220 | 1897–1996 | 0.39 | Pietrzak (2002) |
| Podhale | | 189,000 | 1931–1986 | 0.77 | Górz (1994) |

[a] Includes partially Slovakia.
[b] Główny Urząd Statystyczny 2007.

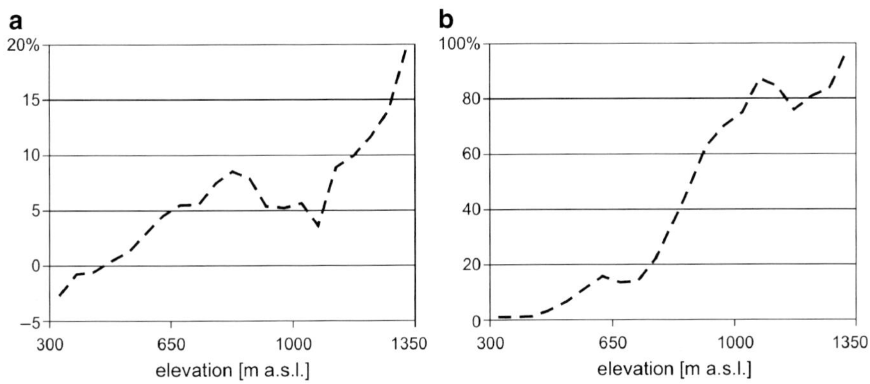


**Fig. 11.3** Forest cover changes in the Western Beskidy mountains in relation to elevation, (**a**) net forest cover change, 1933–1995, (**b**) proportion of 1933 open land converted to forests by 1995 (after Kozak 2005)

2004) and in the Ukrainian Carpathians (Troll and Sitko 2006). In addition, pastures and meadows located in the higher part of the forest belt have been rapidly disappearing in all mountain ranges in the Polish Carpathians (Wężyk and Pyrkosz 1999; Kozak et al. 1999; Ciurzycki 2004a; Paterek and Olędzki 2005). Only in the Bieszczady mountains have the anthropogenic upper timberline and high mountain pastures been stable since the mid-nineteenth century, probably due to several biophysical factors preventing forest expansion (Kucharzyk 2004; Wolski 2007).

The forest-agriculture boundary advanced to lower elevation, thereby decreasing the open area designated for agricultural activities. This process was slow, and frequently counteracted by the expansion of settlements. In areas of depopulation, the forest-agriculture boundary was depressed by 150–200 m in the twentieth century (Lach 2005); in other areas the shift was less significant, for example 100 m in the Orawa region since the 1820s (Kozak 2003).

## 11.5 Forest Cover Change After 1989

The political and economic breakthrough in 1989 completely changed the economy of the former communist states (Turnock 2003; Chapter 6). While a market-oriented economy was introduced, at the same time, political and socio-economic changes accelerated. The recent accession of Poland to the European Union has triggered a completely new set of drivers related to the implementation of EU policies, in particular the Common Agricultural Policy (CAP). The profound transformation of the past two decades has affected forests in a variety of ways, both directly and indirectly.

### *11.5.1 Drivers*

In the Polish Carpathians, there was little change in land ownership after 1989: private lands (both forested and agricultural) remained private, and state forests remained state property. Only the state-owned farms, which developed in the post-war period in areas of re-settlement and depopulation, were re-privatised. A much more important factor in the economic transformation was the exposure of agriculture to a market-oriented economy at not only the national but also the global scale. In addition, several economic opportunities since 1989 have led to a relative decline in the attractiveness of mountain agriculture. One example is the development of local tourism services, which have become common in the mountainous areas of Poland, particularly in the Carpathians (Kurek 1996; Bański and Stola 2002; Górz 2002).

Overall, the transformation of the 1990s had a significant effect on agriculture and rural areas in Poland (Bański 2003). One of the consequences was a rapid increase in the amount of fallow land throughout Poland, changing from 1% of the arable land in 1990 to 18% of the arable land and 14% of the total agricultural land in 2002 (Główny Urząd Statystyczny 2003; Kulikowski 2004). In Poland, fallow land is typical either in suburban areas, where fallowing precedes a formal request to change agricultural land to other uses (Główny Urząd Statystyczny 2003; Bański 2005), or in areas less favourable for agriculture, where it typically precedes land abandonment, forest succession or afforestation. In fact, early forest succession stages in Poland were estimated to occupy 900,000 ha at the end of the 1990s (Gorzelak 2006).

In relation to changes in agriculture, new afforestation programmes were launched in Poland in the 1990s, with an overall goal of increasing the percentage of forested land to 30% by 2020 and 33% by 2050 (Ministerstwo Środowiska 2003), continuing the trends of the post-war period. Although the afforestation rate has recently been decreasing and afforestation plans for 2001–2005 were executed at only 79% of planned levels on a national scale (Państwowe Gospodarstwo Leśne Lasy Państwowe 2006), the afforestation programmes will constitute an important part of rural development policies to be carried out between 2007 and 2013 (Ministerstwo Rolnictwa i Rozwoju Wsi 2006).

With the accession of Poland (and other countries in the region) to the EU, the profitability of mountain agriculture may be at least partially increased by the implementation of CAP measures. However, recent demographic processes in Poland may strengthen the tendencies to abandon agricultural land. Poland has undergone a major demographic shift since the 1990s, with a slow decline in population that is expected to continue in the near future (Węcławowicz et al. 2006). The most significant drivers of population decrease in Poland have been declining fertility and foreign migrations (Węcławowicz et al. 2006). As in most countries of East Central Europe, future population decline will be most evident in rural areas (Turnock 2003). In addition, the expansion of the EU in 2004 facilitated migration to Western Europe, especially to the UK and Ireland. In 2006, the Polish government estimated that there were around 800,000 temporary and permanent migrants (Gazeta 2007), while Węcławowicz et al. (2006) gave a value of 1,000,000 total migrants in 2005, noting however serious difficulties in obtaining reliable estimates. A recent survey (Czapiński and Panek 2007) estimated that the percentage of temporary or permanent workers abroad between 2005 and 2007 was 4.1% of the population aged 16 or older, which would be equivalent to roughly 1,300,000 people. Such a migration has had an impact on regional labour markets, including agriculture, and may have increased the tendency to abandon agricultural land.

Population decline has not yet affected the Polish Carpathians, which in general witnessed an increase in population between 1950 and 1998 (Soja 2004) and between 1989 and 2004, except in the easternmost part (Węcławowicz et al. 2006). Significant foreign migration has however frequently been reported in this area; for example, for the commune of Zarszyn in the eastern part of the Polish Carpathians, Guzik (2003) estimated that 5% of the total working population was working abroad in the early 2000s. At the same time, Bański (2005) observed a similar phenomenon in the commune of Zawoja.

### *11.5.2 Forest Cover at the Beginning of the Twenty First Century, Current Change Estimates and Future Trends*

Forests in the Polish Carpathians covered slightly more than 40% of the total mountain area in 2000. One estimate, formed on the basis of official statistics for communes located in the Polish Carpathians, was 41.0% in 2000 (Kozak 2005). A similar value was given by Niemtur (2007) and can also be found on the basis of both Corine Land Cover 2000 (40.6%) and a forest map of the Carpathians published recently (Kozak et al. 2008; 40.5%).

Forest cover in the region increased slowly over the past two decades. Using a bi-temporal comparison of Landsat data, Kozak et al. (2007b) noted an increase in forest area between 1987 and 2000 in the central part of the Western Carpathians (Poland and Slovakia), with an annual change rate of 0.38%. This value is higher than the values derived from official statistics. For example, the annual forest cover change rate was found to be 0.13% between 1995 and 2000 for communes located in the Polish Carpathians (Kozak 2005) and 0.09% between 2000 and 2005 for three provinces covering the Polish Carpathians and their foreland (Śląskie, Małopolskie and Podkarpackie) (based on data from Główny Urząd Statystyczny 2007).

Several processes currently affect net forest cover change in the Polish Carpathians. Data from the recent agricultural census show that leaving agricultural land fallow is a common practice in the Polish Carpathians: in a majority of communes, more than 20% of the total agricultural area is left fallow (Główny Urząd Statystyczny 2003). Indeed, land abandonment and early stages of forest succession (Fig. 11.4) can be found practically everywhere in the Polish Carpathians. For example, German and Sadowski (2005) estimated that approximately 5–7% of the commune of Pcim underwent re-naturalization (succession on abandoned fields), while Bański (2005) confirmed a high percentage of fallows in the commune of Zawoja in the 2000s. Using Landsat data, signs of abandonment and forest succession between 1985 and 2000 were found on 6–7% of meadows and pastures in the western and central part of the Polish Carpathians (Kozak 2005). Widespread land abandonment and forest succession since the 1980s were found in the eastern part of the Polish Carpathians by Kuemmerle et al. (2008), amounting to 13.9% of the farmland used in the 1980s in the Polish part of the mountains. A similar number was found in the Ukrainian Carpathians, and significantly more was found in the Slovakian Carpathians.

Forest succession on abandoned agricultural land typically leads to a decline of biodiversity, numerous evidence found in the Polish Carpathians was convincingly summarized by Michalik (1990). Another consequence of land abandonment and forest succession in the Polish Carpathians is loss of cultural landscapes (Angelstam et al. 2003; Wolski 2007). Therefore, forest succession on abandoned agricultural land poses currently a major management challenge in nature conservation areas in the Polish Carpathians as their managers attempt to maintain valuable, diverse, semi-natural open areas below the climatic timberline (Ciurzycki 2004b). While losses of

**Fig. 11.4** Early stages of forest succession on an abandoned meadow, Zubrzyca, Orawa region

biodiversity and cultural values resulting from land abandonment and forest succession are well documented in the European context (MacDonald et al. 2000), Weiss (2004) – referring to various concepts of mountain forest restoration – reasonably claims that "*restoration concepts are not only objective categories and determined by natural science knowledge but depend to a great extent on the value systems and interests of the actors involved as well as on the institutional settings of the programmes*". It seems that a similar statement may apply to the valuation of forest succession on abandoned agricultural land in mountains of Central Europe.

Afforestation (Fig. 11.5) does not play an important part in the Polish Carpathians, especially when compared to the area of fallowed or abandoned land (Kozak 2005). In three provinces comprising the Polish Carpathians, afforestation occurred in 10,600 ha between 1995 and 2000 (Ministerstwo Środowiska 2003), and 8,100 ha between 2001 and 2005 (based on data from Główny Urząd Statystyczny 2007).

Deforestation, defined as a relatively permanent removal of forest cover and alteration of forest use of land, occurs only sporadically, and is well controlled by regional and forest authorities. Construction of recreation areas for skiing provides one instance of deforestation in the Carpathians. Forest degradation is most frequently related to forest decline phenomena and has been observed in the region in several locations (Troll 1995; Kozak 1996; Widacki 1999; Niemtur 2007). In the early 1990s, changes in the regulations of private forest management induced temporarily excessive timber extraction and degradation of private forests in Poland (Broda 2000). Local forest degradation related to changing forest regulations simultaneously occurred elsewhere in the Carpathians (Turnock 2002; Haigh et al. 2004;

**Fig. 11.5** Afforestation at the forest-agricultural boundary, Beskid Niski

Kuemmerle et al. 2007), but negative impacts on forests were much smaller in the Polish Carpathians than in the neighbouring Carpathian countries, especially Ukraine (Kuemmerle et al. 2007).

Forest succession on abandoned fields is doubtless the most widespread process contributing to forest cover change in the Polish Carpathians. Unlike afforestation or deforestation, which are documented in land records, forest succession is particularly difficult to measure, being very slow, spatially dispersed and not typically reported in land surveys. For example, Kozak et al. (2004) found that most agricultural areas in the village of Ponikiewka in the Beskid Mały mountains that had been planned to be converted to forests were in fact partially forested before the plans were made. One of the difficulties in obtaining accurate estimates of forest succession is the small average farm size in the Polish Carpathians. Using recent agricultural census data, Bański (2005) notes that average farm size in three Carpathian provinces is currently between 2.6 and 3.2 ha; in addition, farms in southeastern Poland are typically fragmented.

According to Szwagrzyk (2004), difficulties in monitoring the process of fallowing and secondary forest succession may result in a severe underestimate of future forest cover increases. Since secondary succession is not taken into account in afforestation plans, it seems quite likely that target levels of forest cover in Poland will be reached much sooner than planned (Szwagrzyk 2004). Afforestation activities planned for the period between 2001 and 2020 in three provinces of the Polish

Carpathians should give rise to approximately 100,000 ha of new forests (Ministerstwo Środowiska 2003), provided they are successful. Still, this covers at most 20–40% of the land estimated to be fallow in this area in 2001 (Kozak 2005). The remaining 60–80% demonstrates a huge uncertainty related to the future area of forests in the Polish Carpathians.

## 11.6 Concluding Remarks

The areal extent of mountain forests in the Polish Carpathians became relatively stable as early as the nineteenth century. A significant growth in forest cover started in the second half of the twentieth century and continues today. The long-term trends of forest cover changes in the Polish Carpathians have thus followed the typical path of 'forest transition' described by Mather (1992) and are similar to trends observed in other parts of the Carpathians, most mountain areas in Europe and many mountainous regions throughout the world. A specific driver of forest cover change in the Polish Carpathians was a post-war depopulation in the eastern part of the mountains that subsequently triggered extensive and rapid forest expansion.

Today, afforestation adds every year to an overall forest cover increase in the Polish Carpathians. However, it will not be the primary process determining future forest cover in the Polish Carpathians. Much more important are changes in mountain agriculture, possible land abandonment and subsequent forest succession.

While an increase in forest cover in the Polish Carpathians seems inevitable, the period since 1989 is still too short to be used to form decisive conclusions about forest cover trends. Small-scale farming and the fine resolution of the conversion of agricultural land into forests complicate reliable estimates of the intensity of the land cover change processes. Forecasts of the future rate of forest cover increase are uncertain because of a number of direct and indirect impacts on land, especially those related to the demographical situation and labour market in rural areas.

**Acknowledgments** I would like to thank the editors and two anonymous reviewers for their valuable comments and suggestions.

## References

Adamczyk B, Gerlach T, Obrębska-Starklowa B, Starkel L (1980) Zonal and azonal aspects of the forest-agricultural boundary in the Polish Carpathians. Geogr Pol 40:71–84

Angelstam P, Boresjö-Bronge L, Mikusiński G, Sporrong U, Wästfelt A (2003) Asessing village authenticity with satellite images: a method to identify intact cultural landscapes in Europe. Ambio 32:594–604

Augustyn M (2000) Antropogeniczne zmiany środowiska przyrodniczego na obszarze dawnej wsi Ustrzyki Górne w świetle źródeł historycznych. Roczn Bieszcz 9:237–262

Balon J, German K, Kozak J, Malara A, Widacki W, Ziaja W (1995) Regiony fizycznogeograficzne Polskich Karpat. In: Warszyńska J (ed) Karpaty Polskie. Uniwersytet Jagielloński, Kraków

Bański J (2003) Selected aspects of present-day changes in Polish rural space. Geogr Pol 76(1):73–96
Bański J (2005) Przestrzenny wymiar współczesnych procesów na wsi. Stud Obsz Wiej 9:3–149
Bański J, Stola W (2002) Przemiany struktury przestrzennej i funkcjonalnej obszarów wiejskich w Polsce. Stud Obsz Wiej 3:5–111
Broda J (1985) Proces wylesienia na ziemiach polskich od czasów najdawniejszych. Czas Geogr 56:151–173
Broda J (2000) Historia leśnictwa w Polsce. Wydawnictwo Akademii Rolniczej w Poznaniu, Poznań
Canadell JG, Raupach MR (2008) Managing forests for climate change mitigation. Science 320:1456–1457
Ciurzycki W (2004a) Struktura przestrzenna naturalnych odnowień świerkowych na górnoreglowych polanach popasterskich w Tatrach Polskich. Sylwan 7:20–30
Ciurzycki W (2004b) Wtórna sukcesja lasu na polanach górskich wyłączonych z gospodarki pasterskiej. Sylwan 11:59–66
Czapiński J, Panek T (eds) (2007) Diagnoza społeczna 2007. Warunki i jakość życia Polaków. Rada Monitoringu Społecznego, Warszawa. http://www.diagnoza.com. Cited June 2008
Długosz Z, Soja M (1995) Ludność. In: Warszyńska J (ed) Karpaty Polskie. Uniwersytet Jagielloński, Kraków
Dobrowolska M (1985) Procesy osadnicze w dorzeczu Wisłoki i Białej Dunajcowej w tysiącleciu. Wydawnictwo Naukowe WSP w Krakowie, Kraków
Galarowski T, Kostuch R (1965) Próba analizy przebiegu granicy rolno-leśnej na Pogórzu i w Beskidach Zachodnich. Probl Zagospod Ziem Górs 9:165–192
Gawlikowski Z, Zabierowski K (1979) Problemy gospodarcze w rolnictwie górskim ze szczególnym uwzględnieniem regionu muszyńskiego. Zesz Nauk UJ Pr Hist 60:81–109
Gawryszewski A (2005) Ludność Polski w XX wieku. Polska Akademia Nauk, Instytut Geografii i Przestrzennego Zagospodarowania, Warszawa
Gazeta Wyborcza (2007) On-line archive. http://szukaj.gazeta.pl/archiwum/0,0.html. Cited 8th May 2007
German K, Sadowski P (2005) Wpływ renaturalizacji środowiska przyrodniczego na zmiany granicy rolno-leśnej w gminie Pcim. Probl Zagospod Ziem Górs 51:71–80
Główny Urząd Statystyczny (1971) Rocznik Statystyczny 1971. Główny Urząd Statystyczny, Warszawa
Główny Urząd Statystyczny (2003) Charakterystyka rolniczej przestrzeni produkcyjnej Polski. Główny Urząd Statystyczny, Warszawa
Główny Urząd Statystyczny (2007) Bank Danych Regionalnych. http://www.stat.gov.pl/bdr/bdrap.strona.indeks. Cited June 2007
Górka Z (1995) Osadnictwo. In: Warszyńska J (ed) Karpaty Polskie. Uniwersytet Jagielloński, Kraków
Górz B (1994) Rolnictwo Podhala. In: Górz B (ed) Studia nad przemianami Podhala. Wyższa Szkoła Pedagogiczna w Krakowie Kraków
Górz B (2002) Współczesne przemiany na obszarach wiejskich Podhala. Prz Geogr 74:451–468
Gorzelak A (2006) Zalesianie i zadrzewianie kraju. In: Szujecki A (ed) Z dziejów lasów państwowych i leśnictwa polskiego 1924–2004, vol 3, Lata powojenne i współczesność. Centrum Informacyjne Lasów Państwowych, Warszawa
Grainger A (1995) The forest transition: an alternative approach. Area 27:242–251
Guzik C (2003) Społeczny kontekst przemian na wsi polskiej u progu XXI wieku–na przykładzie gminy Zarszyn. Stud Obsz Wiej 4:147–162
Haigh MJ, Jansky L, Hellin J (2004) Headwater deforestation: a challenge for environmental management. Global Environ Change 14:51–61
Janicki R (2004) Specific features of the landscape succession processes in Przemyskie Foothills. In: Kirchner K, Wojtanowicz J (eds) Cultural landscapes. Regiograph, Brno
Jostowa W (1972) Pasterstwo na polskiej Orawie. Ossolineum, Wrocław
Kaim D (2007) Zastosowanie GIS i teledetekcji do badania zmian użytkowania ziemi i pokrycia terenu w regionach transgranicznych na przykładzie Małych Pienin. MSc thesis in the Institute of Geography and Spatial Management, Jagiellonian University, Kraków

Kardaś P (2000) Zastosowanie telegeoinformacji w badaniach zmian pokrycia terenu Magurskiego Parku Narodowego w latach 1935–1999. Fotointerpret Geogr Probl Telegeoinformacji 31:82–90
Kondracki J (1989) Karpaty. Wydawnictwa Szkolne i Pedagogiczne, 2nd edn. Wydawnictwo Rolnicze i Leśne Warszawa
Kozak J (1996) Przestrzenny model degradacji lasów Beskidu Śląskiego. Biul KPZK 174:511–538
Kozak J (2003) Forest cover change in the Western Carpathians in the past 180 years. A case study in the Orawa region in Poland. Mt Res Dev 23:369–375
Kozak J (2005) Zmiany powierzchni lasów w Karpatach Polskich na tle innych gór świata. Wydawnictwo Uniwersytetu Jagiellońskiego, Kraków
Kozak J, Estreguil C, Ostapowicz K (2008) European forest cover mapping with high resolution satellite data: the Carpathians case study. Int J Appl Earth Obs Geoinf 10:44–55
Kozak J, Estreguil C, Troll M (2007a) Forest cover changes in the northern Carpathians in the 20th century: a slow transition. J Land Use Sci 2:127–146
Kozak J, Estreguil C, Vogt P (2007b) Forest cover and pattern changes in the Carpathians over the last decades. Eur J For Res 126:77–90
Kozak J, Ostapowicz K, Szablowska-Midor A, Widacki W (2004) Land abandonment in the Western Beskidy Mts. and its environmental background. Ekol Bratisl 23(1):116–126
Kozak J, Troll M, Widacki W (1999) Semi-natural landscapes of the Western Beskidy Mts. Ekol Bratisl 18(1):53–62
Krakowska A (1994) Zmiany w zaludnieniu Podhala w latach 1931–1988. In: Górz B (ed) Studia nad przemianami Podhala. Wyższa Szkoła Pedagogiczna w Krakowie Kraków
Kucharzyk S (2004) Zmiany przebiegu górnej granicy lasu w Paśmie Szerokiego Wierchu w Bieszczadzkim Parku Narodowym. Roczn Bieszcz 12:81–102
Kuemmerle T, Hostert P, Radeloff VC, Perzanowski K, Kruhlov I (2007) Post-socialist forest disturbance in the Carpathian border region of Poland, Slovakia, and Ukraine. Ecol Appl 17:1279–1295
Kuemmerle T, Hostert P, Radeloff VC, van der Linden S, Perzanowski K, Kruhlov I (2008) Cross-border comparison of post-socialist farmland abandonment in the Carpathians. Ecosystems 11:614–628
Kulig L (1956) Zalesienia, dolesienia i zadrzewienia karpackich terenów górskich. Sylwan 100(10):58–70
Kulikowski R (2004) Struktura przestrzenna rolnictwa Polski w świetle wyników Powszechnego Spisu Rolnego 2002 r. Biul KPZK 213:40–65
Kurek W (1996) Agriculture versus tourism in rural areas of the Polish Carpathians. GeoJournal 38:191–196
Lach J (2005) Przyrodnicze skutki zmian granicy rolno-leśnej w Beskidzie Niskim. Pr Zagospod Ziem Górs 51:17–29
Lach J, Wyżga B (2002) Channel incision and flow increase of the upper Wisłoka river, southern Poland, subsequent to the reafforestation of its catchment. Earth Surf Process Landf 27:445–462
Lichtenberger E (1978) The crisis of rural settlement and farming in the high mountain region of continental Europe. Geogr Pol 38:181–187
MacDonald D, Crabtree JR, Wiesinger G, Dax T, Stamou N, Fleury P, Lazpita JG, Gibon A (2000) Agricultural abandonment in mountain areas of Europe: environmental consequences and policy response. J Environ Manag 59:47–69
Maciejowski W (2001) Zmiany użytkowania ziemi i ich wpływ na funkcjonowanie środowiska przyrodniczego w zlewni górnej Wilszni (Beskid Niski) w okresie 1920–2000. In: German K, Balon J (eds) Przemiany środowiska przyrodniczego Polski a jego funkcjonowanie. Instytut Geografii i Gospodarki Przestrzennej Uniwersytetu Jagiellońskiego, Kraków
Maggi M (2005) Which indicators for monitoring forests within Alpine Natura 2000 sites? Ph.D. thesis, the Ecole Nationale Du Génie Rural, des Eaux et des Forêts, Paris
Mather AS (1992) The forest transition. Area 24:367–379
Mather AS, Fairbairn J (2000) From floods to reforestation: The forest transition in Switzerland. Environ Hist 6:399–421

Mather AS, Fairbairn J, Needle CL (1999) The course and drivers of forest transition: the case of France. J Rural Stud 15:65–90
Mather AS, Needle CL (1998) The forest transition: a theoretical basis. Area 30:117–124
Michalik S (1990) Sukcesja wtórna i problemy aktywnej ochrony biocenoz półnaturalnych w parkach narodowych i rezerwatach przyrody. Prądnik Pr Muz Szafera 2:175–198
Ministerstwo Rolnictwa i Rozwoju Wsi (2006) Projekt Programu Rozwoju Obszarów Wiejskich 2007–2013. http://www.minrol.gov.pl/DesktopDefault.aspx?TabOrgId = 1469&LangId = 0. Cited June 2007
Ministerstwo Środowiska (2003) Krajowy Program Zwiększenia Lesistości. http://www.mos.gov.pl/artykul/326_lesnictwo/296_krajowy_program_zwiekszania_lesistosci.html. Cited June 2007
Mrazkówna M, Kubijowicz W (1925) Ze studiów nad osadnictwem Babiej Góry. Czas Geogr 3:45–71
Nanowski J (1974) Stan i perspektywy osadnictwa w Zawoi. Zesz Nauk UJ, Pr Geogr 38:95–111
Niemtur S (2007) Lasy karpackie: stan aktualny i próba prognozy. Probl Zagospod Ziem Górs 54:31–44
Opieliński J (1989) Zalesienia terenów połemkowskich w granicach byłego powiatu nowosądeckiego. Sylwan 133(7):31–44
Ostafin K (2003) Zmiany użytkowania ziemi w krajobrazie wsi Trzebunia w Beskidzie Średnim od połowy XIX w. In: Michalczyk Z (ed) Badania geograficzne w poznawaniu środowiska. UMCS, Lublin
Ostafin K (2008) Zmiany granicy rolno-leśnej w środkowej części Beskidu Średniego od połowy XIX w. do 2005 r. Ph.D. thesis, Institute of Geography and Spatial Management, Jagiellonian University, Kraków
Ostapowicz K (2007) Modelowanie zmian pokrycia terenu w przestrzeni i w czasie. Ph.D. thesis in the Institute of Geography and Spatial Management, Jagiellonian University, Kraków
Państwowe Gospodarstwo Leśne Lasy Państwowe (2006) Raport o stanie lasów w Polsce 2005. http://www.lp.gov.pl/media/biblioteka/raporty. Cited June 2007
Paterek A, Olędzki JR (2005) Zmiany w zasięgu pięter roślinnych w Tatrach w latach 1977–1999. Teledetekcja Śr 36:106–118
Pietrzak M (2002) Geomorfologiczne skutki zmian użytkowania ziemi na Pogórzu Wiśnickim. Instytut Geografii i Gospodarki Przestrzennej Uniwersytetu Jagiellońskiego, Kraków
Piussi P (2000) Expansion of European mountain forests. In: Price MF, Butt N (eds) Forests in sustainable mountain development: a state of knowledge report for 2000. IUFRO Research Series 5, CABI Publishing, Wallingford, New York
Podraza A (1981) L'agriculture dans Les Carpates Polonaises aux XVIIIe–XXe siécles. Zesz Nauk UJ, Pr Hist 69:75–89
Polskie Towarzystwo Leśne (1965) Dzieje lasów, leśnictwa i drzewnictwa w Polsce. Państwowe Wydawnictwo Rolnicze i Leśne, Warszawa
Puyravaud J-P (2003) Standardizing the calculation of the annual rate of deforestation. For Ecol Manag 177:593–596
Reiser A (2006) Przemiany użytkowania ziemi na tle przekształceń demograficznych we wsi Hańczowa w latach 1933–2003. Studium metodyczne. M.Sc. thesis, Institute of Geography and Spatial Management, Jagiellonian University, Kraków
Rudel TK, Coomes OT, Moran E, Achard F, Angelsen A, Xu J, Lambin E (2005) Forest transitions: towards a global understanding of land use change. Global Environ Change 15:23–31
Smykała J (1990) Historia, rozmiar i rozmieszczenie zalesień gruntów porolnych w Polsce w latach 1945–1987. Sylwan 134(3–12):1–7
Snyder T (1999) To resolve the Ukrainian Problem Once and for All. The ethnic cleansing of Ukrainians in Poland, 1943–1947. J Cold War Stud 1:86–120
Soja M (2001a) Rozwój ludnościowy a zmiany użytkowania ziemi w Beskidzie Niskim w XIX i XX wieku. In: German K, Balon J (eds) Przemiany środowiska przyrodniczego Polski a jego funkcjonowanie. Instytut Geografii i Gospodarki Przestrzennej Uniwersytetu Jagiellońskiego, Kraków

Soja M (2001b) Zmiany zaludnienia Łemkowszczyzny w latach 1869–1998. In: Kortus B (ed) Człowiek i przestrzeń. Instytut Geografii i Gospodarki Przestrzennej Uniwersytetu Jagiellońskiego, Kraków
Soja M (2004) La situation demographique des Carpates au tournant des XXe et XXIe siecles. Zesz Nauk UJ, Pr Geogr 113:185–192
Sokołowski M (1928) O górnej granicy lasu w Tatrach. Wydawnictwo Fundacji "Zakłady Kórnickie". Wydawnictwo Rolnicze i Leśne Kraków
Szwagrzyk J (2004) Sukcesja leśna na gruntach porolnych; stan obecny, prognozy i wątpliwości. Sylwan 4:53–59
Środoń A (1948) Górna granica lasu na Czarnohorze i w Górach Czywczyńskich. Rozprawy Wydziału Matematyczno-Przyrodniczego B(72)(7), Polska Akademia Umiejętności, Kraków
Towpasz K, Zemanek B (1995) Szata roślinna. In: Warszyńska J (ed) Karpaty Polskie. Uniwersytet Jagielloński, Kraków
Troll M (1995) Degradacja lasów masywu Skrzycznego (Beskid Śląski) i jej związek z orografią terenu. Sylwan 7:87–91
Troll M, Sitko I (2006) Pasterstwo w Zachodniej Czarnohorze (Karpaty Ukraińskie) w ujęciu przestrzenno-czasowym. In: Troll M (ed) Czarnohora. Przyroda i człowiek. Instytut Geografii i Gospodarki Przestrzennej Uniwersytetu Jagiellońskiego, Kraków
Turnock D (2002) Ecoregion-based conservation in the Carpathians and the land use implications. Land Use Policy 19:47–63
Turnock D (2003) The human geography of East Central Europe. Routledge, London and New York
Valde-Nowak P (1988) Etapy i strefy zasiedlenia Karpat polskich w neolicie i na początku epoki brązu. Ossolineum, Wrocław
Warcholik W (2005) Rejestracja różnic w przebiegu granicy rolno-leśnej w Beskidzie Niskim na obszarze Polski i Słowacji (1933–1975) z wykorzystaniem GIS. Probl Zagospod Ziem Górs 51:59–69
Warszyńska J (1995) Wstęp. In: Warszyńska J (ed) Karpaty Polskie. Uniwersytet Jagielloński, Kraków
Weiss G (2004) The political practice of mountain forest restoration–comparing restoration concepts in four European countries. For Ecol Manag 195:1–13
Węcławowicz G, Bański J, Degórski M, Komornicki T, Korcelli P, Śleszyński P (2006) Przestrzenne zagospodarowanie Polski na początku XXI wieku. Polska Akademia Nauk, Instytut Geografii i Przestrzennego Zagospodarowania, Warszawa
Wężyk P, Guzik M (2004) Techniki geoinformatyczne w badaniach czasowo-przestrzennych zmian szaty roślinnej na przykładzie rejonu Kasprowego Wierchu w Tatrach. Teledetekcja Śr 33:58–67
Wężyk P, Pyrkosz R (1999) Użytkowanie polan w Gorcach na przestrzeni lat 1954–1997 na podstawie fotogrametrycznego opracowania zdjęć lotniczych. Arch Fotogr Kartogr Teledetekcji 9:223–232
Whited TL (2000) Extinguishing disaster in Alpine France: The fate of reforestation as technocratic debacle. GeoJournal 51:263–270
Wiącko E (1956) Lasy i gospodarstwo leśne na ziemiach polskich od połowy XIX wieku. Sylwan 100(7):88–135
Widacki W (ed) (1999) Przemiany środowiska przyrodniczego zachodniej części Beskidów pod wpływem antropopresji. Instytut Geografii Uniwersytetu Jagiellońskiego, Kraków
Wolski J (2001) Kierunki zmian krajobrazu okolic bieszczadzkiej wsi Caryńskie. Pr Geogr IGiPZ PAN 179:149–167
Wolski J (2007) Przekształcenia krajobrazu wiejskiego Bieszczadów Wysokich w ciągu ostatnich 150 lat. Pr Geogr IGiPZ PAN 214:5–228
Woś B (2005) Zmiany pokrycia terenu w wybranych gminach Beskidów w drugiej połowie XX w. na podstawie analizy zdjęć lotniczych. Teledetekcja Śr 35:1–114
Żabko-Potopowicz A (1956) Lasy i leśnictwo na ziemiach polskich w pierwszej połowie XIX wieku. Sylwan 100(7):58–87

# Chapter 12
# Parks as a Mechanism to Maintain and Facilitate Recovery of Forest Cover: Examining Reforestation, Forest Maintenance and Productivity in Uganda

**Joel Hartter, Jane Southworth, and Michael Binford**

J. Hartter (✉)
Department of Geography, University of New Hampshire, NH, USA
e-mail: joel.hartter@unh.edu

J. Southworth and M. Binford
Department of Geography and Land Use and Environmental Change Institute (LUECI), University of Florida, FL, USA

H. Nagendra and J. Southworth (eds.), *Reforesting Landscapes: Linking Pattern and Process*, Landscape Series 10,
DOI 10.1007/978-1-4020-9656-3_12, © Springer Science+Business Media B.V. 2010

## 12.1 Introduction

Tropical forests are among the world's most productive ecosystems, providing important social and environmental benefits. However, they are increasingly threatened by accelerating rates of forest conversion and degradation (Brown and Lugo 1990). The Food and Agriculture Organization estimated that 13 million ha of forest are converted annually to agriculture and pasture in developing countries (FAO 2006). Forest loss and fragmentation of tropical forests has been called the single greatest threat to global biological diversity (Turner and Corlett 1996; Laurance and Bierregaard 1997). Hill and Curran (2003) assert that fragmentation negatively impacts species composition due to a reduction in forest area and an isolation of the remaining forest fragments.

As tropical deforestation continues to threaten biodiversity and forest-based livelihoods, the regeneration of cut forests and the conservation/protection of those that remain is an increasingly important component of land-cover change in many tropical regions. Biologists have called for increased protection as a primary means to sustain these natural ecosystems (Struhsaker 1987). Traditionally, conservation efforts have concentrated on the establishment of protected areas, which have become the cornerstones of most national strategies to conserve biological diversity (Howard et al. 2000; Wilkie et al. 2006). McNeely (1990) sees protected areas as making fundamental contributions to sustaining human societies by preserving places of cultural and biologic heritage that contain critical habitat and biodiversity, while also generating cash to local and national economies and providing employment and improved infrastructure (McNeely 1994; Struhsaker 2005). In the century and a quarter since the formation of Yellowstone National Park, the world's first national park, the number of national parks, reserves, and other protected areas around the world has grown substantially. There are now more than 100,000 parks, reserves, and other protected areas around the world, covering well over 10% of the earth's land area (Hayes 2006).

Parks (we use the term "parks" to refer to protected areas of all sorts – areas where land use is restricted mostly to wildlife and/or preservation of "natural" existing habitat) can be effective mechanisms for maintaining and even improving forest cover despite growing agricultural populations outside their boundaries (Southworth 2004; Perz 2007), a condition facing many parks in developing countries today. Only 11% of the world's forests are currently designated for conservation of biological diversity (FAO 2006). By controlling access, resource extraction, settlement, and other activities, parks not only provide large continuous tracts of forest that can provide important ecosystem services such as carbon sequestration, nutrient uptake, water level maintenance, and pollutant filtration, but they are most effective in conserving a full array of biodiversity compared to unprotected lands (Lwanga 2003; Struhsaker et al. 2005).

In addition to the important ecosystem services, forest parks in the tropics are important for reforestation programs because they provide natural conditions with minimal human disturbance that promote natural processes and pathways to forest succession and recovery. Generally reforestation strategies within parks differ from those of afforestation programs and plantation forests, in which only a few species (which are predominantly exotic) are planted (Condit et al. 1993), in turn creating even-aged monocultures. In natural ecosystems such as parks, where species richness and diversity is higher than in plantation forests, forests are less vulnerable to perturbations and can provide a larger and more diverse array of ecosystem functions. Though some ecosystem services can be achieved in plantation forests of exotic species, biodiversity is a critical component to forest maintenance and recovery. Lwanga (2003) notes the best way to successfully integrate biodiversity is with indigenous trees. Frugivores and aviaries that reside or feed within the park play critical roles to forest succession in seed dispersal (Duncan and Chapman 2002; Lwanga 2003). Seed dispersal by frugivores is vital to the survival of fruiting tree populations in tropical forests, which in turn supports other species and the services they provide (Chapman and Chapman 1995). Under dry conditions, human intervention in reforestation efforts may be required, but in areas that receive abundant rainfall such as moist tropical forests, trees can promote seed dispersal by attracting dispersers, leading to increased seed germination and establishment of endemic tree species (Lugo 1992).

Outside parks, the situation often is different. The surrounding landscape contains forest fragments (remnants of once larger tracts of forest) that are important lifelines to neighboring communities, providing subsistence-based resources. These forests represent reservoirs of land (e.g., land that could be converted later for farming or pasture), resources (e.g., fuelwood, building poles, thatching), and economic opportunity for people (e.g., tourism, sale of forest products), while at the same time are often viewed as buffers for parks by managers (Schonewald-Cox and Bayless 1986). In unprotected forests, where levels of human disturbance and degradation tend to be higher than those inside a park, limited numbers or even a complete lack of seed dispersal mechanisms is considered one of the primary obstacles to forest regrowth (Nepstad et al. 1991). Whereas human populations around savanna parks usually are limited by low and sporadic rainfall, which strongly constrains crop agriculture, forest parks in the tropics, particularly those within a broad range of mid-altitudes, often occupy and/or are surrounded by land that is highly suitable for agriculture (Goldman et al. 2008). As such historically forested landscapes have been rapidly converted to farmland around parks in East and Central Africa, creating a mosaic of interacting natural and human-influenced patches.

Park establishment is one mechanism to mitigate the impacts of forest fragmentation in tropical forest landscapes. Where parks have hard-edged boundaries ("fortress conservation" narrative (Hulme and Murphree 2001)) and can effectively prevent large-scale incursion and limit or even prohibit consumptive uses within park boundaries, forests can be allowed to regrow in absence of human disturbance. Despite social pressures on forests outside parks to provide resources or to be converted to support rural livelihoods, forests under protection can allow natural processes to persist; thereby maintaining and improving existing forest while also assisting in natural regeneration.

The viability and sustainability of forested ecosystems depends not only on successful park establishment and management, but also on the ability to monitor their change in spatial extent and productivity. Satellite remote sensing techniques can be advantageous to efficiently observe and monitor changes that occur in park landscapes within different land-cover types, over multiple dates, at multiple locales, and over large areas. This chapter addresses changes in area and productivity of forest in the park as compared to the forest outside the park and between 1984 and 2003 using satellite imagery. Kibale National Park in western Uganda exhibits attributes of fortress conservation (strict exclusion of humans, the prevention of consumptive use, and the minimization of other forms of human impact [Hulme and Murphree 2001]). Forest is an important land cover in this region due to the population's exclusion from the park, where timber, building poles and firewood were previously obtained. We would expect that with increasing population pressure, and conversion of land to agriculture, forests outside the park would decrease in size and productivity over time, thus harming their ability to provide resources, while the forest in the park has sustained little incursion and has been maintained and improved in some areas over time. Kibale is a good example to illustrate the role of parks for forest maintenance, regeneration and regrowth within a densely settled agricultural landscape. Kibale will be used to highlight the use of parks conceptually as a tool for reforestation and forest maintenance. We will also discuss methodological approaches and the limitations with current techniques to study land cover change.

### *12.1.1 Case Study: Kibale National Park, Uganda*

Sub-Saharan Africa continues to lose an estimated 4 million ha of forest cover annually (FAO 2006). Forest planting, landscape restoration, and natural expansion of forests have significantly reduced the net loss of the forest area, but still only 16.4% (74,585 ha) of forest area on the continent is designated primarily for conservation. Most deforested areas, especially in the tropics, are converted to other land uses, mainly agriculture and pasture (Brown and Lugo 1994). Forests in Uganda are widespread and complex, representing one of the most dominant forms of land-cover in Uganda, covering approximately 4.9 million ha of the country's total land surface (24.1 million ha NEMA 2001). Nearly three million hectares of forests and woodlands remain unprotected and are used by nearby communities for various purposes (NEMA 2001). They are important to the livelihoods of neighboring communities in several ways. They provide thatch, handcraft materials, medicines, food, fuel, building poles, timber, and other resources (Hartter 2007).

As is the case in many countries in Sub-Saharan Africa, Uganda's forests are under continuous threat of conversion due to population growth, in-migration, and intensive agriculture. Over 80% of the land in Uganda is used for small-scale farming (Mukiibi 2001) and continued population growth leads to added pressure on the land. Unprotected forests are vulnerable to exploitation and agricultural encroachment.

Forest conversion into other land uses such as cultivating crops, growing fuelwood, and expanding pasture is common. Although estimates vary, Uganda continues to lose between 0.8% (NEMA 2001) and 3% of closed-canopy forest annually. Nationwide, unsustainable domestic tree harvesting for firewood and non-timber forest products continues. Since 95% of the wood consumed in Uganda is used for fuelwood, and two-thirds of this amount is used at the household level, pressure continues to mount on remaining forests (Kayanja and Byarugaba 2001).

Kibale National Park (Kibale) in western Uganda (Fig. 12.1) is illustrative of the agricultural expansion and intensification surrounding parks. Kibale is a medium-altitude tropical moist forest covering about 795 km$^2$ (Hartter 2007). Kibale Forest was demarcated in 1932 as a Crown Forest and a Central Forest Reserve in 1948 and elevated to national park status in 1993 (Struhsaker 1997). While officially Kibale now includes a "game corridor" that connects the southern portion of Kibale

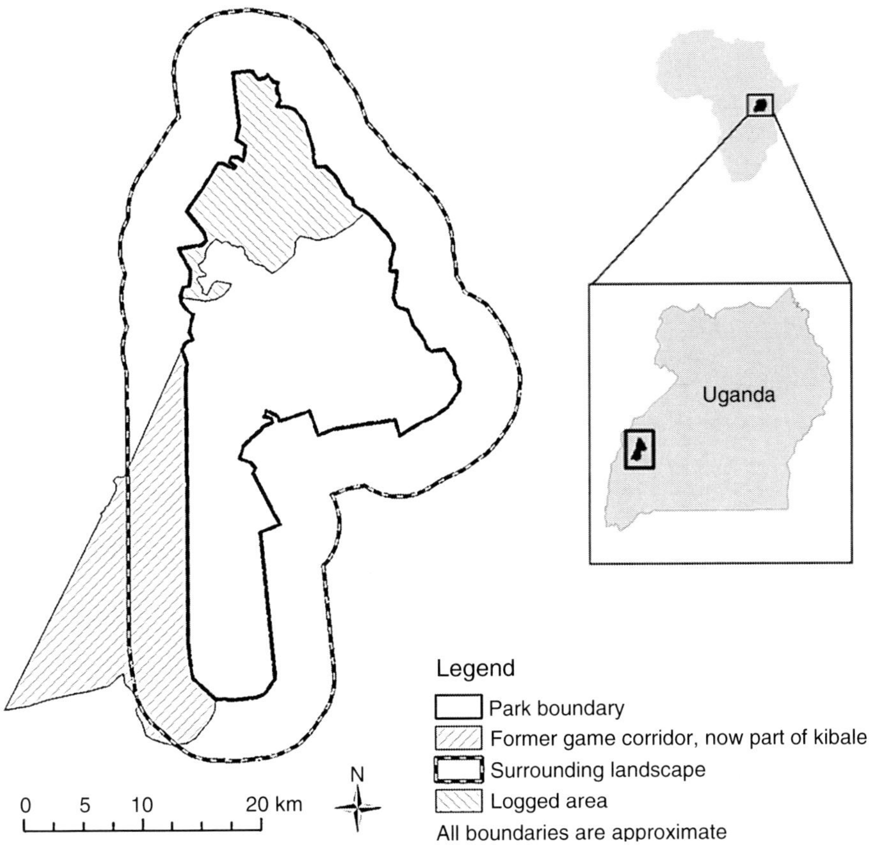


**Fig. 12.1** Kibale park boundary, game corridor and surrounding landscape buffer area, and the location of the study area in Uganda as inset

to Queen Elizabeth National Park, this corridor was formally gazetted as part of Kibale in 1993 (at which time human settlement was prohibited and those living in the corridor were expelled). Soon after, efforts began (and progress continues) to reforest the mixed agricultural land. Therefore, the former corridor was not classified as park in this analysis as it was not treated as a park area on the ground during this study period. By doing so, our analysis is conservative since the addition of the corridor would tend to lessen the difference between park and non-park. Excluding the game corridor, the Kibale Forest in this analysis contains 561 km$^2$.

This transitional forest (between lowland rainforest and montane forest) is at an average elevation of 1,110–1,590 m and is a remnant of a previously larger mid-altitude forest region (Struhsaker 1997). Although the amount of rainfall and length of season change, the average annual rainfall for the region is 1,552 mm for the period 1903–2008 (although average rainfall has risen to 1,698 mm for the period 1990–2008) (C. Chapman unpublished data). The climate is warm throughout the year, with an average range of 15–23°C (Struhsaker 1997).

As a Crown Forest Reserve, the main objective was to provide a sustained production of hardwood timber (Struhsaker 1997). Large-scale harvest began in 1950 for hardwood timber production, but was elevated between 1968 and 1975 to open the canopy by 50% with the harvest of trees greater than 1.52 m in girth with mechanized felling which occurred in the northern and central parts of Kibale (Skorupa 1988). Logging was conducted at variable intensities (low-high) (Fig. 12.1), resulting in large forest gaps and variable forest disturbance. Harvested areas were allowed to regenerate naturally. During Uganda's political upheaval in the 1970s and 1980s, the plantations were not managed (Chapman et al. 2002). Management plans changed when Kibale became a national park in 1993. Plantations started to be harvested and the harvested pine areas were left to regenerate to native forest. From 1993 until 2006, the planted exotics in Kibale have been extracted (using manual pit saws and portable sawmills) to facilitate forest regeneration with indigenous species (Chapman et al. 2002; C. Chapman unpublished data).

In the Kibale landscape, the vast majority of the population sustains their livelihoods through agricultural-based activities. Agriculturalists in the area belong primarily to two ethnic groups: the Batoro (west side of Kibale) and the Bakiga (east side of Kibale) (Fig. 12.1). The Batoro are the largest ethnic group in the area (~52% of population). But the immigrant Bakiga, who came to the region from southwestern Uganda beginning in the 1950s (Naughton-Treves 1998), and individuals of other ethnic groups have greatly contributed to population growth and the demand for agricultural land and resources (NEMA 2001). This region is one of the most densely populated areas on the African continent (Lepp and Holland 2006), and the population around Kibale has more than tripled between 1959 and 1990 (Naughton-Treves 1998). Population (2006) in the study area is estimated at 262 individuals/km$^2$ on the west side of the park and 335 individuals/km$^2$ on the east side of the park (Hartter 2007).

Kabarole District in western Uganda, where the western portion of Kibale lies, is a good example of the pressure on forest resources and subsequent landscape fragmentation facing protected areas in developing countries. The park is surrounded by a growing agricultural population, large tea estates, and a vast network of wetlands and bottomland forest fragments (Fig. 12.2). Nearly all of the forests found on potentially arable lands have been converted to small-scale agriculture, tea, or pasture and soil degradation is a major concern (NEMA 2001). The remaining forests typically occur in valley bottoms or on the steep rims of crater lakes (Gillespie and Chapman 2006; Hartter pers. obs.). Nearly 43% of the land within a 5 km periphery of the park is under cultivation or pasture. Tea dominates much of the landscape bordering the northwest portion of Kibale. Farm sizes on average are less than five hectares. Forest product consumption is at an all-time high as rural communities depend on them to sustain their livelihoods, yet available forested areas are decreasing in area and accessibility. These areas serve as important resource bases for food, thatching, craft materials, medicines, building poles, timber, firewood, and water. Forest degradation and forest fragment extinction is common in this landscape because of the heavy reliance on these areas (Hartter 2007).

### *12.1.2 Monitoring Land Cover Change in and Around a Forest Park*

The importance of tropical forests has long been recognized (Brown and Lugo 1994), but quantifying forest change in spatial extent and quality is essential for monitoring and management. Typical vegetation surveys (e.g., plots, transects) can be costly, time-consuming, spatially and temporally limited, and politically sensitive. The integration of remote sensing into vegetation studies can be a valuable addition to research and can be conducted at a larger spatial scale and over time (Ramsey and Jensen 1996).

Remote sensing using satellites provides robust techniques to efficiently observe and monitor the changes that occur within different land cover types over multiple dates at multiple locales over a large area. However, the introduction of satellite imagery into forest research is not without associated errors. Landscapes are nearly never homogeneous and therefore any discrete classification will incorrectly characterize certain portions of the landscape. In addition, remote sensing data analyses are typically of relatively coarse spatial resolution.

The most frequent use for the mapping of changes in landscapes is by analysis of satellite data, which can generate land-cover maps over greater spatial extents and more frequent time steps than is possible with detailed field studies (Jensen 2000). Despite the abundant use and relative success of classification methods used in landscape analysis, the use of these methods alone may not be optimal.

Alternatively, forested landscapes can be monitored using continuous data, because data are not broken or segregated in discrete elements but rather can represent the gradients on a landscape. Continuous data analyses can help provide more revealing spatial analyses and focus more on biophysical indicators. Acquiring continuous data allows for within-class differentiation, and not just across-class conversion monitoring (Southworth 2004). Discrete classification can be used to identify the presence or absence of certain land-cover types, but continuous methods also provide a measure of productivity.

The Normalized Difference Vegetation Index (NDVI) is widely used and has been shown to be a robust and reliable estimator of vegetation trends and status (Jensen et al. 1991; Ehrlich et al. 1994), including forest vegetation (Foody and Curran 1994, Rey-Benayas and Pope 1995, Sader et al. 2001). A healthy, vigorous, dense tract of forest will often have higher NDVI values than many other types of vegetation, although some forms of intensive agriculture for example, sugarcane, tea, rice fields, will often have the highest NDVI values.

## 12.2 Methods

A land-cover analysis was used to determine the extent and spatial distribution of temporal forest change and to provide an initial analysis of forest cover change both within and around the park. As such, we can examine not only absence and presence of forest and conversion to different land-cover classes (e.g., conversion from forest to field crops), but also the size and location in the landscape of patches of different land covers at each date.

### *12.2.1 Discrete Data Analysis – Land Cover Classification and Change Detection*

Landsat imagery was chosen because it offers the best combination of spatial, spectral, temporal and radiometric resolutions. Three dry-season images were acquired: May 26, 1984, January 17, 1995, and January 31, 2003 (path 132, row 060). Images were geometrically registered to 1:50,000 scale survey topographic maps of the region, followed by radiometric calibration and atmospheric correction to correct for atmospheric haze, sensor bias, and differences in sun angle. The first image provided baseline data prior to park establishment, the second captures conditions at park establishment, and the third represents current conditions.

During 2004 and 2005, 180 training sites (sites that are representative of land cover classes of interest and used to "train" the satellite image to detect classes of interest) were collected and used to group areas of similar spectral reflectance value together in a supervised classification to delineate land cover types. Five classes were used in the classification: (1) forest, (2) wetland vegetation (papyrus (*Cyperus papyrus* L.) and elephant grass (*Pennisetum purpureum*)), (3) water, (4) tea, and (5) other (bare soil, field crops, short grasses). The final classification of the 2003 composite had an accuracy of 89% and a kappa statistic of 0.87.

Land-cover maps were derived for each date by independent supervised classification of the Landsat scenes using a Gaussian maximum likelihood classifier.

Additional ground data control points were collected by interpretation of 1:31,000 scale aerial photography acquired in December 1988.

Land-cover classifications were then used to create change trajectories, that is, sequences of successive changes in land cover types (Petit et al. 2001). We restricted this analysis to those change-trajectory classes that highlighted changes in forest cover, since that is the focus of this chapter. Land-cover classes examined here are areas forested on all three image dates (forest), areas of wetland and grasses on all three image dates (wetland vegetation), and areas of agriculture on all three image dates (crops or tea) which represents the three stable land cover classes in the region. Next we have two conversion classes of interest: reforestation (when any non forest class [crops or wetland vegetation] on date one or date two, is forested by date three and deforestation [any area of forest] on date one or date two, which is non-forest [wetland vegetation, tea or crops] on date three). Water was excluded from this analysis as the crater lakes predominant in this region do not vary in area.

### *12.2.2 Continuous Data Analysis*

To remove the effect of phenological and potential precipitation differences and emphasize those areas that changed, both positively and negatively, more than the average change which may due to patterns of precipitation occurrences and other seasonal not inter-annual effects, we calculated surfaces free of these influences for the continuous analyses. First we created NDVI surfaces for each scene. We then calculated the standard normal deviate (Z-score) scene-by-scene, which transforms the mean NDVI in an image to 0, and the index value in each pixel to an expression of the number of standard deviations from the mean. Then NDVI-change trajectory images were generated by subtracting the 1995 NDVI image from the 1984 image, and the 2003 image from the 1995 image. A positive number indicates a reduction in normalized NDVI from one time to the next, and a negative number indicates an increase in the normalized NDVI. If anything, this normalization process will underestimate any changes in the longer time frame, due to shifts in precipitation patterns or long term changes.

## 12.3 Results

### *12.3.1 Land Cover Classification and Change Detection*

Forest is the dominant land cover type, with over 79% of Kibale National Park being in stable forest cover (Table 12.1, Fig. 12.2). Since 1984, the park boundaries have remained relatively intact, and only 4% of the area has experienced any deforestation during this time period, in contrast with over 11% reforested. While there is 1% in agriculture (crops, tea) in the park and almost 4% in wetland vegetation/elephant grass, there is no evidence of active large-scale incursion into the park.

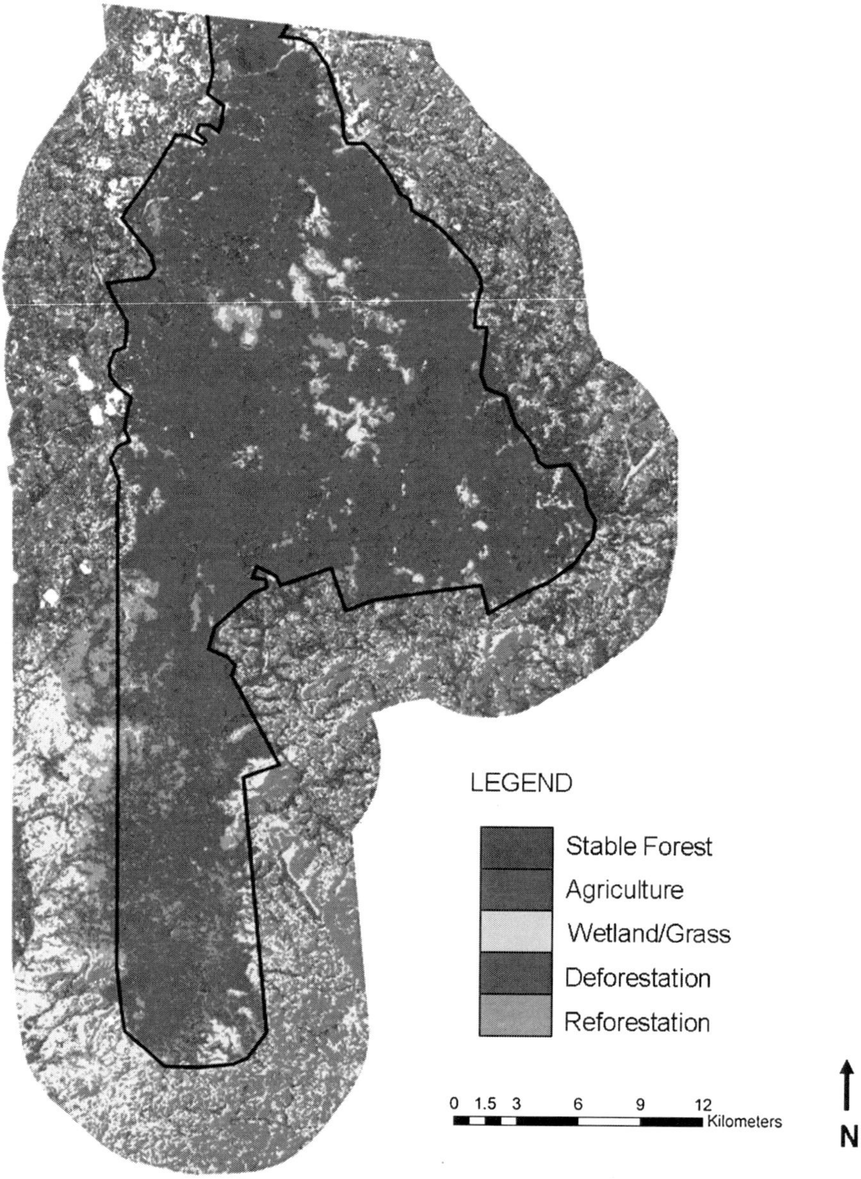


**Fig. 12.2** Trajectories of land cover change occurring within the park boundaries and surrounding landscape from 1984 to 1995 to 2003 (*see Color Plates*)

Forest cover within Kibale has become consolidated over time, with a total of over 90% area in forest cover by 2003 (stable forest plus reforested). There are fewer forest patches and the patches that remain have become larger and/or are linked to other forest patches (Fig. 12.2).

**Table 12.1** Land cover/use trajectory results for the study region for 2003, shown as within Kibale or outside park (within 5 km around the park), for the five trajectories of interest: stable forest cover, stable agriculture, stable wetland/grass, reforestation or deforestation, in terms of the percent of the landscape

| | Kibale | Outside Park |
|---|---|---|
| Trajectory | (% area) | (% area) |
| Stable Forest | 79.6 | 9.9 |
| Stable Agriculture | 1.2 | 36.4 |
| Stable Wetland | 3.8 | 17.1 |
| Reforestation | 11.3 | 11.5 |
| Deforestation | 4.1 | 25.0 |

The landscape outside Kibale presents a stark contrast to the in-park scenario, where conversion of forests continues. In contrast to inside Kibale where the dominant land cover is forest in the surrounding landscape less than 10% is in stable forest cover. Rather, there are two main land covers in the surrounding landscape – wetland vegetation/elephant grass at over 17% and crops at over 36%. Since 1984, there are fewer forest patches and those patches are smaller in size, as shown by the conversion classes, with over 25% deforestation and only 11% reforestation. The forests that remain have become smaller and are more isolated within the surrounding matrix of agricultural land, with an overall total of just over 20% in forest cover by 2003.

## 12.3.2 *Continuous Data Analysis*

### 12.3.2.1 NDVI Description

NDVI is highest in tea plantations, then in forests, wetland vegetation/elephant grass, and crops, in order (Fig. 12.3). The higher value in tea plantations is probably a function of evenness and density of the canopy of tea bushes. Tea is grown with an even canopy (the youngest tea leaves are harvested from the very ends of the branches) about 1 m high throughout the stand. Forest canopies are more diverse and have openings, and are thus more heterogeneous at the scale of Landsat imagery, resulting in more shadows and open areas and lower NDVI. Other than tea, NDVI distribution follows the general pattern seen with land-cover classes: areas with higher NDVI are mostly forest and the fine-grained mosaic of forest patches and dendritic bottomland forests in the surrounding landscape. There is a subtle difference in the park between the areas in the northern section that were logged (logged forest NDVI $0.67 \pm 0.05$, $0.65 \pm 0.06$, $0.48 \pm 0.06$ in 1984, 1995, and 2003, respectively) and the rest of the park, which has not been logged ($0.60 \pm 0.08$,

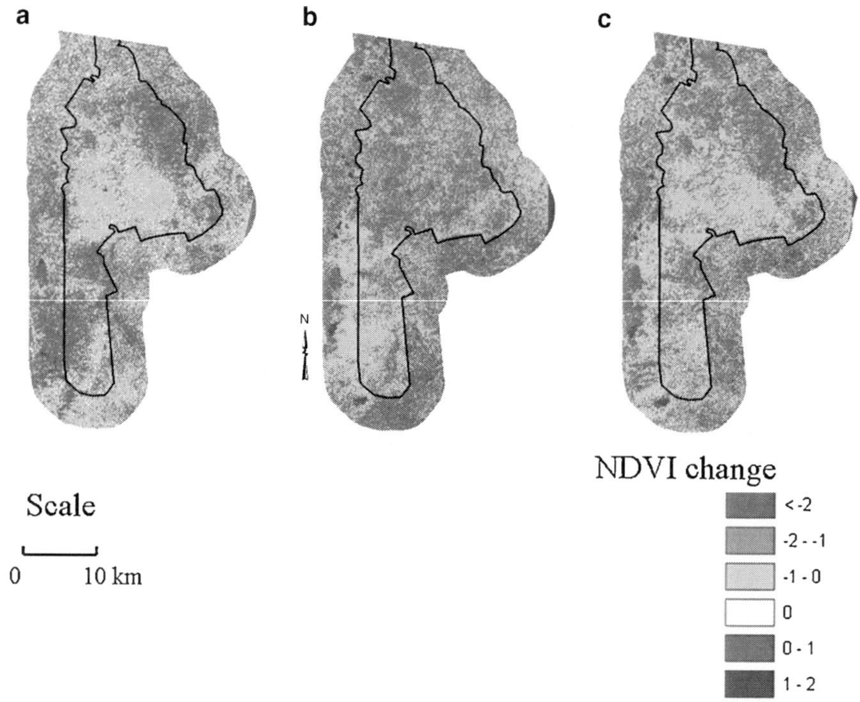


**Fig. 12.3** NDVI change trajectories showing the NDVI difference based on the individual image date NDVI standard normal deviates, for the three different time-steps, (**a**) 1984–1995, (**b**) 1995–2003, and (**c**) 1984–2003, where negative values indicates an increase in NDVI over time (*see Color Plates*)

0.59±0.06, and 0.44±0.06). There is also an increasing gradient of NDVI in the park from the lower elevations in the east and south to the higher northern forests, which may also be related to the logged-unlogged areas.

#### 12.3.2.2 NDVI Change Trajectories

The general trend over the three time periods is a decrease of NDVI (Fig. 12.3). In spatial terms, there was a slight increase in NDVI from 1984 to 1995 in the central part of the park with major decreases in the northeast and southern parts of the park. The NDVI decrease was widespread outside the park, with only two small areas, one south of the park and the other on the east side, showing increases. The 1995–2003 change shows more general NDVI decline within the park, except for the southern region which had a slight increase (Fig. 12.3). The park boundary seems to have had a stronger role in NDVI change, especially on the southeast and southern boundaries, where a larger and more widespread decrease occurred outside the park than inside. So while the park has maintained and increased forest cover the quality or productivity of that forest cover has actually declined over time.

### 12.3.3 Spatial Patterns of Land Cover Change

The most evident spatial pattern of the land-cover trajectories of the Kibale system is the difference between the continuous forest within the park boundaries and the fragmented or thinly dendritic forested areas in the surrounding landscape (5 km buffer around the park) (Fig. 12.3). These figures show the NDVI of areas classified only as forests, and starkly illustrate the "island" nature of the park itself. The vast majority of area classified as forest outside the park is the riparian forest in the bottoms of valleys where the combination of thick vegetation and wet soils have discouraged clearing and agriculture and some of these bottomland wetlands have been recently been protected by a local community-based conservation organization (Kibale Association for Rural and Economic Development). The forest fragments are denser on the northeast side of the park, where there are fewer roads and therefore less access to land. The southeast side of the park has both the most intensive agriculture and the sparsest forest cover.

Mean NDVI values in park and in the surrounding landscape drop from 1984 and 1995, but there is a much larger drop between 1995 and 2003 (Table 12.2). The maximum and minimum NDVI values as well as the standard deviation values are greatest in 1984 and at their least in 2003. What is surprising about these results is not only that there is a drop in overall productivity (mean NDVI values), but from the land cover analysis, forest and wetland inside Kibale have experienced little change. In 1984, Kibale forest has much higher mean NDVI, but there is higher variability in NDVI values, meaning that forest in the park may be in various stages of re-growth or succession. However, in 2003, when NDVI values for forest are lowest, NDVI values have less variability. This decrease in mean NDVI is illustrated in Fig. 12.3, which shows that the drop in NDVI is not attributed to a particular sector of the park, rather the majority of Kibale has decreased substantially in productivity.

**Table 12.2** Forest cover and NDVI values for Kibale and outside the park

| | Year | Forest (% area) | Mean NDVI | NDVI Standard Deviation | NDVI minimum | NDVI maximum |
|---|---|---|---|---|---|---|
| Kibale | 1984 | 86.6 | 0.591 | 0.078 | 0.315 | 0.805 |
| | 1995 | 86.9 | 0.548 | 0.055 | 0.143 | 0.770 |
| | 2003 | 90.6 | 0.360 | 0.052 | 0.133 | 0.559 |
| Outside Park | 1984 | 31.9 | 0.644 | 0.063 | 0.338 | 0.799 |
| | 1995 | 34.3 | 0.580 | 0.044 | 0.333 | 0.778 |
| | 2003 | 29.2 | 0.398 | 0.049 | 0.043 | 0.586 |

## 12.4 Discussion

### *12.4.1 Forest Change in and Around Kibale and Implications of Those Changes*

By many different physical measures from simple visual examination of composite satellite images to the comparison of various landscape indices, Kibale is an island of forest surrounded by intensively used agricultural land. This agricultural matrix is characterized by highly fragmented and rapidly changing land covers (and associated land uses). Although legislatively the park excludes human influence, the park itself is not unchanging, as there is also natural variability within the system. Clearly, there is evidence to support the relative success of Kibale in forest maintenance and reforestation. Our results indicate an increase in forest cover inside the park since 1984. The majority of Kibale's forest has remained intact, with only a small proportion lost to deforestation in 20 years in contrast to what has happened in the landscape surrounding the park. One of Struhsaker's (2002) indicators of park success is a decrease in illegal activities. Field interviews suggest large-scale encroachment into the park has virtually halted since formal park establishment in 1993 and boundaries are well understood and maintained, which is also supported by these research findings.

However, forest productivity and health (as represented by NDVI) has decreased over time both within the park and in the forest fragments outside the park. While the latter is expected due to the increased use of fragments by local communities, as a function of exclusion from the park resource, the decrease in NDVI values within the park boundaries, where forest maintenance and regrowth dominates, was unexpected, especially across the full extent of the park area.

Examining forest productivity through NDVI provides another layer to assist in understanding the importance of reforestation and forest cover maintenance. The decreased NDVI value is puzzling, and no doubt troubling. There is no clear explanation for the decline and further analysis with increased time steps should address this issue. However, the impacts of decreased productivity should not be ignored. Lower forest productivity has notable ecological implications and may be a precursor to future changes in forests. A degraded forest (one with lower productivity) will produce less above- and below-ground biomass, thus limiting the ability of nutrient uptake and erosion abatement. Decreased productivity may have negative impacts on species richness (Kay et al. 1997), which may correlate to changes in phenology, seasonal fruiting, and nutritional content of foliage.

Outside the park, the situation is different. Deterred by fear of punitive measures against illegal activity (Struhsaker et al. 2005), many households have turned to the unprotected forests. Growing population has led to intensive resource extraction in unprotected forests and other forests have been cleared and converted to agriculture or pasture land. The clearing of forest is highly variable in each fragment because of local land tenure, individual resource needs, the number of households dependent on the particular forest fragment, human–wildlife interaction, local geography and ecology, and other factors. Increased population and further fragmentation of the landscape implies further use of the fragments until resources have been exhausted.

As the rural population outside Kibale grows and continues to cut forest seeking land and resources and to prevent crop raiding (Hartter 2007), long term detriment is caused to the lands. There is anecdotal evidence to suggest that the increased isolation of the forest fragments within the intensely cultivated landscape is related to trees dying out in the fragments. Even though some of the fragments remain, many farmers have noticed that bigger trees are dying and they say that it is taking longer for the endemic species to grow to even pole size. Fragmentation not only lowers species number, but also alters community composition and their ecosystem processes because of the reduction in size and the change in shape (Hill and Curran 2003). Laurance et al. (2000) found that forest fragment size and tree species diversity were directly related. As in the logged areas of Kibale, regeneration is slower in the fragments due to human disturbance. Continued extraction of smaller trees for fuelwood and building poles decreases the periodicity of disturbance and nutrient removal, but reduces the juvenile class that would eventually grow to larger trees. As a result of continued resource extraction and no effort to replant trees in species or in number, there are fewer seed trees and reduced habitat important for seed-dispersing frugivores and aviaries. Tree seedlings are often outcompeted in the degraded forest interior and at edges. These areas then are inhabited by weedy species that have minimal use for home consumption. Fragment size is important. Higher mortality rates in forest fragments are found in smaller fragments in the Amazon (Laurance et al. 2000) and Laurance and Bierregard (1997) found that the number of rare species increased with fragment size. The disruption of natural succession in these forest fragments leads to a decreased ability to perform ecosystem functions, and in turn the availability of resources to meet subsistence needs.

While the outlook for small forest fragments may look bleak (Chapman et al. 2006), the decrease in availability and access to fuelwood resources for the rural communities outside Kibale may actually be one of the largest drivers of positive forest change. Ninety-five percent of all Uganda's energy needs are met with fuelwood and charcoal (Naughton-Treves et al. 2007). The decline and extinction of nearby forest fragments and their products (particularly fuelwood) has forced households to adapt. Some travel farther and/or seek alternative sources of energy. Many households in the rural communities outside Kibale who have enough land are reforesting bottomland, hilltop, and former agricultural areas to compensate for resource shortages and/or to serve as a security net for risks or future opportunities. The most prolific type planted is eucalyptus. In the 1950s, eucalyptus *(Eucalyptus* var., but mostly *globulus* and *grandis*) was introduced in many parts of Uganda as a solution to wood shortages, because it is fast-growing, coppices, somewhat resistant to rot and termites, burns at a higher BTU than other available softwoods, and can provide poles and timber as early as three and eight years respectively after planting. Agricultural land is also being reforested with other species, such as musizi (*Maesopsis eminii*), oocarpa pine (*Pinus oocarpa*), Caribbean pine (*Pinus caribea*), patula pine (*Pinus patula*), silver oak (*Grevillea robusta*), and markamia

(*Markhamia lutea*) mainly for fuelwood and building poles. Future analyses of land cover change in this region may therefore detect reforestation occurring outside the park boundary.

### 12.4.2 Parks as Mechanisms for Forest Recovery and Maintenance

Parks offer one mechanism for forest maintenance and recovery. However, there is no clear way forward. On one side, parks with hard boundaries should be a core component of conservation strategies as a way to maintain forest cover and the biodiversity contained therein (Bruner et al. 2001). Given high rates of deforestation outside the park and the protection provided within Kibale's boundaries, reforestation efforts have been quite successful. However, the park-people linkage cannot be ignored. Several studies around the world have documented the success of parks in forest maintenance. Roy Chowdhury (2006) found that the reserve did appear to successfully curtail deforestation and contributed to reforestation efforts in Mexico. Outside the reserve, forest fragmentation is much greater than inside. Nagendra et al. (2006) found a similar result in the Tadoba Andhari Tiger Reserve in India. Compared to unprotected lands, federal conservation lands in Brazil suffered fewer impacts of logging and deforestation (Asner et al. 2006). Additionally, forest recovery in parks with hard boundaries could be faster because the human influence has been excluded.

Forest maintenance and improved forest recovery may come at a high price to the surrounding landscape. Byron and Arnold (1999) estimate an enormous proportion of people in the world are dependent on the forests to support their livelihoods in one way or another. Excluding or severely restricting access and extraction forces them to seek alternatives. Often, they turn first to the unprotected forest fragments that remain. As population increases, more resources are consumed, and landscapes surrounding the park is parceled further and resource bases become degraded.

Despite their social importance, best management practices for forest fragments do not exist, or if they do exist, are not practical or are not followed by the rural majority who depend on them to survive. Chapman et al. (2006) conclude that many of the forest fragments surrounding Kibale will disappear in the coming years. Therefore a concerted effort is needed to boost local perception of the value of forests, the benefits they provide, and management options (Kayanja and Byarugaba 2001).

Knowledge of the locations, characteristics, and changes of forests and the surrounding communities is useful to establish protection and management policies. Conservationists must work towards a better understanding of park-people relationships and the impacts of land-use decisions on different taxa within the park and in the surrounding landscape. It may be important first to recognize that it is more important to protect some forests over others because some may be more valuable in socio-economic terms and others that have important conversation or biodiversity values. This has implications on monitoring because it requires the location of forest change to be known and the relationship of forests to other forests and resource bases within the surrounding landscape. In many areas of Uganda, forests are not easily accessible. The budgets and staff for the District Environmental Officer are small and resources are limited, making broad-scale monitoring difficult. Therefore identifying 'hot-spots' through remote sensing is an important phase in allocating limited resources in forest management.

### *12.4.3 Continuous Data Analysis Using NDVI in the Kibale Landscape*

Much of the analysis presented is based on NDVI analysis. NDVI has been proven to be a robust measure of vegetation attributes (net primary productivity, green biomass, and green leaf area index) (Serneels et al. 2001). However, NDVI is well correlated with climate variables, such as evapotranspiration and precipitation (Anyamba et al. 2001). In his meta-analysis of net primary productivity (NPP) and global climate in the wet tropical forests, Schuur (2003) found that net primary productivity becomes less sensitive to mean annual precipitation at high precipitation levels (around 2,445 mm annually), at which point NPP levels off and then declines. Wang et al. (2001) also report the lowest correlations of NDVI to differences in precipitation with forest.

Precipitation in the Kibale region is extremely local. The bi-modal rainfall pattern produces two major rainy seasons. The long rains are between late February and early May and the short rains occur between late August to early December. Each of the images were captured following abundant rainfall during the rainy seasons and all three images are not anomalous (not in peak or trough precipitation months). Mean rainfall totals for the four-month period prior to image acquisition for 1984 and 1995 images were not significantly different than the four month period prior to the 2003 image capture (Dunnett's test, $p > 0.05$). If the area was precipitation limited, we would expect that there would be a distinct seasonal difference. However, the Kibale region, receiving more than 1,500 mm of rainfall annually, is not precipitation limited. Further, Serneels et al. (2001) and Richard and Poccard (1998) report the sensitivity of NDVI to inter-annual rainfall in rainfall limited regions ($<1,200$ and $<900$ mm/yr respectively). There is a distinct dry and wet season in the Kibale region, but local farmers report that these seasons are becoming less and less distinct and increasingly blend into one another. In addition, seasonality has little impact on the Kibale forest signal. In addition, we also used standard normal deviates to examine change from the mean in a given year and not changes in means across years. Although we acknowledge the use of images acquired from non-anniversary dates may introduce some error into subsequent analyses, we can conclude that forest conditions represented in the 1984, 1995, and 2003 Landsat images are analogous, and the use of NDVI was an appropriate measure of long-term vegetation changes and not interannual variation in precipitation.

Remote sensing methods can be advantageous in efficiently observing and monitoring changes that occur in Uganda's forests multiple dates at multiple locales over a large area. Satellite remote sensing is especially appropriate for initial reconnaissance mapping and continued monitoring of forests over large geographic areas. The heterogeneous forest park landscape of Kibale National Park demands a more integrative approach to simple discrete methods. By supplementing the analyses with continuous data sets, we can better identify areas that have changed from one land cover class to the next, but perhaps more importantly those areas with biophysical change. In this study, we focused on the absence/presence and productivity of the forest inside Kibale and in the surrounding landscape. Kibale, a park with hard boundaries that exclude access and resource extracting can be considered a success. We found that Kibale has maintained its forest cover and has contributed to the recovery of forests inside its boundaries. However, forest maintenance may come at a high price as the surrounding landscape has a higher proportion of deforestation and a lower reforestation. The loss of forests outside the park has important direct and indirect implications on both the rural communities and wildlife that depend on them for their survival. In addition, the decrease in productivity in both the park and surrounding landscape is puzzling, and future research efforts will address this important issue. Without Kibale though, the forest would almost certainly face the same fate that has befallen many of the fragments outside it. The continued viability of these ecosystems depends on monitoring their change and managing them in a sustainable fashion.

## 12.5 Conclusion

Synergy of remotely sensed biophysical and human change data at multiple scales is imperative to understand drivers and effects of landscape change. Through a linked methodology to we can build a better understanding of social and ecological interactions, areas where changes are occurring and where more detailed information must be gathered can be identified. Such information can be used by conservation scientists, park managers, and community leaders in order to build an understanding of the importance of forests and work with communities to develop cooperative management agreements or outreach programs in communities outside Kibale. Encouraging recognition of finite resources by local communities and stressing their importance on them may help to foster a stronger sense of conservation and may lead to improved management of resource pools (Johnson and Nelson 2004) outside of the park while maintaining for full range of ecosystem services within the park.

The chapter highlights an important research imperative in the reforestation discourse; to examine the impacts of park establishment and their management on reforestation, forest recovery and forest cover maintenance. The landscape in and around Kibale is emblematic of broader-scale issues (e.g., population and resource pressures and the subsequent islandization of parks) threatening other forest parks and is illustrative of a greater trend that may befall newer parks in Central Africa, including Gabon and the Democratic Republic of Congo. Kibale is in many respects an example of the "fortress conservation" model of park conservation that has often been criticized by social scientists but generally supported by conservationists. Results from this and other studies suggest that, despite its hard-edged "fortress" features, Kibale has been fairly successful both in terms of its wildlife conservation objectives and in reforestation and forest maintenance. Much of the success, in terms of low border incursions for resource extraction and resilience of wildlife populations may be due to the remnant fragments of forest in the external landscape, serving as a "resource buffer", in addition to providing additional natural habitat for forest species.

Despite the success of Kibale in maintaining forest cover, the forests in the surrounding landscaped continue to be heavily exploited, and that heavy reliance and continued resource extraction from forest fragments by local communities will perpetuate their degradation. Since resource use within Kibale is restricted, the remaining forest fragments are in jeopardy and face almost certain eradication (Chapman et al. 2007). The fragments farther from Kibale may currently be displacing that pressure. However, once remnant forests surrounding Kibale are destroyed or degraded beyond use, then pressure on Kibale may increase as it has with other protected areas (Chhetri et al. 2003).

Combining discrete and continuous data analysis techniques with satellite imagery can provide valuable information to aid in conservation of these invaluable resources. A new potential pathway to forest transitions (see Rudel, Chapter 3) could be one of conservation and protection. More studies linking parks and conservation to reforestation are needed, as well as the inclusion of studies both of forest area (classification/ discrete methods) and quality (e.g., NDVI, continuous methods). By protecting forest parks, we not only maintain forests, but allow natural processes to drive forest recovery and preserve pathways of forest succession. Moreover, the incorporation of social and ecological data with remotely sensed datasets and analyses can identify areas where changes are occurring and where more detailed information must be gathered. The continued viability of these ecosystems depends on monitoring reforestation, understanding the changes in forest cover, and managing those forests protected in parks through appropriate and sustainable management policies.

**Acknowledgments** This research was supported by a National Science Foundation grant (#0352008), a Working Forests in the Tropics Field Research Grant, and the Center for African Studies David L. Niddrie Memorial Scholarship. Permission to conduct this research was granted by the Uganda National Council for Science and Technology, Uganda Wildlife Authority, and local council leaders and their representatives in Kabarole and Kibale districts.

# References

Anyamba A, Tucker CJ, Eastman JR (2001) NDVI anomaly patterns over Africa during the 1997/98 ENSO warm event. Int J Remote Sens 22:1847–1859

Asner GP, Broadbent EN, Oliveira PJC, Keller M, Knapp DE, Silva JNM (2006) Condition and fate of logged forests in the Brazilian Amazon. PNAS 103:12947–12950

Brown S, Lugo AE (1990) Tropical secondary forests. J Trop Ecol 6:1–32

Brown S, Lugo AE (1994) Rehabilitation of tropical lands: a key to sustaining development. Restor Ecol 2:97–111

Byron N, Arnold M (1999) What futures the people of the tropical forests? World Dev 27:789–805

Bruner AG, Gullison RE, Rice RE, da Fonseca GA (2001) Effectiveness of parks in protecting tropical biodiversity. Science 191:125–128

Chapman CA, Chapman LJ (1995) Survival without dispersers: seedling recruitment under parents. Conserv Biol 9:675–678

Chapman CA, Chapman LJ, Zanne A, Burgess A (2002) Does weeding promote regeneration of an indigenous tree community in felled pine plantations in Uganda? Restor Ecol 10:408–415

Chapman CA, Wasserman MD, Gillespie TR, Speirs ML, Lawes MJ, Saj TL, Zieglers TE (2006) Do food availability, parasitism, and stress have synergistic effects on red colobus populations living in forest fragments? Am J Phys Anthropol 131:525

Chhetri P, Mugisha A, White S (2003) Community resources use in Kibale and Mt. Elgon National Parks, Uganda. Parks 13:28–49

Chapman CA, Naughton-Treves L, Lawes MJ, Wasserman MD, Gillespie TR (2007) The conservation value of forest fragments: explanations for population decline of the colobus of western Uganda. International Journal of Primatology. 28(3):513–528

Condit R, Hubbell SP, Foster RB (1993) Identifying fast-growing native tre4es from the neotropics using data from a large, permanent census plot. For Ecol Manage 62:123–144

Duncan RS, Chapman C (2002) Limitations of animal seed dispersal for enhancing forest succession on degraded lands. In: Levey DJ, Silva WR, Galetti M (eds) Seed dispersal and frugivory: ecology, evolution and conservation. CABI International, Wallingford, UK, pp 437–450

Ehrlich D, Estes JE, Singh A (1994) Applications of NOAA-AVHRR 1 km data for environmental monitoring. Int J Remote Sens 15:145–161

Food and Agriculture Organization (2006) Global Forest Resources Assessment 2005. Food and Agriculture Organization of the United Nations, Rome

Foody GM, Curran PJ (1994) Estimation of tropical forest extent and regenerative stage using remotely-sensed data. J Biogeogr 21:223–244

Gillespie TR, Chapman CA (2006) Prediction of parasite infection dynamics in primate metapopulations based on attributes of forest fragmentation. Conserv Biol 20:441–448

Goldman A, Hartter J, Southworth J, Binford M (2008) Studying the human landscape around Kibale National Park. In: Wrangham R, Ross E (eds) Science and conservation in a Ugandan forest: How long-term research can help habitat management. Cambridge University Press, Cambridge, UK, pp 129–144

Hartter J (2007) Landscape change around Kibale National Park, Uganda: impacts on land cover, land use, and livelihoods. Ph.D. Dissertation, University of Florida, Gainesville, FL

Hayes TM (2006) Parks, people, and forest protection: an institutional assessment of the effectiveness of protected areas. World Dev 34:2064–2075

Hill J, Curran P (2003) Area, shape and isolation of tropical forest fragments: effects on tree species diversity and implications for conservation. J Biogeogr 30:1391–1403

Howard P, Davenport T, Kigenyi F, Viskanic P, Baltzer M, Dickinson C, Lwanga J, Matthews R, Mupada E (2000) Protected area planning in the tropics: Uganda's national system of forest nature reserves. Conserv Biol 14:858–875

Hulme D, Murphree M (eds) (2001) African wildlife and livelihoods: the promise and performance of community conservation. James Curry, Oxford

Jensen JR (2000) Introductory digital image processing. Prentice Hall, Upper Saddle River, USA

Jensen JR, Lin H, Yang X, Ramsey EW III, Davis BA, Thoemke CW (1991) The measurement of mangrove characteristics in southwest Florida using SPOT multispectral data. Geocarto Intl 2:13–21
Johnson KA, Nelson KC (2004) Common property and conservation: the potential for effective communal forest management within a national park in Mexico. Human Ecology 32(6): 703–733.
Kay RF, Madden RH, van Schaik C, Higdon D (1997) Primate species richness is determined by plant productivity: implications for conservation. Proc Natl Acad Sci U S A 94:13023–13027
Kayanja FIB, Byarugaba D (2001) Disappearing forests of Uganda: the way forward. Curr Sci 81:936–947
Laurance WF, Bierregaard RO Jr (eds) (1997) Tropical forest remnants–ecology, management, and conservation of fragmented communities. The University of Chicago Press, Chicago, IL, USA
Laurance WF, Delamônica P, Laurance SG, Vasconcelos HL, Lovejoy TE (2000) Rainforest fragmentation kills big trees. Nature 404:836
Lepp A, Holland S (2006) A comparison of attitudes toward state-led conservation and community-based conservation in the village of Bigodi, Uganda. Soc Natl Res 19:609–623
Lugo AE (1992) Tree plantations for rehabilitating damaged forest lands in the tropics. In: Wali MK (ed) Ecosystem rehabilitation, vol 2. Ecosystem analysis and synthesis. SPB Academic Publishing, The Hague, The Netherlands, pp 247–255
Lwanga JS (2003) Forest sucesión in Kibale National Park, Uganda: implications for forest restoration and Management. Afr J Ecol 41:9–22
McNeely JA (1990) The future of national parks. Environment 32(1):16–41
McNeely J (1994) Protected areas for the 21st century: working to provide benefits to society. Biodiv Conserv 3:390–405
Mukiibi JK (ed) (2001) Agriculture in Uganda. General information, vol 1. Fountain Publishers, Kampala, Uganda
Nagendra H, Pareeth S, Ghate R (2006) People within parks – forest villages, land-cover change and landscape fragmentation in the Tadoba Andhari Tiger Reserve, India. Appl Geogr 26:96–112
National Environment Management Authority (2001) State of Environment Report for Uganda 2000/2001. National Environment Management Authority, Kampala, Uganda
Naughton-Treves L (1998) Predicting patterns of crop damage by wildlife around Kibale National Park, Uganda. Conserv Biol 12:156–168
Naughton-Treves L, Kammen DM, Chapman C (2007) Burning biodiversity: woody biomass use by commercial and subsistence groups in western Uganda's forests. Biol Conserv 134:232–241
Nepstad DC, Uhl C, Serreo AS (1991) Recuperation of a degraded Amazonian landscape: forest recovery and agriculture restoration. Ambio 20:248–255
Perz SG (2007) Grand theory and context-specificity in the study of forest dynamics: forest transition theory and other directions. Prof Geogr 59:105–114
Petit C, Scudder T, Lambin E (2001) Quantifying processes of land-cover change by remote sensing: resettlement and rapid land-cover changes in south-eastern Zambia. Int J Remote Sens 22:3435–3456
Ramsey EW, Jensen JR (1996) Remote sensing of mangrove wetlands: relating canopy spectra to site-specific data. PERS 62:939–948
Rey-Benayas JM, Pope KO (1995) Landscape ecology and diversity patterns in the seasonal tropics from Landsat TM imagery. Ecol Appl 5:386–394
Richard Y, Poccard I (1998) A statistical study of NDVI sensitivity to seasonal and inter-annual rainfall variations in southern Africa. Int J Remote Sens 19:2907–2920
Roy Chowdhury R (2006) Landscape change in the Calakmul Biosphere Reserve, Mexico: modeling the driving forces of smallholder deforestation in land parcels. Appl Geogr 26:129–152
Sader SA, Hayes DJ, Hepinstall JA, Coan M, Soza C (2001) Forest change monitoring of a remote biosphere reserve. Int J Remote Sens 22:1937–1950
Schonewald-Cox CM, Bayless JW (1986) The boundary model: a geographic analysis of design and conservation of nature reserves. Biol Conserv 38:305–322

Schuur EAG (2003) Productivity and global climate revisited: the sensitivity of tropical forest growth to precipitation. Ecology 84:1165–1170

Serneels S, Said MY, Lambin EF (2001) Land cover changes around a major East African wildlife reserve: The Mara ecosystem (Kenya). Int J Remote Sens 22:3397–3420

Skorupa JP (1988) The effects of selective timber harvesting on rain-forest primates in Kibale Forest, Uganda. Ph.D. Dissertation, University of Florida, Gainesville, FL

Southworth J (2004) An assessment of Landsat TM band 6 thermal data for analysing land cover in tropical dry forest regions. Int J Remote Sens 25:68–706

Struhsaker T (1987) Forestry issues and conservation in Uganda. Biol Conserv 39:209–234

Struhsaker TT (1997) Ecology of an African rain forest: logging in Kibale and the conflict between conservation and exploitation. University Press of Florida, Gainesville, FL

Struhsaker TT (2002) Strategies for conserving forest national parks in Africa with a case study from Uganda. In: Terborgh J, Van Schaik C, Davenport L, Rao M (eds) Making parks work: strategies for preserving tropical nature. Island Press, Washington, DC, pp 97–111

Struhsaker TT (2005) Conservation of red colobus and their habitats. Int J Primatol 26:525–538

Struhsaker TT, Struhsaker PJ, Siex KS (2005) Conserving Africa's rain forests: problems in protected areas and possible solutions. Biol Conserv 123:5–54

Turner I, Corlett R (1996) The conservation value of small, isolated fragments of lowland tropical rain forest. TREE 11:330–333

Wang J, Price KP, Rich PM (2001) Spatial patterns of NDVI in response to precipitation and temperature in the central Great Plains. Int J Remote Sens 22:3827–3844

Wilkie DS, Morelli GA, Demmer J, Starkey M, Telfer P, Steil M (2006) Parks and people: assessing the human welfare effects of establishing protected areas for biodiversity conservation. Conserv Biol 20:247–249

# Chapter 13
# Spontaneous Regeneration of Tropical Dry Forest in Madagascar: The Social–Ecological Dimension

**Thomas Elmqvist, Markku Pyykönen, and Maria Tengö**

T. Elmqvist (✉), M. Pyykönen, and M. Tengö
Department of Systems Ecology, Stockholm University, Stockholm, Sweden and
Stockholm Resilience Centre, Stockholm, Sweden
e-mail: thomase@ecology.su.se

H. Nagendra and J. Southworth (eds.), *Reforesting Landscapes: Linking Pattern and Process*, Landscape Series 10,
DOI 10.1007/978-1-4020-9656-3_13, © Springer Science+Business Media B.V. 2010

## 13.1 Introduction

Loss of tropical forests and changes in land use/land cover affect climate and environmental change at global scales and are of growing concern worldwide (e.g. Achard et al. 2002; Lambin et al. 2003). Although most focus had been on deforestation, regrowth of tropical forest may be widespread. Achard et al. (2002) estimated that at a global level, the annual regrowth area of humid tropical forest is 1 million hectares compared with the annual forest loss of 5.8 million hectares. In other words, annual regeneration may correspond to roughly 20% of the total area of deforestation (see also Chapter 2 in this volume).

In spite of this, we know surprisingly little about regeneration in terms of functional aspects of biodiversity and generation of ecosystem services for local and regional human consumption and use (e.g. Grau et al. 2003; Dunn 2004). This is particularly true for dry tropical forests (see review by Vieira and Scariot 2006; Sánchez-Azofeifa et al. 2005). Further, the role of the institutional context is increasingly emphasized in disentangling drivers of forest dynamics (Lambin et al. 2001; Dietz et al. 2003). Although we have some knowledge about the institutional context in which tropical forest loss is embedded (Gibson et al. 2000; Lambin et al. 2001, 2003), we know very little about the role social institutions may have in influencing rates of tropical forest regeneration, especially at the local scale (Tucker et al. 2005; Nagendra 2007).

In Madagascar, the southern dry forest harbors the highest level of plant endemism with 48% of the genera and 95% of the species endemic (Koechlin 1972; Rabesandratana 1984), and is listed as one of the 200 most important ecological regions in the world (Olson and Dinerstein 2002). The dry forest has long been considered to be the most intact vegetation type in all of Madagascar (Du Puy and Moat 1996) and it has also been suggested that deforestation here may have much more severe consequences as compared to moist forests due to a limited regeneration capacity (Green and Sussman 1990; Du Puy and Moat 1996). Arid conditions have resulted in a historically less intensive slash and burn agriculture and natural fires are infrequent in this system (Koechlin 1972). Despite global recognition of the value of the southern dry forest, there have been surprisingly few studies on forest cover changes or effects of anthropogenic impacts (Elmqvist 2004, but see Elmqvist et al. 2007).

In contrast to other types of forests in Madagascar there are only a few, very small areas formally under protection (Fenn 2003). Informal institutions, however, play an important role in southern Madagascar in protecting forest ecosystems and maintaining their capacity to generate valuable ecosystem services (Tengö et al. 2007). Other studies have shown the value of existing institutions and customary authority and values for the success of conservation (Lingard et al. 2003; Horning 2003a, 2008; McConnell and Sweeney 2005; Schachenmann 2006). Few studies have, however, linked the existence of a social capital (sensu Pretty 2003) related to forest management with spatial analysis of forest dynamics.

In this study, we used Landsat images from southern Madagascar from three different years (1986, 1993, and 2000) for a time series analysis of dry forest cover change.

We performed field surveys, interviews and analyses of local institutions to interpret the observed forest cover changes in a social institutional context.

## 13.2 Study Area and Methodology

The Androy region is situated in the southernmost part of Madagascar between Lat 24° 13′ and 25° 24′ S and Lon 45° 20′ and 46° 26′ E (Fig. 13.1). The area is characterized by semi-arid climatic conditions with irregular rainfall averaging less than 500 mm per year. The annual rainfall declines from north to south and from northeast to southwest (Battistini and Richard-Vindard 1972). The dry season usually lasts eight to nine months, between March–October/November, but locally it can extend over several years (Dewar and Wallis 1999; Richard et al. 2002). The mean temperature is generally between 23°C and 26°C but the daily amplitude may be as large as 22°C during the cold season, May–October.

The dry forest of southern Madagascar is characterized by drought tolerant woody species of Didiereaceae and Euphorbiaceae (Rabesandratana 1984). The most common species of the Didiereaceae, *Alluaudia procera,* dominates forest stands in central and northern Androy, while in southern Androy, forest stands are dominated by Euphorbiaceae, mainly *Euphorbia decorsei.* The southernmost part of Androy is a sandy area with paleodunes while the northern part is a hilly upland on Precambrian crystalline bedrock (Battistini and Richard-Vindard 1972). Since the early 1970s, the dry forest cover has been reported to be in decline, principally due to clearing for agriculture, cattle herding, timber harvest and charcoal production (Sussman et al. 1994; Sussman and Rakotozafy 1994; Sussman et al. 2003).

Technically, all non-private forested land in Madagascar is state property (Kull 2004).The traditional land claims inherited from the ancestors *(tanin-drazana)* related to clans and lineages are however still effective in Androy, representing a common property regime with collective owners that organize to exclude or regulate non-owners and their use of resources (Hanna et al. 1996). Customary institutions are however often challenged or weakened by top down interventions, the local formal government and modernization through institutions such as schools and churches that contradict authority based on lineage and traditional belief systems (Gezon 2006; Tengö et al. 2007). A recent legislation scheme of Madagascar, the GELOSE (Gestion Locale Sécurisée or secured local management) from 1996, allows for decentralization of management rights of renewable natural resources, including forest (Antona et al. 2002; Kull 2004), but this is not yet effective in Androy.

### *13.2.1 Analyses of Landsat Images*

Satellite images from three different years were used for time series analysis of forest cover changes. The images from 25 June 1984, 15 April 1993, and 28 May

2000 respectively, were all dry season synoptic views from path/row 159/77 with 30 m resolution. The 1984 and 1993 scenes were Landsat 5 TM and the 2000 scene was a Landsat 7 ETM+. The software ERDAS 9.0 was utilised for all processing of the satellite images.

Since a classification of a whole Landsat scene covering 180 by 180 km would introduce considerable error and thereby cause misinterpretations, a subset was used from each scene, approximately 87 by 63 km (Lat 24° 47′ and 25° 24′ S, Lon 45° 36′ and 46° 26′E) (Fig. 13.1). These sub-scenes were rectified to one pixel accuracy relative to each other. An unsupervised classification containing 250 classes was performed for each of the images including the 6 bands with 30 m resolution. A preliminary reclassification was based on initial fieldwork in 2001, while the final classification was based on surveys carried out in May 2002 and January 2004. From the classified images, the class for dense/mature forest was extracted for further analysis.

A change detection function, utilizing ArcGIS (ESRI), was applied on the three classified forest images to show changes that had occurred between 1984, 1993, and 2000. Within the study area of 243,600 ha, we calculated the areas covered by the following classes:

Loss of forest cover:
(a) Forest 84/--/--Forest in 1984 only (loss of forest after 1984)
(b) Forest 84/93/--Forest in 1984 and 1993 (loss of forest after 1993)

Stable forest cover:
Forest 84/93/00 Forest during 1984, 1993, and 2000

Increased forest cover:

(a) Forest --/93/00 Forest in 1993 and 2000 (increase of forest cover after 1984)
(b) Forest --/--/00 Forest in 2000 only (increase of forest cover after 1993)

### *13.2.2 Ground-Truthing of Forest Classification*

Surveys to verify forest classification were carried out in May 2002 and January 2004. We selected four areas corresponding to the three forest classes for detailed field investigations (Fig. 13.1) DEF = loss of forest cover (10,000 ha), REG = regenerating forest (15,000 ha), STF = stable forest (22,000 ha), FAD = forest protected by fady (taboo)(13,000 ha). Criteria for selection were: (1) large and contiguous representations of each of the forest classes and (2) accessibility of these areas. The four finally selected areas were mapped using a mobile GIS system, ArcPad from ESRI, run on handheld computers/GPS. In the analyses here, we focus on stable and regenerating forests in Northern and Central Androy (areas REG and STF in Fig. 13.1). The detailed analyses of forest cover loss in area DEF and stable forest in area FAD are given in Elmqvist et al. (2007) and Tengö et al. (2007). In REG and STF we used 20 × 20 m plots to verify the classes of dense stable forest and increasing

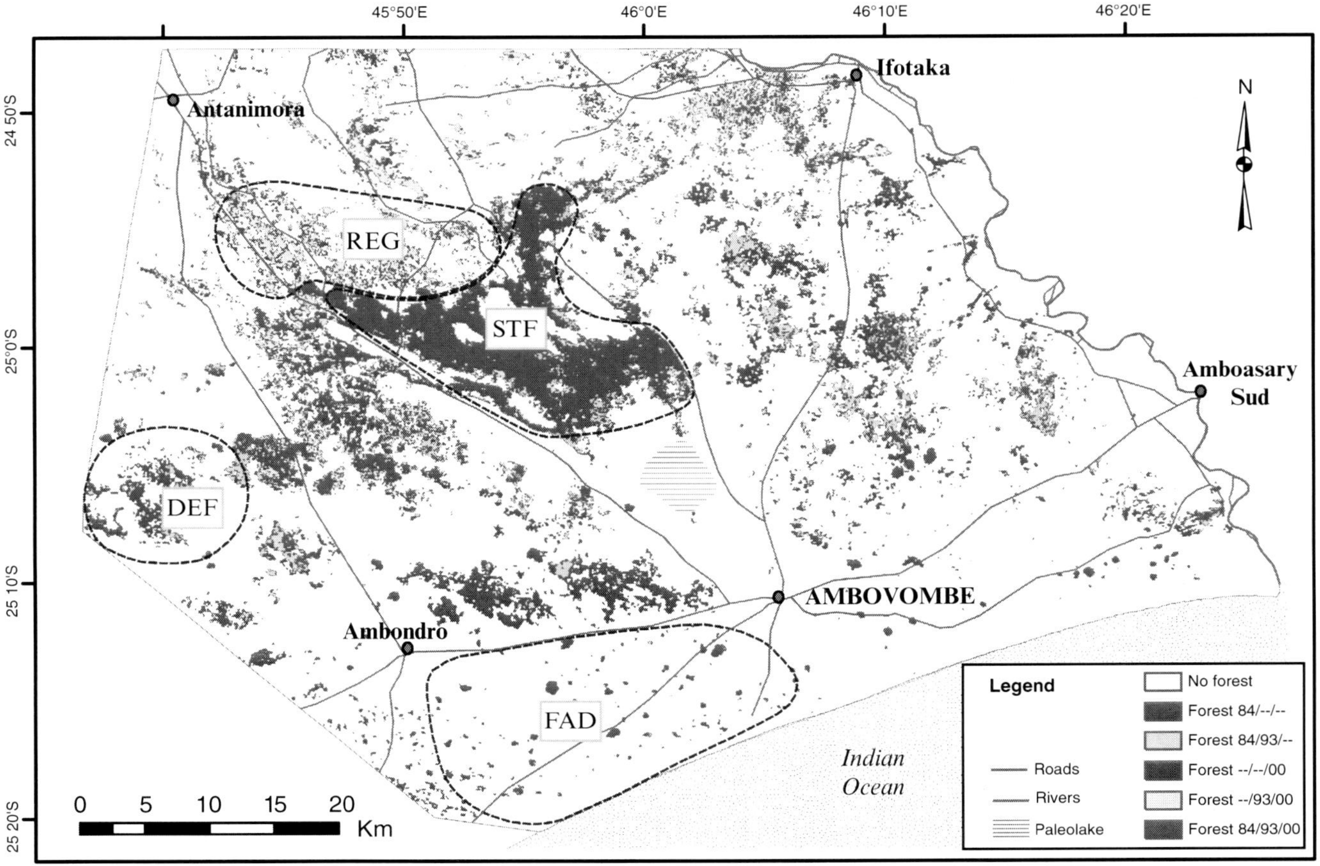


**Fig. 13.1** Time-series analysis of changes in forest cover based on satellite images (dry season synoptic views from 25 June 1984 (Landsat 5 TM), 15 April 1993 (Landsat TM) and 28 May 2000 (Landsat 7 ETM+) from Androy, southern Madagascar. Four areas were used for ecological and social surveys REG=regenerating forest, STF=stable forest, FAD=forest patches protected through taboos, DEF=deforestation (modified after Elmqvist et al. 2007) (*see Color Plates*)

forest cover. Five plots were located in the area classified as regenerating and four plots in the area classified as stable and dense forest cover. Plots were located randomly within stands of *Alluaudia procera*.

In all plots, individual trees and shrubs were identified to the species level. Each individual's height was estimated using a metered stick. Diameter at breast height (dbh = 130 cm) was measured for all individuals over 150 cm height. Tests of statistically significant differences in densities, height, and dbh were performed using non-parametric tests (Mann–Whitney *U*-test). Analyses of species richness in relation to sampling effort using the Chao index revealed that sampling efforts were adequate for the regenerating forest, but larger samples were needed in the stable forest where variation between plots was larger.

### 13.2.3 Analyses of Trends in Aridity

Trends in precipitation as a driver of forest change were analyzed using a national dataset (annual resolution, 1963–2005) from Direction de la Meteorologie Nationale et de l'Hydrologie in Madagascar. In order to capture the ecophysiological effects of variable precipitation, we analyzed trends in aridity expressed as no. of consecutive days with rainfall <1 mm. Rainfall patterns in the study area are extremely variable with locally high precipitation over short periods of time and, unfortunately, reliable local precipitation data were unavailable for the time period 1963–2005. We compiled available data on severe arid conditions in southern Madagascar using EM-DAT: The OFDA/CRED International Disaster Database www.em-dat.net.

### 13.2.4 Social Surveys

Information on local views on forest cover change, drivers, and local institutions, for example the rules-in-use including property right schemes and enforcement characteristics, was obtained through interviews in all four areas. To avoid applying preconceived ideas of local institutions, we used a qualitative interview approach described by, for example, Kvale (1996) rather than a predefined questionnaire, and held semi-structured open ended interviews with a checklist. The checklists included the following questions: (a) who has access to forest resources, (b) which rules regulate access, (c) which authority is responsible for rule enforcement, and (d) to what extent are the rules actually followed and enforced. Further, we also discussed with all informants their view on forest cover changes, drivers of change and how people have responded to temporary drought conditions.

Informants included the forest officials active in Ambovombe and Antanimora, and key informants as well as other villagers in the four areas. Key informants were persons with authority in relation to forest resources at the village level, either as representatives of the official local government, *fanjakana*, village presidents and counselors, or of customary authority, the *fokonolona*, village elders and clan leaders. These were all men of various ages. Official representatives were generally younger than persons representing the customary authority.

To supplement and triangulate their opinion, interviews were also held with women and younger persons in the villages. In total, 26 informants were interviewed in 10 local communities (DEF: Mareñy, Lahabe, Bemonzola $n=8$; REG: Manave and Mitsoriake $n=8$; STF: Ankilivalo and Belaza $n=3$; FAD: Amboanaivo, Ambazoa, Marolamainty $n=7$). The interviews with the forest officials had a broader geographical focus compared to the community interviews, but the foresters were also asked site specific questions for triangulation of the interview information from the local key informants.

## 13.3 Ecological and Social Dimensions of Forest Cover Change

### *13.3.1 The Ecological Dimension*

The overall results of the remote sensing analysis are shown in Fig. 13.1. Overall there was a total decrease of 7% of forest cover between 1984 and 2000 but during 1993–2000 forest cover increased by 4%. The REG area had a high density of juvenile (<2 m height) *A. procera* and *Cedrelopsis grevei* (Ptaeroxylaceae) and significantly higher densities than in the STF area (Mann–Whitney *U*-test, $P<0.02$, Fig. 13.2).

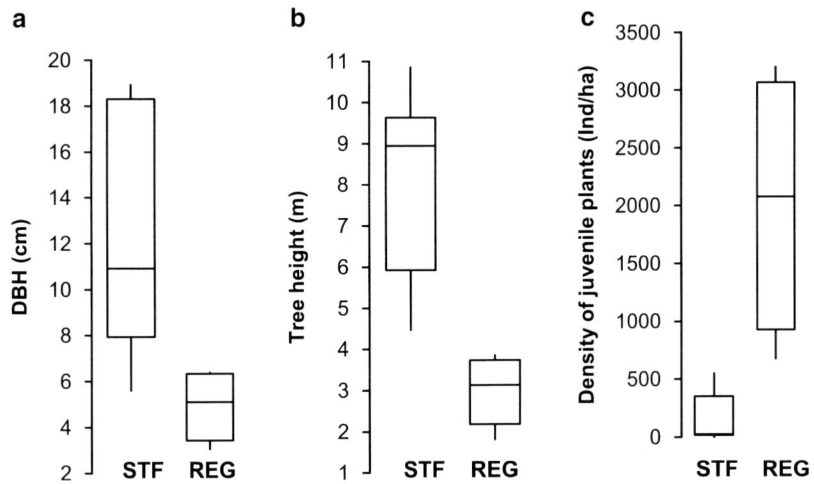


**Fig. 13.2** Analyses of (**a**) DBH, (**b**) tree height, and (**c**) density of juvenile plants in plots (20×20 m) in REG=regenerating forest ($n=5$ plots) and STF=stable forest cover ($n=4$ plots). Box plots – 95% confidence interval

Mean height and diameter (dbh) of *A. procera* was significantly larger in STF than in the REG area (Mann–Whitney *U*-test, $P < 0.02$ and $P < 0.05$ respectively) (Fig. 13.2). Species number of woody plants >1 m height were similar in the two areas but variation in the regeneration forests was less (range 17–24, $n=5$ plots) compared to the forest with stable cover (range 8–19, $n=4$ plots). In the DEF area, deforestation seems to have occurred mostly in the period 1984–1993 with 1,177 ha lost and only 253 ha in the period 1994–2000 (Fig. 13.3). The pattern of increase in forest cover in REG was a more continuous process throughout the whole period (Fig. 13.3).

Previous theoretical analyses of arid and semi-arid ecosystems and degradation have emphasized that human population growth and over-grazing have led to a process of degradation, moving these ecosystems away from a natural single equilibrium state (e.g. Swift 1996), as a consequence of dry forests having a low regeneration potential (e.g. Green and Sussman 1990). This view has been increasingly challenged and recent studies have emphasized that semi-arid systems exhibit large spatial and temporal variations and are far better described in terms of non-equilibrium systems (e.g. Leach et al. 1999; Holmgren and Scheffer 2001) with different alternative states (Holmgren and Scheffer 2001). Rather than gradual responses to changing conditions, semi-arid systems may experience sudden transitions from one state to another triggered for instance by management and climatic conditions (Westoby et al. 1989; Holmgren and Scheffer 2001). Rapid recovery of vegetation in semi-arid areas, as observed in this study, has been observed elsewhere when grazing pressure falls below a low critical value (Holmgren and Scheffer 2001).

In Androy, grazing pressure has decreased during the last decades due to droughts (Casse et al. 2004). Since the 1960s there is a trend of increased aridity in Madagascar (Fig. 13.4). In semi-arid Androy, aridity is likely to be even more pronounced and the region has experienced declining precipitation since the 1970s and recurrent drought conditions which have almost become chronic (Casse et al. 2004). Severe droughts in southern Madagascar have been reported in 1981, 1988–1992, 2000, and 2003. The one in 1981 affected one million people, while in 1992, 950,000 people and in 2000, 230,000 people were affected (EM-DAT: The OFDA/CRED International Disaster Database www.em-dat.net) (Fig. 13.4).

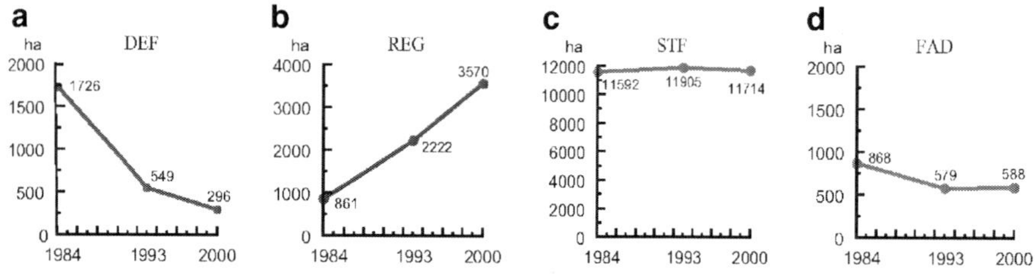


**Fig. 13.3** Changes in forest cover (ha) in the four surveyed areas (**a**) DEF=loss of forest cover, (**b**) REG-increase in forest cover, (**c**) STF-stable and (**d**) FAD –stable forest cover. Based on analyses of Landsat images from 1984, 1993, and 2000 and ground-truthing (modified after Elmqvist et al. 2007)

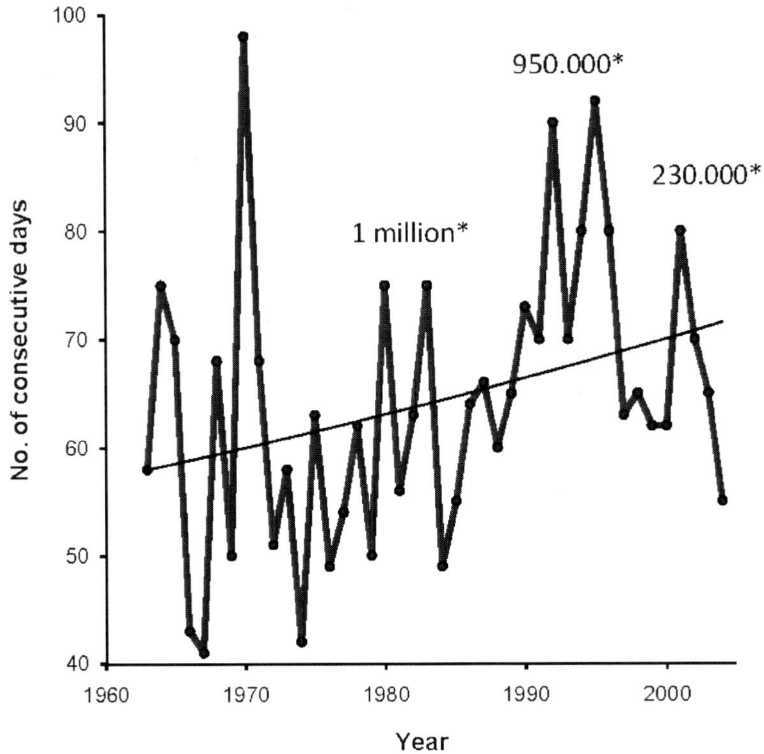


**Fig. 13.4** Number of consecutive days/year with precipitation < 1 mm 1963–2005. National data from Direction de la Meteorologie Nationale et de l'Hydrologie in Madagascar. Smoothed trend-line. * denote numbers of people affected by severe drought in southern Madagascar according to EM-DAT: The OFDA/CRED International Disaster Database www.em-dat.net

Frequent droughts have resulted in: (1) decline in grazing intensity due to direct reductions in livestock numbers (cf. Raharison 1997) and (2) farming becoming a less reliable source of livelihood. As a consequence, migration to areas and urban centers outside Androy has increased during the last decades (Casse et al. 2004). The resulting decrease in pressure on vegetation from grazing cattle and from people harvesting wood, has provided favourable conditions for vegetation regrowth. Large scale regeneration of forests has also in other areas been found to be related to migration and abandonment of land (e.g. Pascarella et al. 1999; Grau et al. 2003, see also Chapter 3 in this volume).

The dominant tree species in the regenerating stands, *A. procera* and *C. grevei* have small drought resistant wind-dispersed seeds, typical of tropical dry forests. Vieira and Scariot (2006) suggested that this may often result in a high regeneration potential, since such species are less affected by fragmentation. Nonetheless, the majority of regenerating stands were located adjacent to mature stable forests

(Fig. 13.1) but it is notable that the regenerating forest was as species rich, or richer, than the mature forests. The latter suggests that the regeneration process was not severely limited by seed dispersal.

Large-scale regeneration of tropical forests has been argued to be greatly overlooked (Grainger 2008), but represents an important economic potential as well as a potential for conservation of biodiversity (e.g. Chazdon 1998; Bawa and Siedler 1998) and carbon sink services (Ramirez et al. 2002). Vieira and Scariot (2006) concluded in their review that contrary to the common perceptions about dry forests there is a high potential for regeneration due to: (i) high proportion of small seeded wind-dispersed species, (ii) high ability of sprouting after disturbance and (iii) relatively simple community diversity and structure. Whether or not an area will regenerate may perhaps therefore, in many areas, be less a question of ecological constraints, and instead determined by social variables.

### 13.3.2 The Social Dimension

The four areas are within 50 km of each other but still display considerable difference in terms of rules-in-use. The DEF area, characterized by loss of forest cover was found to have insecure property rights and failing rule enforcement whereas the REG, STF and FAD areas with stable or regenerating forest cover all were found to have well defined property rights (Table 13.1). Across all four areas, the formal national rules regarding land clearing or tree cutting were not efficiently enforced. This was repeatedly confirmed by the forest officers, stating lack of funding as the main reason. The most strict and well reinforced rule was the informal taboo against harvesting in certain portions of forest in all four areas, which applied to all forest patches in southern Androy (see Table 13.1).

In the DEF area in western Androy, local informants and the forest officer in charge described a situation where the transition from forest to cleared or partly cleared land had occurred as a response to insecure property rights. Customary rules were not enforced as the *tompon-tany,* the land owners by ancestral laws, had migrated from the area during early nineteenth century (Heurtebize 1986). The present inhabitants settled around the 1950s, and in an analysis of aerial photos from this period, the forest was found to as regenerate in abandoned fields. Following the more limited governmental forest control after independence 1960 (see Duffy 2006), forest clearing was used during the 1980s to secure individual land rights as the collective rights enforced by customary authority was no longer in place. Several recent clearings and signs of extensive cutting were observed in May 2003 and January 2004.

In REG and STF in northern and central Androy, settlements are currently small and scattered. The forest is utilized as a seasonal resource for cattle herding by people from the more densely populated south and southwest, and informants claim well defined property rights (Table 13.1). For example, seasonal dwellers need to pay respect to the *tompon-tany*, and generally do so and the procedure is well

**Table 13.1** Population density, social institutions and land use in the four surveyed areas (modified after Elmqvist et al. 2007)

| Type of forest cover change | Area surveyed ($km^2$) | Ind./$km^2$ [a] | Institutional characteristics | Social–ecological interactions – effects on forest cover |
|---|---|---|---|---|
| Regeneration (REG), Northern Androy | 150 | <10 | Well defined customary land rights. Limited capacity for monitoring and enforcement due to low population density | Decline in grazing pressure due to permanent and temporal outmigration of people and decline in seasonal grazing in the area. Decline in land clearings for agriculture |
| Stable forest (STF), Central Androy | 220 | <10 | Well defined customary land rights. Limited capacity for monitoring and enforcement due to low population density | Decline in grazing pressure due to permanent and temporal outmigration of people and decline in seasonal grazing in the area |
| Loss of forest cover (DEF), Western Androy | 100 | 20 | Neither formal nor customary tenure enforced. Rules for forest access and utilization | Open access conditions leading to land clearings to acquire land coinciding with increasing aridity and recurrent droughts |
| Stable forest cover (FAD), Southern Androy | 130 | >150 | Well defined customary land rights, strengthened by taboos preventing forest resource extraction, that are very well monitored and enforced | Changes in seasonal migration with cattle (longer duration, more distant) |

[a] *Source*: LandScan 2001 Global Population Database (Oakridge, TN: Oak Ridge National Laboratory http://www.ornl.gov/gist/)

recognized by the formal village authority. However, the informants claimed rule enforcement was challenging as the area is vast and scarcely populated, and potentially some undetected logging could occur. Both of the dominant species in the regeneration area, *A. alluaudia* and *C. grevei*, have a high economic value mainly as source for charcoal production. There is therefore a high risk of future loss of forest cover if the current social institutions fail in rule enforcement, and outsiders start logging in the area.

### 13.3.3 The Social–Ecological Dimension

This volume clearly shows that insights into the social and ecological context in which reforestation may occur are required to enhance forest management and human livelihoods. We found in this study that local communities and the existence of social capital, expressed as shared norms and rules within communities, can contribute to favorable conditions for forest regeneration. Human presence is often routinely seen by external observers as a threat to forest ecosystems and biodiversity. While this may be true in a general sense, the view that separates people and ecosystems provides no constructive means to move forward and also overlooks essential dynamics that underlies the patterns and trends of biodiversity and resource use that emerge in a landscape.

Patterns of forest loss and regeneration have long been shaped by human activities in Madagascar (Kull 2004). In the south, Dawson and Ingram (2008) demonstrate that environmental change in south-eastern Madagascar is highly variable in space and time, and that cover, quality, and composition of the littoral forest is related to human activities such as timber harvest, coppicing for fuel wood, and harvest of medicinal plants and indirect drivers such as population density. Kaufmann and Tsirahamba (2006) provide a detailed account of how the Mahafaly pastoralists view the dry forests of the southwest as essential reserves for browse and fodder in times of drought. The herders open up areas of pasture in the forest by grazing animals, and plant and encourage species that provide sources of water and fodder for cattle, such as the endemic *Euphorbia stenoclada*.

We find that a key factor driving changes in forest cover patterns and dynamics are human responses to social and ecological drivers, as mediated by local institutions. Corroborating this, several authors describe spiritual values assigned to forests by local people in Madagascar, values that in some instances have been formally recognized as an asset for conservation (Horning 2003b; McConnell and Sweeney 2005; Schachenmann 2006; Tengö et al. 2007). For example, the detailed study of the patchy forest landscape in southern Androy reveal that the scattered patches are created and maintained by cultural practices and well enforced religious taboos (Tengö et al. 2007). Thus, we argue that forest ecosystems where human presence is significant are best described and understood as social–ecological systems, where social and ecological processes interact and reinforce each other, creating situations favoring forest loss, stable cover or regeneration (Berkes and Folke 1998).

A key insight from the studies mentioned above is that not only do we need to see human activities as intrinsic to ecosystem dynamics; we also need to acknowledge that they are not always detrimental for goals such as high biodiversity or capacity of forest ecosystems to regenerate. Some practices may enhance the capacity of an ecosystem to generate services that human rely upon (Tengö M 2004) while enhancing diversity and building resilience in the social–ecological system (Berkes and Folke 1998; Berkes et al. 2003). Fairhead and Leach (1996) showed how farmers' everyday activities in the forest-savanna transition zone of West Africa increased the density of woody vegetation and positively influenced the formation of forests in fallow and savanna areas. By controlling fire, and influencing soil fertility and tree regeneration, farmers managed to move the system across thresholds between vegetation states of grass or woody species dominance.

The taboo forests in southern as well as other parts of Androy, ranging in size up to ~ 100 ha, are often old growth forests that may serve as important propagule sources for dispersal and colonization of the surrounding areas (cf. Chazdon 2003). Even though our data cannot show a clear linkage between the areas of forests protected by taboos and forest regeneration, these areas are likely to add to the capacity of the system to regenerate through providing (a) propagule sources for dispersal and colonization of the other areas, and (b) undisturbed habitat for "mobile links" (Lundberg and Moberg 2003) such as lemurs (i.e. *Lemur catta)* and forest birds that act as vectors for seed dispersal (see Bodin et al. 2006). Dawson and Ingram (2008) show empirically that harvest intensity of forest products in southeastern Madagascar are not necessarily associated with lower diversity or conservation values. The adaptive and seasonal use of forest resources by the pastoralist Mahafaly west of Androy (Kaufmann 2008) may create opportunity for regeneration, and introduces small scale disturbances that add to the heterogeneity of the forest, thus increasing resilience (cf. Bengtsson et al. 2003).

The relationship between human presence and density, and biodiversity and potential for regeneration is not necessarily negative but clearly complex. In general, change may be driven by external factors but manifests itself locally and the scale of agency – the direct causation of action – is often intrinsically localized (Wilbanks and Kates 1999). As a consequence, processes that are of a similar type can lead to different outcomes in areas separated by only short geographical distances. In Androy, human responses to increasing aridity had very different consequences. In the north, outmigration, decreased grazing pressure and stability of property regimes, contributed to conditions for forest regeneration. However, in Western Androy, previous migration and abandonment contributed to a situation with insecure property rights, insufficient monitoring and enforcement of local rules, and degrading forest (cf. Sheperd 1992; Seddon et al. 2000; Horning 2003b; Casse et al. 2004). There is thus an urgent need for a greater understanding of the positive and negative feedback interactions in an interlinked social–ecological system.

The role of social institutions in mediating human responses to change has proven to be a useful framework for understanding the outcome of social–ecological interactions and the consequences for forest ecosystems and the services they generate (Gibson et al. 2000; Berkes and Folke 1998; Lambin et al. 2001).

Corresponding to the findings of this study, Nagendra (2007) reports from an analysis of 55 forests in Nepal, that type of land tenure and degree of monitoring were the most important factors for understanding forest regeneration. In particular, monitoring and rule enforcement have been identified as key aspects of institutions for sustainable resource use (Ostrom 1990; Gibson et al. 2005). Hayes (2006) recently argued that rather than formal legal definitions of forest protection (e.g. parks), it was the informal rules acknowledged and made by forest users, that influenced forest conditions in both parks and formally unprotected forests. More research is urgently needed regarding how legitimate norms and rules develop in relation to local role of local ecological knowledge and understanding, how such rules may be adaptive and responsive to ecosystem dynamics, and in what contexts they create conditions beneficial for forest regeneration (cf. Sánchez-Azofeifa et al. 2005).

In conclusion, we want to emphasize that interpretation of patterns of forest regeneration need to include simultaneous studies of changing environmental factors and existing and changing characteristics of social institutions. Our study points to the large, but often neglected, capacity of a semi-arid system to spontaneously regenerate given a window triggered by changing precipitation and grazing pressure. The outcome however, is critically dependent on functioning social institutions that develop along with an understanding of the local ecosystem.

**Acknowledgments** We thank Chantal Radimilahy, Elisabeth Rabakonandrianina, Paul Sinclair and Jean-Aimé Rakotoarisoa for help, advice and support throughout this study. We thank Jean-Aimé Andriamaherilala for fieldwork guidance and translation. We also thank Mark Fenn, WWF Madagascar, and the villagers in Manave, Mitsoriake, Mareñy, Lahabe and Amboanaivo for assistance during fieldwork. The study was supported by grants from Sida/Sarec and Formas. The analyses presented here builds on and is an extension of analyses published by the authors in PLoSONE (Elmqvist et al. 2007)

## References

Achard FHD, Eva HJ, Stibig P et al (2002) Determination of deforestation rates of the world's humid tropical forests. Science 297:999–1002

Antona M, Motte E, Salles J-M et al (2002) Property rights transfer in Madagascar biodiversity policies. Paper presented at BioEcon Rome meeting, Rome, 30–31 May 2002

Battistini R, Richard-Vindard G (1972) Biogeography and ecology in Madagascar. Dr. W Junk B.V. Publishers, Hague

Bawa KS, Siedler R (1998) Natural forest management and conservation of tropical biodiversity. Conserv Biol 12:46–55

Bodin Ö, Tengö M, Norman A et al (2006) The value of small size: loss of forest patches and thresholds in southern Madagascar. Ecol Appl 16:440–451

Bengtsson J, Angelstam P, Elmqvist T et al (2003) Reserves, resilience, and dynamic landscapes. Ambio 32:389–396

Berkes F, Colding J, Folke C, eds. (2003) Navigating Social-Ecological Systems: Building Resilience for Complexity and Change. Cambridge: Cambridge University Press

Berkes F, Folke C, eds. (1998) Linking Social and Ecological Systems. Management Practices and Social Mechanisms for Building Resilience. Cambridge: Cambridge University Press.
Casse T, Milhoj A, Ranaivoson S et al (2004) Causes of deforestation in southwestern Madagascar: what do we know? Forest Policy Econ 6:33–48
Chazdon RL (1998) Tropical forests – log 'em or leave 'em? Science 281:1295–1296
Chazdon RL (2003) Tropical forest recovery: legacies of human impact and natural disturbances. Perspect Plant Evol Syst 6:51–71
Dewar RE, Wallis JR (1999) Geographical patterning of interannual rainfall variability in the tropics and near tropics: an L-moments approach. J Clim 12:3457–3466
Dietz T, Ostrom E, Stern PC (2003) The struggle to govern the commons. Science 302:1907–1912
Dunn RR (2004) Recovery of faunal communities. Conservation Biology 18:302–309.
Du Puy D, Moat J (1996) A refined classification of the primary vegetation in Madagascar based on the underlying geology: using GIS to map its distribution and to assess its conservation status. In: Lourenco WR (ed) Biogéographe de Madagascar. Editions ORSTROM, Paris
Duffy R (2006) Non-governmental organisations and governance states: The impact of transnational environmental management networks in Madagascar. Environ Polit 15:731–749
Elmqvist T (2004) The forgotten dry forest of southern Madagascar. Plant Talk 35:29–31
Elmqvist T, Pyykönen M, Tengö M et al (2007) Patterns of loss and regeneration of tropical dry forest in Madagascar: The social institutional context. PLoS ONE 2(5):e402. doi:10.1371/journal.pone.0000402
Fairhead J, Leach M (1996) Misreading the African landscape; Society and ecology in a forest-savanna mosaic. Cambridge University Press, Cambridge
Fenn M (2003) The spiny forest ecoregion. In: Goodman SM, Benstead JP (eds) The natural history of Madagascar. University of Chicago Press, Chicago, IL
Gezon L (2006) Global visions, local landscapes. A political ecology of conservation, conflict, and control in Northern Madagascar. Altamira Press, Lanham, MD
Gibson C, McKean MA, Ostrom E (2000) People and forests. Communities, institutions and governance. The MIT Press, Cambridge, MA
Gibson CC, Williams JT, Ostrom E (2005) Local enforcement and better forests. World Development 33:273–284
Grainger A (2008) Difficulties in tracking the long-term global trend in tropical forest area. Proc Natl Acad Sci U S A, 10.1073/pnas.0703015105
Grau HR, AideTM ZJK et al (2003) The ecological consequences of socioeconomic and land-use changes in postagriculture Puerto Rico. Bioscience 53:1159–1168
Green GM, Sussman RW (1990) Deforestation history of eastern rain forests of Madagascar from satellite images. Science 248:212–215
Hanna S, Folke C, Mäler K-G (1996) Rights to nature. Island Press, Washington, DC
Hayes TM (2006) Parks, people, and forest protection: an institutional assessment of the effectiveness of protected areas. World Dev 34:2064–2075
Heurtebize G (1986) Histoire des Afomarolahy (Extrême-Sud de Madagascar). CNRS, Paris
Holmgren M, Scheffer M (2001) El Niño as a window of opportunity for the restoration of degraded arid ecosystems. Ecosystems 4:151–159
Horning NR (2003a) How rules affect conservation outcomes. In: Goodman SM, Benstead JP (eds) The natural history of Madagascar. Chicago University Press, Chicago, IL, pp 146–153
Horning NR (2003b) The cost of ignoring rules: how Madagascar's biodiversity and rural livelihoods have suffered from institutional shortcomings. Paper presented at The International Conference on Rural Livelihoods, Forests and Biodiversity, Bonn, Germany, 19–23 May 2003
Horning NR (2008) Madagascar's biodiversity conservation challenge: from local- to national-level dynamics. Environ Sci 5:109–128
Kaufmann JC (2008) The non-modern constitution of famines in Madagascar's spiny forests: "water-food" plants, cattle and Mahafale landscape praxis. Environ Sci 5:73–89
Kaufmann JC, Tsirahamba S (2006) Forests and thorns: conditions of change affecting Mahafale pastoralists in Southwestern Madagascar. Conserv Soc 4:231–261

Koechlin J (1972) Flora and vegetation of Madagascar. In: Battistini R, Richard-Vindard G (eds) Biogeography and ecology in Madagascar. Dr. W Junk B.V. Publishers, Hague

Kull CA (2004) Isle of fire. The political ecology of landscape burning in Madagascar. The University of Chicago Press, Chicago, IL

Kvale S (1996) Interviews: an introduction to qualitative research interviewing. Sage, Thousand Oaks, CA

Lambin E, Geist HJ, Lepers E (2003) Dynamics of land-use and land-cover change in tropical regions. Annu Rev Environ Res 28:205–241

Lambin EF, Turner BL, Geist HJ et al (2001) The causes of land-use and land-cover change: moving beyond the myths. Global Environ Change-Hum Policy Dimens 11:261–269

Leach M, Mearns R, Scoones I (1999) Environmental entitlements: dynamics and institutions in community-based natural resource management. World Dev 27:225–247

Lingard M, Raharison N, Rabakonandrianina E et al (2003) The role of local taboos in conservation and management of species: the Sokake or radiated tortoise (*Geochelone radiata*) in southern Madagascar. Conserv Soc 1:223–246

Lundberg J, Moberg F (2003) Mobile link organisms and ecosystem functioning: implications for ecosystem resilience and management. Ecosystems 6:87–98

McConnell WJ, Sweeney SP (2005) Challenges of forest governance in Madagascar. Geogr J 171:223–238

Nagendra H (2007) Drivers of reforestation in human-dominated forests. PNAS 104:15218–15223

Olson DM, Dinerstein E (2002) The global 200: priority ecoregions for global conservation. Ann Missouri Bot Gard 89:199–224

Ostrom E (1990) Governing the commons. The evolution of institutions for collective action. Cambridge University Press, Cambridge

Pascarella JB, Aide TM, Serrano MI et al (1999) Land use history and forest regeneration in the Cayey mountains, Puerto Rico. Ecosystems 3:217–228

Pretty J (2003) Social capital and the collective management of resources. Science 302:1912–1914

Rabesandratana R (1984) Flora of the Malagasy southwest. In: Jolly A, Oberlé P, Albignac R (eds) Madagascar. Pergamon Press, Oxford, pp 55–74

Raharison L-JR (1997) Effets et conséquences sur le régime des aquifères des anomalies climatiques dans l'extreme sud de Madagascar: cas du bassin endoréique d'Ambovombe. Cachiers Sécheresse 8:129–137

Ramirez OA, Carpio CE, Ortiz R (2002) Economic value of the carbon sink services of tropical secondary forests and its management implications. Environ Res Econ 21:23–46

Richard AF, Dewar RE, Schwartz M et al (2002) Life in the slow lane? Demography and life histories of male and female sifaka (Propithecus verreauxi verreauxi). J Zool 256:421–436

Sánchez-Azofeifa G et al (2005) Research priorities for neotropical dry forests. Biotropica 37:477–485

Schachenmann P (2006) Spiritual values in Madagascar – the starting point for endogenous conservation initiatives. Mount Res Dev 26:323–327

Seddon N, Tobias J, Yount JW et al (2000) Conservation issues and priorities in the Mikea forest of south-west Madagascar. Oryx 34:287–304

Sheperd G (1992) Managing Africa's tropical dry forests. A review of indigenous methods. Overseas Development Institute, London

Sussman RW, Rakotozafy A (1994) Plant diversity and structural-analysis of a tropical dry forest in Southwestern Madagascar. Biotropica 26:241–254

Sussman RW, Green GM, Sussman LK (1994) Satellite imagery, human-ecology, anthropology, and deforestation in Madagascar. Hum Ecol 22:333–354

Sussman RW, Green GM, Porton I et al (2003) A survey of the habitat of Lemur catta in Southwestern and Southern Madagascar. Primate Conserv 19:32–57

Swift J (1996) Desertification, narratives, winners and losers. In: Leach M, Mearns R (eds) The lie of the land: challenging received wisdom in African environmental change. James Currey/International African Institute, London, pp 73–90

Tengö M (2004) Management practices for dealing with uncertainty and change-Social-ecological systems in Tanzania and Madagascar [PhD-thesis]. Stockholm, Sweden: Stockholm University

Tengö M, Johansson K, Rakotondrasoa F et al (2007) Taboos and forest governance: informal protection of hot spot dry forest in southern Madagascar. Ambio 36:683–691

Tucker CM, Munroe DK, Nagendra H et al (2005) Comparative spatial analyses of forest conservation and change in Honduras and Guatemala. Conserv Soc 3:174–200

Vieira D, Scariot A (2006) Principles of natural regeneration of tropical dry forests for restoration. Restor Ecol 14:11–20

Westoby M, Walker B, Noy-Meir I (1989) Opportunistic management for rangelands not at equilibrium. J Range Manag 42:266–274

Wilbanks TJ, Kates RW (1999) Global change in local places: how scale matters. Clim Change 43:601–628

# Chapter 14
# Forest Transition in Vietnam and Bhutan: Causes and Environmental Impacts

**Patrick Meyfroidt and Eric F. Lambin**

## 14.1 Introduction

Prospects for a forest transition in Tropical Asia would have wide implications for global environmental change. Data from the latest Forest Resources Assessment (FAO 2006) showed a net increase in forest cover between 1990 and 2005 in four Asian countries: Bhutan, India, China, and Vietnam (Kauppi et al. 2006; Mather 2007, Chapters 7 and 15). Vietnam had been severely deforested during the previous

P. Meyfroidt (✉) and E.F. Lambin
Department of Geography, University of Louvain, Louvain-La-Neuve, Belgium
e-mail: patrick.meyfroidt@uclouvain.be

H. Nagendra and J. Southworth (eds.), *Reforesting Landscapes: Linking Pattern and Process*, Landscape Series 10,
DOI 10.1007/978-1-4020-9656-3_14, 

decades (FAO 1993). Since the mid-1990s, however, the deforestation trend has been largely reversed and forest cover has increased notably, although not everywhere in the country.

This reforestation was accompanied by political and economic changes in favour of decentralisation and liberalisation (i.e., the Doi Moi reforms initiated in the 1980s, as a response to the economic stagnation of the country). These changes induced rapid economic growth and the development of the industrial and service sectors, and also strongly affected the agricultural and forestry sectors (Kerkvliet and Porter 1995; Pingali et al. 1997). Forestry policies and reforestation programs were also the government's response to the increasing scarcity of forests. Farmers also responded to land degradation and scarcity by changing their agricultural practices.

In this chapter we describe the forest transition that took place in Vietnam. We then analyse it by reference to the main pathways of reforestation proposed in the emerging theory of forest transition (Rudel et al. 2005, Chapter 3). We also evaluate the impacts of this forest transition on two of the main environmental attributes of forests: biomass carbon stock, and changes in habitat quality and fragmentation. Finally, we present more briefly what is known on the forest transition in Bhutan.

## 14.2 Methods

### *14.2.1 Study Areas*

#### 14.2.1.1 Vietnam

Vietnam is characterised by a large range of biophysical conditions due to its longitudinal extension, topography and location at the southeastern edge of mainland Asia (Sterling et al. 2006). It is therefore divided in highly diverse ecoregions (Wikramanayake et al. 2001). One fourth of the country lies below 20 m of elevation, corresponding mainly to the deltas of the two major rivers, the Mekong and the Red River (Fig. 14.1). These two regions are densely populated and mostly covered by agriculture. Another fourth of the country lies above 625 m, while the remaining half is covered by hills, lower slopes, and plateaus.

The northern mountains – the largest mountainous area – are characterized by a relatively colder and more seasonal climate than the rest of Vietnam. This region marks the transition between the northern distributional limit of tropical plants and the southern distributional limit of subtropical and temperate ones. The dominant formation is evergreen forests of broad-leaved and coniferous trees, with pockets of semi-evergreen forests and specific communities of limestone forests. The western part – the Northern Indochina subtropical forests ecoregion – has a more pronounced dry season due to a winter westerly wind. The cool winter temperatures and higher elevation bring a montane flora with a noticeable Himalayan component. In the

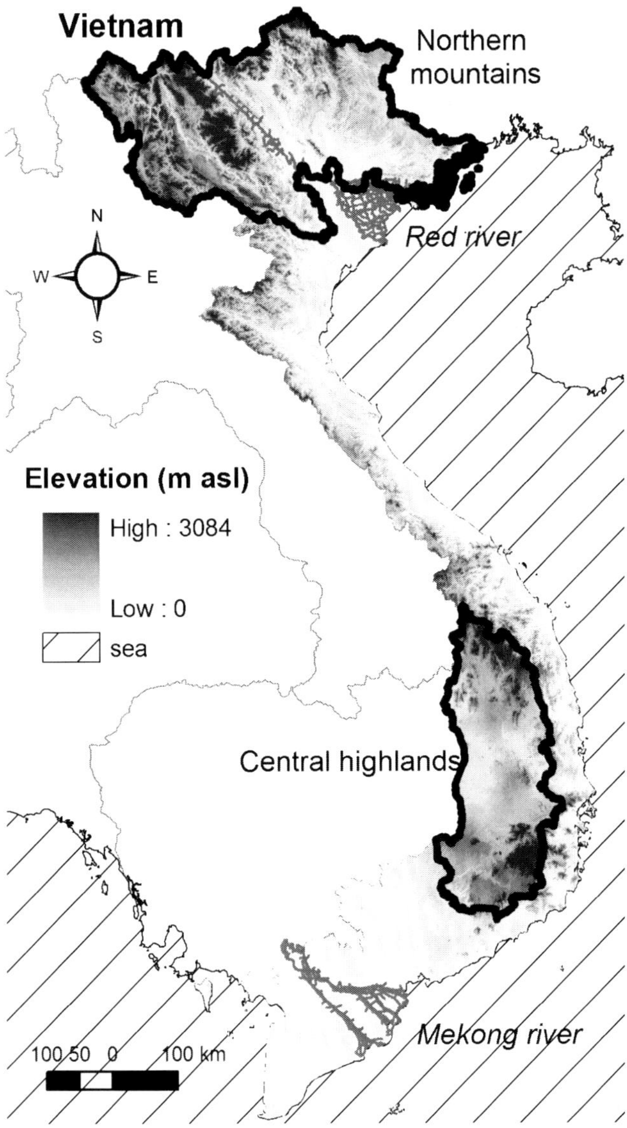


**Fig. 14.1** The main geographical features of Vietnam
(*Source*: authors)

eastern part – the South China–Vietnam subtropical evergreen forests ecoregion – there are fewer months without rainfall, with sometimes discontinuous rainfall all year round. This region is dominated by mountains of lower elevation. Karstic and coniferous forests are dominant in this region.

The Annamite Mountains (or Truong Son Range) are a string of mountains extending along the central coast of Vietnam. In the southwest, these mountains extend into a series of high plateaus composing a region usually called the central highlands. The Annamite Mountains and central highlands have a generally wetter and hotter climate than the north, especially as one moves southward. The forests are mainly evergreen with tropical families, although temperate families begin to dominate as elevation increases. Coniferous trees are an important component of these forests. In the south, lower elevations and especially eastern slopes in the rain shadow of the mountains are drier and characterised by semi-evergreen or deciduous forests – the Southeastern Indochina dry evergreen forests ecoregion. The central region also comprises, at the border with Cambodia, a dry plain of deciduous dipterocarp forests – the Central Indochina dry forests ecoregion.

After decades of war, Vietnam was officially reunified under socialist rule in 1976. A socialist economic model, including collectivised farming, was implemented in the south of Vietnam as it had been already the case in the north. However, from the 1980s, the economic and food crisis, and the spread of an informal economy triggered reforms towards liberalisation, de-collectivisation and decentralisation (Kerkvliet 1995; Irvin 1995).

This study covers roughly the whole country, although some data and statements apply primarily to the main forested regions – that is the upland regions described above.

#### 14.2.1.2 Bhutan

Bhutan is a Buddhist kingdom that lies in the eastern Himalayas, at an altitude ranging from 200 m to almost 8,000 m. The country is dominated in the north by the Himalayas. Most of the population is concentrated in isolated valleys in the highlands. This region is characterised by many rivers and large forests. The extreme south of the country consists mostly of tropical plains where agricultural land is concentrated. Centuries of isolationism, the Buddhist culture, a small population, and the rugged topography have helped Bhutan to preserve rather intact ecosystems. The country harbors a very high richness of species per unit area. Bhutan has enjoyed political stability over the long term. Government policies have pursued sustainable development objectives for several decades. Bhutan was only opened to foreigners in the 1970s, and to technologies allowing communication with the outside world even more recently.

### *14.2.2 Data and Methods*

#### 14.2.2.1 Vietnam

Meyfroidt and Lambin (2008b) assembled all available land cover maps of Vietnam derived by remote sensing, and homogenised their thematic content and spatial details. These maps were compared with the official forest cover maps from the government of Vietnam for the early and late 1990s. The spatial patterns of these maps were compared using Fuzzy Kappa, Kappa Location and other indicators (Hagen-Zanker et al. 2005). The most consistent maps were then compiled with other available statistics of forest cover to build a time series of forest cover for Vietnam for the last few decades.

The causes of the reforestation during the 1990s were assessed using statistical analyses (Meyfroidt and Lambin 2008a). Two multivariate spatial lag regressions were performed, using data at the district level over the whole country, to explain respectively the changes in natural and planted forest cover. All analyses of the pathways of forest transition (Section 14.3.3) were based on these results, supported and interpreted by a review of local case studies (Meyfroidt and Lambin 2008a).

Finally, the environmental impacts of the forest transition were assessed. Meyfroidt and Lambin (2008b) evaluated the changes in forest fragmentation for the different ecoregions of the country. The Vegetation Continuous Fields data of 1992–1993 and 2000–2001, produced by University of Maryland (DeFries et al. 2000; Hansen et al. 2003), which were among the most consistent datasets identified during the reconstruction of time series of forest cover, were used (Fig. 14.2a). These were the only pair of data produced by the same methodology, thus minimizing spurious effects on landscape indices. Ecoregion maps were adapted from Olson et al. (2001) (Fig. 14.2b).

Forest cover data from the Forest Inventory and Planning Institute of Vietnam (FIPI), available for 1980, 1992–1993, 1995, and 2005, were also among the most consistent datasets. Estimates of biomass carbon density (aboveground and belowground) for the different forest types and densities present in these data were based on the literature. Based on these parameters, Meyfroidt and Lambin (2008b) estimated the total biomass carbon stock and average values of carbon density for different years and forest categories.

#### 14.2.2.2 Bhutan

No formal data analysis was conducted for Bhutan as we only offer here a preliminary and more superficial analysis compared to Vietnam. We relied on a review of the literature and data from the successive FAO Forest Resource Assessments. A more in-depth study is forthcoming.

## 14.3 Results

### *14.3.1 The Pattern of Forest Transition in Vietnam*

The reconstructed time series confirm unambiguously that a forest transition took place in Vietnam (Fig. 14.3). Vietnam experienced several decades of deforestation at least since the 1930s, with the highest rates – averaging 1.4% per year – reached

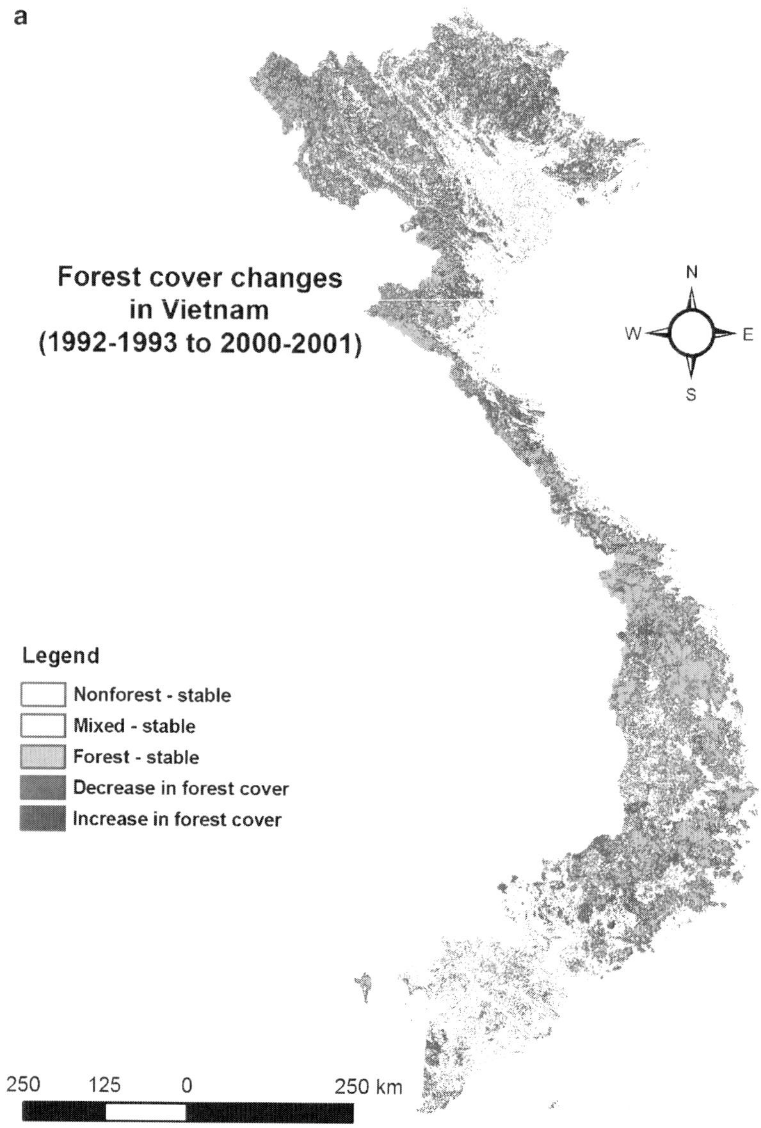


**Fig. 14.2** Forest cover changes and ecoregions in Vietnam (adapted from Meyfroidt and Lambin 2008b). (**a**) Forest cover changes were derived from the University of Maryland VCF maps. (*see Color Plates*)

during the 1970s and 1980s. The forest cover was reduced to around 25–31% of the total land area of the country in 1991–1993, the turning point of the forest transition. Deforestation was especially high in the northern mountains, where only 17% of land remained covered by forests in the early 1990s.

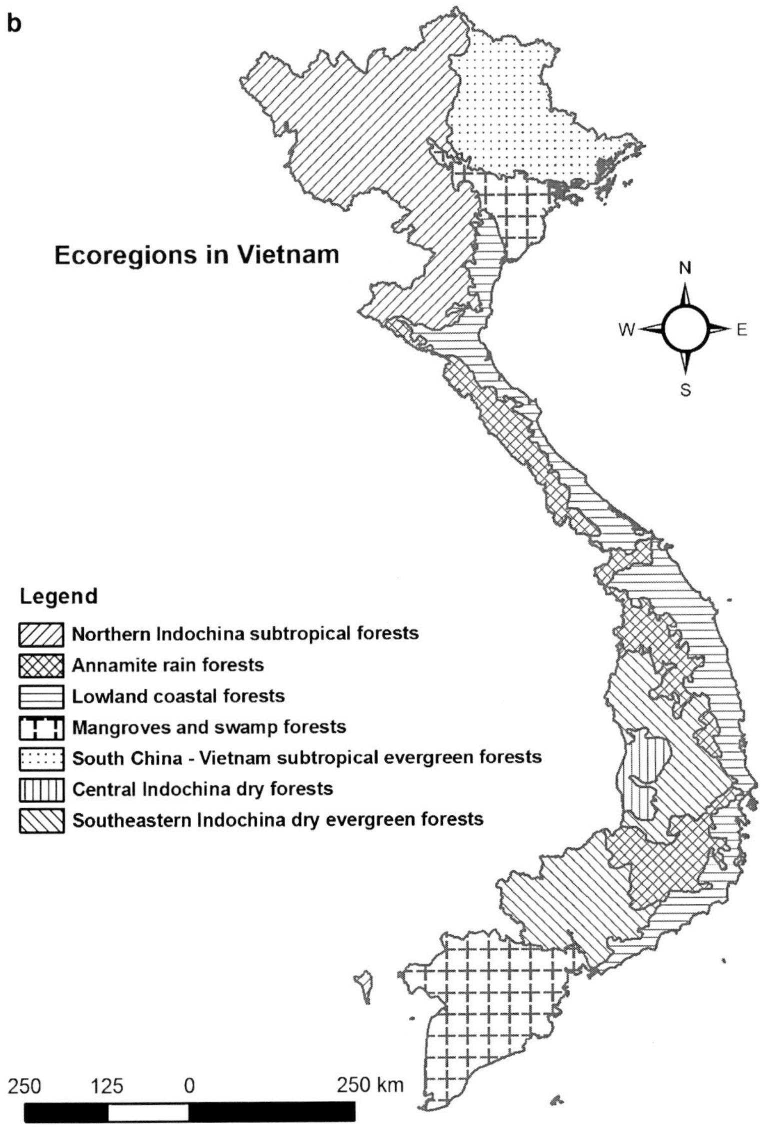


**Fig. 14.2b** Ecoregions were adapted from Olson et al. (2001)

Since then the forest cover increased at a national scale, to around 32–37% in 1999–2001 and 38% in 2005, which represents an average rate of reforestation of 3.1% per year between 1995 and 2005. This high rate of reforestation – more than twice as high as the rate of deforestation in the previous decades – is a unique feature of the forest transition in Vietnam. In other forest transitions, forest recovery

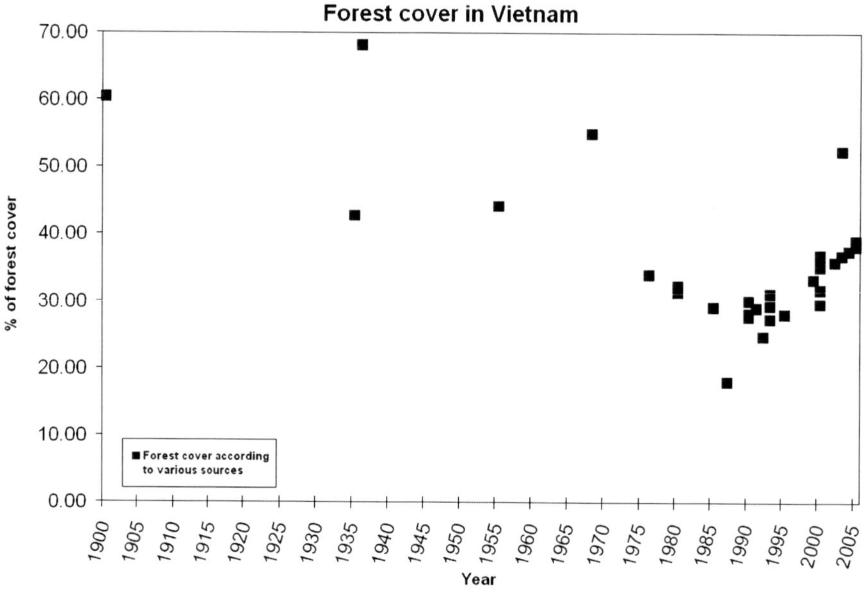


**Fig. 14.3** Forest cover in Vietnam at several dates according to different sources (adapted from Meyfroidt and Lambin 2008b)

is slower than the earlier deforestation (Mather and Needle 1998; Rudel et al. 2005) (see Satake and Rudel 2007 for a discussion of possible implications).

The forest cover at the turning point was higher than for nineteenth or early twentieth centuries forest transitions in Europe (between 0% and 25%), which corresponds to what was hypothesized by Rudel et al. (2005). This reforestation was due in similar proportions to natural forest regeneration and to planted forests. Planted forests were almost non-existent until the 1980s, and then increased continuously to reach 7–8% of the total land area in 2005.

Forest cover changes were not homogeneous geographically (Fig. 14.2). Most of the forest cover remaining in 1991–1993 was located in the Annamite mountains (along the central coast of Vietnam) and the central highlands. The regions with the largest net reforestation were the highly deforested northeastern mountains, the northern Annamite range and, to a lesser extent, the northwestern mountains, mainly due to natural regeneration. In the central highlands, reforestation occurred in some places but the region as a whole experienced a net deforestation. In the lowlands, few forests remained in the early 1990s and reforestation was small: a few scattered plantations and mangrove rehabilitation in the south.

Tree plantations were more scattered geographically than natural forest regeneration and located mainly in midlands and along coasts, where the road network is more developed. At the local scale, changes in forest cover were also not homogenous. Reforestation mostly took place on mountain slopes while valley bottoms and lower slopes were increasingly converted to permanent agriculture.

### 14.3.2 The Causes of Deforestation in Vietnam

Although there are few comprehensive studies of deforestation in Vietnam, evidence suggests that deforestation in the 1970s and 1980s was mainly caused by agricultural expansion and wood exploitation in the uplands (De Koninck 1999). These processes were driven mainly by population growth, due to natural growth and migration from the lowlands, which was encouraged by policies to relieve pressure on the already densely populated river deltas (Lundberg 2004). Timber demand for urban and industrial needs also contributed to forest clearing, especially during the post-war reconstruction of the country (McElwee 2004). Traditional shifting cultivation practiced by ethnic minorities has been and is still often blamed by the Vietnamese government for causing deforestation, but the evidence for this is not convincing (Do 1994; Fox et al. 2000).

During the 1980s, agricultural policies also caused rapid deforestation in the mountains (Castella and Dang 2002). Cooperatives had been established since 1954 in the north and 1975 in the south of the country. Paddy lands were managed by the cooperatives and, as work on these collective farms was mandatory, little labour was left available for upland shifting cultivation. The productivity in the paddy lands was low due to an inefficient management of the cooperatives – for example, the fertiliser distribution networks – and because farmers had few incentives to intensify production (Pingali et al. 1997).

In 1981, the policy known as "Decree 100" introduced land contracts between farmers and cooperatives, to increase incentives for increasing production (Kerkvliet and Porter 1995; Pingali et al. 1997). The farmers were allowed to keep surpluses above a fixed contracted quantity. However, the short-term contracts encouraged unsustainable land exploitation and the investments were still managed by the cooperatives. Furthermore, the land was distributed according to the number of mouths to feed in the household rather than according to the agricultural labour force. Some families had therefore to manage areas too large for their labour capacity. Yields in the paddy lands thus remained low and food shortages were severe in the mountains.

Throughout the 1980s, as the cooperatives were progressively dismantled and farmers were increasingly free to allocate their labour, families with little paddy lands but abundant labour turned to slash-and-burn cultivation on hillsides to maximise labour returns while growing their own food (Castella et al. 2002; Sikor and Dao 2002). During this period, forests disappeared in the mountains and hillsides were increasingly eroded. Several authors warned of the deepening environmental and development crisis in Vietnam's mountains (Jamieson et al. 1998).

During the 1990s and 2000s, deforestation was high in the central highlands (De Koninck 1999; Muller 2003) due to the large-scale development of perennial crops, mainly coffee but also tea, rubber and other crops. This was spurred by market growth (urban markets in Vietnam and external trade following the liberalisation of the economy), road building, and government-supported colonisation (Hardy 2000; D'haeze et al. 2005). In this region, perennial crops are often the main crops cultivated by households, contrary to the other regions of Vietnam where perennial crops are part of a diversification strategy.

## *14.3.3 The Forest Transition Pathways in Vietnam*

### 14.3.3.1 Economic Reforms and Agricultural Development

A forest transition according to the economic development path occurs when, after a period of deforestation, large areas of land marginally suitable for agriculture are abandoned and left to forest regeneration (Rudel et al. 2005). Because of economic expansion, labour force is driven from agriculture to other economic sectors and from rural to urban areas. At the same time, due to market development, agricultural intensification is concentrated in the most suitable regions.

This pathway was partially at work in Vietnam during the 1990s. Agricultural development occurred when input and output markets were progressively liberalised and land was allocated to households (Kerkvliet and Porter 1995; Pingali et al. 1997). The "Resolution 10" policy in 1988 was a correction to "Decree 100". It further liberalised rice and input prices, land rights and crop choices. Households were also allowed to own all their production after subtracting taxes and charges (Kerkvliet and Porter 1995). The Land Law of 1993 introduced long-term rights to use, transfer, exchange, inherit, rent and mortgage land (Do and Iyer 2003).

These changes created incentives for the farmers to intensify agriculture and invest more labour in paddy lands (Fatoux et al. 2002), invest in land improvements (Do and Iyer 2003), and produce surpluses for new market outlets (Minot and Goletti 2000; Luu 2003). Yield-increasing inputs such as improved seed varieties and fertilisers also became available to farmers with the emergence of market networks (Pingali et al. 1997; Tran and Kajisa 2006). The intensification occurred mainly for paddy rice and maize. Maize cultivation developed following the introduction of high-yielding varieties and as demand for maize, used as fodder for livestock, increased (Sikor 2001; Wezel et al. 2002; Minot 2003; Sikor and Pham 2005). Maize development was particularly important in the northwest of Vietnam.

Farmers also increasingly developed perennial crops (fruit trees, tea, and other), although this remained a marginal and often unsuccessful activity in the mountains. Livestock raising activities, mainly cattle and pigs, also developed.

According to the statistical analyses by Meyfroidt and Lambin (2008a), few of these changes contributed to the regeneration of natural forests, except for maize intensification and, for a small part, livestock development. During the 1990s, maize increasingly replaced upland rice and cassava on hillsides. Households that had poor access to paddies after agricultural land allocation often adopted maize on short fallow fields or on permanent fields with fertilisers to maintain a viable upland cultivation system (Sikor 2001; Castella and Erout 2002). As this intensive maize cultivation replaced upland rice that used to be cultivated without fertilisers and with a long fallow cycle, forest encroachment decreased (Sikor 2006).

As livestock and upland crops compete for space (Eguienta et al. 2002; Sikor 2006; Clement and Amezaga 2008a), the development of livestock may have led locally to the abandonment of cultivated land and regeneration of forest on extensively-managed pastures. However, animal grazing has also had a deteriorating effect on the quality of forests.

The increase in rice yields being mainly dependent on capital rather than labour inputs, it did not affect directly the capacity of households to engage in shifting cultivation in the uplands. Perennial crop plantations were usually created on degraded upland fields. These changes therefore did not contribute to natural forest restoration.

Nevertheless, increases in rice yields and development of perennial crops were associated with and contributed to the increase in planted forests. The conditions that allowed farmers to modernise agriculture – accessibility to market for inputs and outputs, surplus of capital and/or government subsidies – were the same conditions that allowed them to invest in forest plantations activities (Sikor 2001; Tachibana et al. 2001; Muller 2003; Castella et al. 2005). Farmers in Vietnam are often reluctant to diversify their activities before they achieve food self-sufficiency, so increases in rice yields may have been a pre-condition to the development of perennial crops and forest plantations (Fatoux et al. 2002; Castella and Erout 2002; Alther et al. 2002). Furthermore, capital gained in any one of these activities could be invested in the other, so that these activities reinforced each other (Fatoux et al. 2002; Minot 2003; Sikor and Pham 2005). On the contrary, increases in cattle were associated with a decrease in forest plantations (Meyfroidt and Lambin 2008a), suggesting that – as for upland crops – cattle competes with forestry.

There was no rural exodus from the mountains to the cities during the start of the forest transition, in the 1990s. During this period, the rural population density in the mountains increased from 70 to 74 inhabitants per $km^2$, and from 67 to 73 inhabitants per $km^2$ in the northern mountains, where reforestation was the greatest. The share of rural households primarily dependent on farm activities in the mountains remained constant – 87% of the households.

Outmigration from the uplands was low because of its high cost (Sikor and Pham 2005; Castella et al. 2006) and of policies restricting free migration that were only progressively relaxed during the 1990s (Hardy 2000; Lundberg 2004). Despite rapid economic growth, the demand for labour in the industrial sector remained low (Jenkins 2004). The relaxation of constraints on migrations increased rural to urban migration, but these were mostly drawn from the lowlands around the cities (Adger 2002; Dang et al. 1997). According to the 1999 Population and Housing Census (General Statistical Office 2001), only around 10% of the migrants to Hanoi between 1994 and 1999 came from northern mountains.

Government-supported or spontaneous migration to the central highlands could have contributed to the decrease in population growth rates in the other mountainous regions, but this has not affected forest regrowth according to statistical analyses. During the 1990s, no depopulation and/or de-agrarianisation of marginal regions took place. These processes could not have contributed to the increase in forest area.

Evidence in support of the economic development path is therefore ambiguous. The economic growth rate in Vietnam was one of the highest in the world during the 1990s (Fritzen 2002). However, there were geographic disparities in this growth rate, and the income gap between the upland and other regions has widened (Poverty Working Group 1999). The population of the reforested uplands benefited less from this economic growth. Urbanisation and industrialisation did not significantly attract manpower away from agriculture in these areas. There is no evidence either that agricultural intensification in the high potential regions or increasing trade of agricultural products from mountain regions (fruits, livestock) led to the abandonment of cropland and forest regeneration, except for maize. Economic growth did however encourage the development of forest plantations.

In the future, and possibly already since 2000, the economic development path of forest transition may become more important. A geographic redistribution of land use was initiated at a regional scale with the development of maize for fodder and of timber plantations in suitable regions. Perennial crops and livestock are likely to increase in mountains, while rice from the lowlands is likely to reach upland markets. In recent years, urban migration greatly increased and may start to affect populations from the mountain regions. The non-farm sector is playing an increasing role in the economy of mountain regions (Rigg 2001; Sikor and Pham 2005; Castella et al. 2005), notably by providing capital to farmers for agricultural improvements (Reardon et al. 1994; Fatoux et al. 2002).

#### 14.3.3.2 Policy Responses to Forest Scarcity

The forest scarcity path to forest transition posits that political and economic changes will arise as a response to the growing scarcity of forest products and decrease in the provision of ecosystem services following deforestation. In Vietnam many policy changes indeed emerged during the 1990s to halt deforestation and promote forest regrowth. Because of a rather poorly organised civil society in Vietnam, the main actors in the forestry arena during the transition were the government and administrations at different levels, the State Forest Enterprises (the formerly monopolistic forest exploitation organisations), and the households (Clement and Amezaga 2008b).

The Forest Protection and Development Law of 1991 and the Land Law of 1993 introduced the zoning of land according to its current or planned purpose. Forestry land was defined as land covered by forest or planned for forestry uses. Three subdivisions were also introduced: protection forests (for protection of watersheds and from desertification), production forests, and special-use forest (forests of high biological or cultural value). These policies were aimed at protecting forests and restricting shifting cultivation in the uplands.

The 1993 Land Law also introduced a system for the allocation of forestry land to households (Sikor 2001). The purpose of this policy was to restrict slash-and-burn cultivation on forestry lands, and to provide incentives and responsibilities for protection and a sound management of allocated land by households. The implementation of this policy varied locally. The rights and duties of households depended on the status and planning of land use (Nguyen et al. 2004a). Households had a responsibility to preserve existing forests and to manage forest regrowth on bare lands. They had some rights to cultivate crops on bare lands and production forests, and to exploit wood and forest products under some restrictions. According to several programs, they also received small cash payments in return for their commitment to preserve and protect forests, and sometimes to plant trees (Ministry of Agriculture and Rural Development 2001; Sowerwine 2004; McElwee 2004; Sikor 2006). Several authorities (forest management boards, national park administrations, state forest enterprises) also signed forest protection contracts with households, which imposed more restrictions on household rights on forestry land than for allocated land.

Allocation of forestry land to communities had been implemented in several provinces during the 1990s – for example, Son La in the north and Dak Lak in the central highlands (Nguyen 2006a; Clement and Amezaga 2008b). This was, however, given a legal status only through the new Land Law of 2003 and Forest Law of 2004 (Nguyen 2006a). In many villages, areas that were not allocated were *de facto* managed by the communities (Nguyen et al. 2004a). Indeed, some of the payments were given not to households but to villages to preserve natural forests that were not allocated.

Although several case studies have described shortcomings in the enforcement of the forestry land allocation policy (Sikor 2001, 2006), other case studies (Tachibana et al. 2001; Castella et al. 2006; Jakobsen et al. 2007) and a national-level statistical analysis (Meyfroidt and Lambin 2008a) have showed that this policy had a positive impact on natural forest cover. So far, official community-based forest management schemes represents only a small fraction of the allocated land, and there is no conclusive evidence that they contributed significantly to forest cover changes, although under some conditions they may constitute a sustainable way to manage forests (Nguyen et al. 2004a; Nguyen 2006a).

Several tree planting programs were also undertaken by the government, mainly via Decree 327 (in 1992) and its successor, Decree 661 (or the "Five Million Hectare Reforestation Programme", started in 1998) (De Jong et al. 2006). Most of these programs were implemented by the State Forest Enterprises, often with capital from international donors and with contributions of labour provided by farmers. The success of these programs was limited, due to the combination of several factors. In the mountains, farmers lacked labour, capital and technical knowledge about forestry, and the absence of a well-organised market and transportation infrastructure also decreased the profitability of forestry activities (Sunderlin and Huynh 2005; Dinh 2005). The complexity of regulations, frequent and arbitrary changes in regulations, and the long time needed to yield benefits created an uncertain environment for investing in forestry (Dinh 2005). For the same reasons, the forestry land allocation policy was not sufficient to stimulate forest plantations in the mountains during the 1990s. Forest plantations were therefore mainly concentrated in the more accessible midlands and lowlands, where farmers had the production assets to engage in forestry. Reforestation programs, based on monospecific exotic plantations (*Acacia*, *Eucalyptus*, Teak), are criticised for not having beneficial – or even having detrimental – effects on biodiversity, soils and hydrology (Clement and Amezaga 2008a, 2008b).

The Vietnamese government also developed plans to extend protected areas and strengthen their enforcement after signing the Convention on Biological Diversity in 1994 and approving a Biodiversity Action Plan in 1995 (World Bank 2005). Some case studies showed positive effects of protected areas (Muller 2003), while others suggest that they were not effective (Sowerwine et al. 1998; Zingerli et al. 2002). According to Meyfroidt and Lambin (2008a), protected areas did not contribute notably to forest regeneration or protection in terms of the area affected. However, they may have contributed to preserve small ecosystems rich in biodiversity.

The Vietnamese government also enacted a partial ban on logging exports in 1993, modified in 1997, but its impact on forests was likely to be small (McElwee 2004). Information campaigns to increase awareness about the importance of forest protection were also launched. The effects of such measures are difficult to assess.

The forest transition in Vietnam fits several aspects of the forest scarcity pathway. New policies, mostly based on land planning and allocation, were implemented to address the perceived degradation of forest resources. Increasing demand from urban and remote markets also contributed to the growth of the forestry sector and of forest plantations, but only in economically suitable places. The scarcity of natural forests and new policies securing land tenure made investments in forestry more profitable. Meyfroidt and Lambin (2008a) provided national-scale statistical evidence that local scarcity of forests may also have contributed to the reforestation in Vietnam, as has been the case elsewhere (Hyde et al. 1996; Foster and Rosenzweig 2003). However, the perception of this scarcity by rural households remains to be investigated at a local level (Meyfroidt and Lambin, *forthcoming*).

#### 14.3.3.3 Smallholder Agricultural Intensification

The case of Vietnam suggests the existence of a third forest transition path, associated with agricultural intensification *à la* Boserup by smallholders in marginal regions. Rapid deforestation in mountainous regions combined with a high population density led to a reduction in fallows, soil erosion on hillsides and a shortage of land suitable for shifting cultivation (Pandey and Dang 1998; Sikor 2001; Wezel et al. 2002). The relative profitability of slash-and-burn cultivation in the uplands declined in comparison to the cultivation in valleys (Sikor and Dao 2002). Forty percent of the farmers interviewed by Clement and Amezaga (2008a) mention soil degradation as a reason for ending annual upland cultivation.

Farmers therefore intensified paddy rice cultivation by shifting labour to the plots with the highest agro-ecological potential. Mountain farmers progressively added a second (or third) crop on irrigated plots and slowly expanded terraced fields (Castella and Erout 2002; Sikor 2006). Mechanisation being limited in the mountains because of a lack of capital (Henin 2002; Castella et al. 2006), little labour force was left for slash-and-burn agriculture. The upland plots were thus either abandoned or planted with trees, depending on resource endowments and opportunities. In sum, land scarcity triggered agricultural intensification in the paddy lands that, in turn, led to a shortage of labour for upland cultivation. These changes contributed significantly to the increase in natural forest area in mountainous regions of Vietnam (Meyfroidt and Lambin 2008a).

This pathway is more likely to occur in geographic contexts with poor land suitability for agriculture. In such regions, agricultural intensification on plots previously used extensively is costly. It is the case for example in mountains where irrigated agriculture is restricted to valley bottoms. Farmers therefore invest their labour to improve the management of plots already under intensive cultivation and extend them slowly, for example by constructing new terraces (Tachibana et al. 2001; Reid et al. 2006, Chapter 10).

This third forest transition path is consistent with studies by Netting (1993) and others, for example in the Philippines (Shively and Martinez 2001). This pathway is more likely to occur in densely populated regions, such as in many parts of Southeast Asia. In these regions, the initial steps of a forest transition will not result from depopulation and the decline of agriculture in marginal regions, as it would require a massive rural exodus that urban labour markets would not be able to absorb rapidly.

## *14.3.4 Environmental Impacts of the Forest Transition in Vietnam*

### 14.3.4.1 Forest Density and Carbon Stocks

Between 1980 and 2005, the density of forest declined (Meyfroidt and Lambin 2008b) due to the degradation of old-growth forests (Muller 2003; McElwee 2004) and the increasing proportion of young forests regenerating from abandoned fallows. This was particularly significant for broadleaf forests (which, not including dry dipterocarp woodlands, make around 50–60% of the total forest area). Around 43% of broadleaf forests in 1980 were low density compared to around 66% in 2005. The percentage of land covered by medium density and rich broadleaf forests on one hand, and poor forests on the other hand, increased by, respectively, 2% and 7% between 1991–1993 and 2005. Correspondingly, the average biomass carbon stock per hectare of forest declined between 1980 and the 1990s. For broadleaf forests, for which the data are the most accurate, it declined from 147 (125–169) MgC/ha in 1980 to 115 (99–133) MgC/ha in 2005. Most of the rich broadleaf forests remaining in the country are located in the central highlands.

Despite the above, the total carbon stock in forests in Vietnam followed a transition pattern. It decreased from 1314 (993–1716) TgC in 1980 to 903 (770–1307) TgC in 1991–1993, and then increased to 1374 (1058–1744) TgC in 2005. The growth rate was therefore around 2.5% per year since 1995. In agreement with changes in forest area, carbon stocks experienced a net increase in the northern mountains but a net decrease in the central highlands, where deforestation and forest degradation continues.

The fallow area increased from 1980 to 1991–1993 and then declined. The total carbon stock in fallows declined steadily from 321 (273–430) TgC in 1993 to 168 (131–212) TgC in 2005. When forest cover was at its lowest in the early 1990s, fallows were important reservoirs of carbon (26% of the total stock of forests and fallows). Fallows then declined with reforestation and the increase in permanent agriculture, to represent 11% of the total carbon stock of forests and fallows in 2005. Given the high rate of forest plantations, carbon accumulated in planted forests in 2005 may have been equivalent to the carbon added in natural forests between 1991–1993 and 2005 (respectively, 245 (180–323) and around 245 TgC). Although the overall environmental benefits of plantations are disputed by some, these forests made a significant contribution to the total increase in carbon stocks.

Due to the continuing decrease in biomass density, the annual growth rate of total carbon stock between 1995 and 2005 was lower than the rate of change in forest area during the same period. While the forest area in 2005 was 20% above that in 1980, the total carbon stock in 2005 was similar to that of 1980 and the total stock of forests and fallows was lower. Young forests have a low carbon stock but constitute a large annual sink during the first 20 years (Silver et al. 2000; Houghton 2005). Forests in Vietnam constituted a net carbon sink of around 36 TgC per year since the turning point of the forest transition. This sink was, therefore, larger than the annual carbon emissions from fossil fuels of Vietnam (that increased from 6.1 TgC in 1992 to 26.9 TgC in 2004, Marland et al. 2007).

#### 14.3.4.2 Habitat Fragmentation and Biodiversity

The potential impacts of forest cover change on biodiversity vary across regions. Habitat loss is the main factor responsible for the loss of species (Fahrig 2003), and secondary forests can constitute viable habitats for generalist species (Laurance 1997; Dunn 2004; Ewers and Didham 2006). The increase in area of natural forests in Vietnam is thus likely to have a positive effect on biodiversity. As described above, however, the density of forests still declined during the 1990s. In 2005, approximately half of the forests had a low tree density. Primary forests, undisturbed by human activities, were still declining rapidly, from 384,000 ha in 1990 to 187,000 ha in 2000 (FAO 2006). Areas planted with exotic species such as Teak, *Acacia* and *Eucalyptus* do not constitute viable habitats for most animal species.

Changes in spatial pattern of habitats may also constitute an obstacle to rehabilitation of forest biodiversity or even contribute to its degradation. Changes in forest fragmentation during the 1990s differed between regions. In the south China-Vietnam subtropical evergreen forest ecoregion (northeastern Vietnam), the net reforestation during the 1990s was the highest nationwide. Along with the reforestation, forest patches became larger and better connected, with greater core areas, although their shapes increased in complexity. Habitat fragmentation thus decreased during the 1990s. In this region, the trends in landscape change were therefore favourable for biodiversity. However, at the turning point of the forest transition, the forest cover was low and fragmented and, in 2000–2001, forests were still highly fragmented compared to the other mountainous regions of Vietnam. Rehabilitation of biodiversity would thus take time in this ecoregion.

By contrast, in the northern Indochina subtropical forests ecoregion (northwestern Vietnam), reforestation led to an increasingly fragmented landscape, with smaller and more isolated forest patches having more complex shapes and smaller core areas. Despite reforestation, biodiversity might therefore still decrease in this region. Edge effects on birds (Turner 1996; Fahrig 2003) and mammals (Woodroffe and Ginsberg 1998) are particularly worrying for this ecoregion, which has the highest bird and third highest mammal biodiversity of the Indo-Pacific area (Wikramanayake et al. 2001). In these two regions, human disturbance of forests started long ago. Natural successions and species recolonisation is likely to be slow in forest remnants that have been isolated for a long time (Laurance 1997; Chazdon 2003).

In the Annamite ecoregions, forest patches increased in size and connectivity, but also in shape complexity, so that the proportion of core areas remained stable. In the Annamite mountains, forest cover remained large, well connected and with well-preserved core areas. Secondary forests are thus likely to recover more rapidly. In the southern Indochina dry evergreen forests, almost the opposite was observed: patches became smaller and more isolated but their shapes simpler, and the proportion of core areas increased. The central Indochina dry forest area decreased, the proportion of core areas decreased, but the connectivity of patches increased. Dry forests in the highlands of southern Vietnam still hold large unfragmented forests but are threatened by recent deforestation. The lowland forests became fragmented in smaller patches with smaller core areas and increased isolation. The core area of mangroves increased due to a small number of large patches that did not affect the average patch size. The restoration of several areas of mangroves, although insignificant compared to the total forest cover of the country, is important for biodiversity as Indochina mangroves are among the most diverse in Tropical Asia (Wikramanayake et al. 2001).

In the fragmentation analysis, plantations were generally not included. The scattered spatial distribution of plantations decreases their detectability with coarse resolution remote sensing data. With the 60% forest cover threshold used here to identify forest pixels, most plantations were not included in the forest class. Plantations are concentrated in midlands and lowlands where few natural forests remain.

Despite reforestation, forest density declined, and forests in some regions became more fragmented and subjected to edge effects, while the opposite occurred in other areas. The effects of the forest transition on biodiversity are thus likely to vary between ecoregions. To sum up, habitat fragmentation generally decreased in the northeastern region and in the Annamite mountains but increased in the northwestern region and in the lowlands. Because of the long history of human disturbance, the low forest cover at the turning point, the continuing degradation or clearing of old-growth forests, and landscape fragmentation, biodiversity is most likely still threatened in Vietnam despite reforestation in some ecoregions.

### *14.3.5 Forest Transition in Bhutan*

Much less is known on the forest transition in Bhutan compared to that in Vietnam. The vegetation cover of Bhutan has been mapped for the first time based on a Landsat image dating from 1978 combined with field sampling (Sargent et al. 1985). This study estimated the forest cover at 55% of the country area. Closed or partially-closed forests occupied 22% of the country. These forests were undisturbed by human activity and were most probably primary forests. Open forests represented 33% of the land area. Note that the Landsat image only covered 70% of the country and that 20% of the image was at least partially shaded and therefore could not be interpreted. The above forest cover estimates should therefore be taken with caution. At the time of their survey, in 1983, these authors noted pressures to clear forests due to increased commercial falling, overgrazing that caused a failure of forest regeneration, and increasing human population leading to a rise in the demand for fuelwood and building material (Sargent et al. 1985).

FAO's Forest Resource Assessment estimates of forest and other wooded land in Bhutan for 1990 are much larger, at 76.6% of the land area (FAO 1993). Most of this area was classified as forest rather than "other wooded land". Subsequent estimates by FAO for 2000 and 2005 display a clear increase in forest cover, with respectively 79.8% and 81% of the forest and other wooded land category (FAO 2006). Forests represented 64.6%, 66.8%, and 68% of the country area in 1990, 2000, and 2005, respectively. Most of this increase is associated with semi-natural forests. Another national estimate of forest cover for Bhutan for the year 2000 put it at 72.5% of the country area (NEC 2000).

Forest cover increase took place in Bhutan despite an average population growth rate of 3%, with only 21% of the total population living in urban areas in 2003. The rest of the population depends on agriculture, with the agricultural sector still representing 33% of GDP in 2002. Fuelwood accounts for over three-quarters of total energy consumption and nearly all non-commercial energy consumption (Uddin et al. 2007). Private and commercial forest users can extract small amounts of wood from government-owned forests at no charge, which creates a risk of unsustainable forest exploitation. The population also derives a range of non-wood forest products that contribute to the livelihood of rural households.

In 1969, a forestry department was created in Bhutan. Private commercial tree felling was prohibited. The forestry department, who was empowered to release timber for the needs of the local population, controlled all timber extraction (Uddin et al. 2007). Grazing was restricted to certain areas. Free ranging by goats was prohibited but grazing by yak herds was still allowed. Forest plantations were developed. Large areas, mostly in low population density regions, were conserved through the establishment of protected areas. The most significant national park in Bhutan was created as early as in 1962. More than 27% of the country's area was managed within protected areas in 2005.

The forest legislation of Bhutan declares that a minimum forest cover of 60% be maintained permanently at a national scale. Among several Forest Acts, Policies and Royal Decrees, the Forest Policy of Bhutan 1991 and Bhutan Forest and Nature Conservation Act 1995 legislate principles of sustainable management of forests, biodiversity conservation, and social forestry (Uddin et al. 2007). Institutions for community participation in conservation and forestry activities were formally implemented in 2002. These forest policies provide a role for traditional forest users in managing forest activities on publicly-owned land, including community lands. Most forests fall under this category of land ownership rights. Tree planting and forest management is also encouraged on privately-owned land.

Underlying these forest policies is the unique philosophy of development in Bhutan that pursues Gross National Happiness as a main goal. In particular, the promotion of the happiness of the people is viewed as being more important than economic prosperity, in a quest to balance spiritualism and materialism. As part of this policy, preservation and sustainable use of the environment is one of the key objectives consistently pursued by Bhutan over the last 50 years (the other three objectives being economic self-reliance, cultural promotion and good governance). It is inspired by the Buddhist culture of harmony and compassion, and its representation of the place of human beings within the complex web of interdependent relationships between all forms of existence. This underlies a strong ethic of nature conservation that has long predated the international focus on sustainable development.

The path of forest transition in Bhutan is therefore unique, as it is neither caused by a scarcity of forest resources, nor by a modernisation of the economy through market development and rural depopulation. Rather, it is motivated by a development philosophy that assigns a high value to environmental preservation. The policies that implement this Buddhist-inspired worldview are strongly promoted by the central government, through a forest ownership regime of mostly publicly-owned forest land.

## 14.4 Discussion and Conclusion

### *14.4.1 Interactions Between Pathways and Local Diversity*

The three pathways described for Vietnam in Section 14.3.3 are not independent and interact in several ways. Maize intensification was induced not just by the liberalisation of agricultural markets but also by the scarcity of land suitable for rotational shifting cultivation, due to land degradation, high population densities and the forest land allocation policy. Intensification in paddy fields was caused by a combination of "push and pull" factors. On one hand, economic liberalisation, paddy land allocation, and market integration increased profitability of cultivation in lowlands. On the other hand, land degradation and scarcity in hillsides diminished the returns to shifting cultivation. Forestland allocation and zoning policies reinforced land scarcity (Castella et al. 2006; Jakobsen et al. 2007; Clement and Amezaga 2008a) and prohibited newcomers from claiming degraded land abandoned by farmers. So far, the relative contributions of the "push" and "pull" factors remain unclear. The interplay between socio-economic changes and land degradation in agricultural intensification has still to be investigated further (Lambin and Meyfroidt 2008).

Forest policies were not only a response to forest scarcity at a national level, but also the result of broader reforms towards liberalisation and decentralisation in the forestry sector that triggered economic development (Lambin and Meyfroidt 2008). The local effectiveness of forestland allocation policies was also influenced by economic and agricultural changes: where paddy lands were abundant and paddy rice cultivation intensive, farmers had an alternative to shifting cultivation.

These policies were therefore more easily accepted. Where paddy lands were scarce, policy restrictions on shifting cultivation in forestlands were less strongly enforced. Forestland allocation by itself was not sufficient to stimulate forestry activities and forest planting. Market outlets and capital investments were also necessary.

There were local and regional variations in the process of forest transition in Vietnam. Northwestern Vietnam, with soils suitable for maize cultivation but deep valleys with little paddy lands, experienced a large increase of maize cultivation on permanent fields. Northeastern Vietnam, with smaller valleys and less fertile soils, was more affected by intensification in paddy lands. The midlands closer to the deltas were more influenced by forestry activities and forest plantations. There were also regional differences in landscape fragmentation and varying impacts on hydrology and soils.

### *14.4.2 Social and Economic Issues in the Forest Transition*

The forest transition in Vietnam occurred during a rapidly changing socio-economic context of political reforms and transition from socialism to market economy. Changes in land tenure and property rights had large impacts not only on forests but also on people's livelihood.

By restricting the rotation of upland fields, the forestry land allocation policy diminished returns on shifting cultivation. However, poor households lacking access to paddy lands had few other options. They had to continue upland cultivation by reducing the length of fallow and increasing the cultivation frequency of their plots, with detrimental effects on labour time, crop yields and livelihoods (Scott 2000; Nguyen et al. 2004b; Castella et al. 2006; Jakobsen et al. 2007). This affected mainly some minority ethnic groups (e.g., the Hmong and Dao) who had a long tradition of shifting cultivation and had few paddy lands.

The land allocation process was not always equitable, with the well-positioned households (village leaders, commune civil servants) often taking up the most desirable lands before the official land allocation (Dinh 2005). Notwithstanding legal rights, the actual benefits derived from allocated forests also depended on the wealth, political position and labour capacity of households (Nguyen 2006b). A lack of infrastructure and forestry skills, and an uncertain political context also made it difficult for many households to benefit from forestry (Dinh 2005).

When land allocation was initiated, privatisation and individualisation of rights was praised in Vietnam for being the most appropriate way to manage resources. Recently, community-based forest management has emerged as an alternative. Although no comprehensive study of the effectiveness of community forestry exists in Vietnam, some case study evidence shows that, under some conditions, this provides an efficient way to improve livelihoods and manage forests sustainably (Nguyen et al. 2004a; Nguyen 2006a). However, community-based forest management corresponds to a diversity of actual land tenure and land use arrangements, and their actual impact on forest management remains to be investigated.

### 14.4.3 Conclusion

The critical analysis of available forest over data demonstrates unambiguously that a forest transition occurred in Vietnam, with the turning point around 1991–1993 at around 25–31% of land covered by forest. The forest cover then increased to around 32–37% in 1999–2001. The reforestation was due, in similar proportions, to natural forest regeneration and plantations. The rate of reforestation was higher than the rate of deforestation during the previous decades.

According to insights from national-scale statistical modelling and local case studies, forest regrowth in Vietnam was not due to a single process or policy but to a combination of economic and political responses to forest and land scarcity, economic growth, and market integration at the scale of the country. The distribution of forestry land to households, new forest management practices, and food crop intensification were combined in "push and pull" effects to decrease the footprint of agriculture on hillsides. Land scarcity provided strong incentives for agricultural intensification and the sound management of forests. These processes partly correspond to the two pathways described in the emerging forest transition theory – the economic development path and the forest scarcity path – and also to a smallholder agricultural intensification pathway linked to the theory of agricultural intensification of Boserup.

This transition had complex social and economic impacts. The benefits of forestry activities for local land managers, although potentially significant, did not fully materialise so far. Forest issues in Vietnam remain rapidly changing, with the current emergence of community-based forest management and the development of forestry activities by households.

Despite a decreasing density of biomass due to the degradation of forests and the increasing proportion of young trees, the total carbon stock in forests also followed a pattern of transition. Forests in Vietnam represented a net carbon sink of around 36 TgC per year since the beginning of the 1990s. Yet, forests in some regions became more fragmented and subjected to edge effects, while the opposite occurred in other regions. The effects of the forest transition on biodiversity are thus likely to vary between ecoregions. Biodiversity is most likely still threatened in Vietnam. The local scale impacts of different pathways of reforestation on forest composition and spatial pattern, and the associated effects on habitats, carbon stocks, soils and hydrology require more in-depth investigation.

Other forest transitions exist in the region, notably in the Kingdom of Bhutan, where a development philosophy that assigns a high value to environmental preservation promotes forest policies on the mostly publicly-owned forest land. This case demonstrates that net reforestation can take place even in regions with a high rate of population growth and a population living mostly in rural areas. Good governance and a cultural system assigning a high value to nature are essential elements for a successful forest transition. Moreover, contrasting the case of Bhutan with that of Vietnam highlights the diversity of forest transition pathways even within the same world region. Knowledge about the causes, pattern and environmental impacts of the forest transition in these countries is important to understand possible emerging regional trends that would have implications for global environmental change.

**Acknowledgments** The authors thank the anonymous reviewers and the editors for their useful comments.

## References

Adger WN, Kelly PM, Winkels A, Luong QH, Locke C (2002) Migration, Remittances, Livelihood Trajectories, and Social Resilience. Ambio 31 (4): 358–366

Alther C, Castella JC, Novosad P, Rousseau E, Tran TH (2002) Impact of accessibility on the range of livelihood options available to farm households in mountainous areas of northern Vietnam. In: Castella JC, Quang DD (eds) Doi Moi in the mountains land use changes and farmers' livelihood strategies in Bac Kan province, Vietnam. The Agricultural Publishing House, Ha Noi, pp 121–146

Castella JC, Dang DQ (2002) Doi Moi in the mountains: Land use changes and farmer's livelihood strategies in Bac Kan province, Viet Nam. The Agricultural Publishing House, Hanoi

Castella JC, Erout A (2002) Montane paddy rice: the cornerstone of agricultural production systems in Bac Kan Province Vietnam. In: Castella JC, Quang DD (eds) Doi Moi in the mountains land use changes and farmers' livelihood strategies in Bac Kan province, Vietnam. The Agricultural Publishing House, Ha Noi, pp 175–195

Castella JC, Manh PH, Kam SP, Villano L, Tronche NR (2005) Analysis of village accessibility and its impact on land use dynamics in a mountainous province of northern Vietnam. Appl Geogr 25(4):308–326

Castella JC, Boissau S, Hai Thanh N, Novosad P (2006) Impact of forest land allocation on land use in a mountainous Province of Vietnam. Land Use Policy 23(2):147–160

Chazdon RL (2003) Tropical forest recovery: legacies of human impact and natural disturbances. Perspect Plant Ecol Evol Syst 6(1/2):51–71

Clement F, Amezaga JM (2008a) Linking reforestation policies with land use change in northern Vietnam: why local factors matter. Geoforum 39(1):265–277

Clement F, Amezaga JM (2008b) Afforestation and forestry land allocation in northern Vietnam: analysing the gap between policy intentions and outcomes. Land Use Policy, in press.

Dang A, Goldstein S, McNally J (1997) Internal migration and development in Vietnam. Int Migr Rev 31(2):312–337

DeFries RS, Hansen MC, Townshend JRG, Janetos AC, Loveland TR (2000) A new global 1-km dataset of percentage tree cover derived from remote sensing. Global Change Biol 6(2):247–254

De Jong W, Sam DD, Hung TV (2006) Forest rehabilitation in Vietnam: histories, realities and future. Center for International Forestry Research (CIFOR), Bogor

De Koninck R (1999) Deforestation in Vietnam. International Development Research Center (IDRC), Ottawa

D'haeze D, Deckers J, Raes D, Phong TA, Loi HV (2005) Environmental and socio-economic impacts of institutional reforms on the agricultural sector of Vietnam: land suitability assessment for Robusta coffee in the Dak Gan region. Agric Ecosyst Environ 105(1–2):59–76

Dinh DT (2005) Forestry, poverty reduction and rural livelihoods in Vietnam. Labour and Social Affair Publishing House, Hanoi

Do DS (1994) Shifting cultivation in Vietnam: its social, economic and environmental values relative to alternative land use. International Institute for Environment and Development, London

Do QT, Iyer L (2003) Land rights and economic development: evidence from Vietnam — Policy Research Working Paper 3120. Technical Report, World Bank, Washington, DC

Dunn RR (2004) Recovery of faunal communities during tropical forest regeneration. Conserv Biol 18(2):302–309

Ewers RM, Didham RK (2006) Confounding factors in the detection of species responses to habitat fragmentation. Biol Rev 81(1):117–142

Eguienta YK, Martin C, Lecomte P, Husson O, Castella JC (2002) Crop-livestock interactions in northern Vietnam: issues diversity of farmer's responses and alternatives for sustainable integration of animals in upland agricultural systems. In: Castella JC, Quang DD (eds) Doi Moi in the mountains. Land use changes and farmers' livelihood strategies in Bac Kan province, Vietnam. The Agricultural Publishing House, Ha Noi, pp 221–248

Fahrig L (2003) Effects of habitat fragmentation on biodiversity. Annu Rev Ecol Evol Syst 34(1):487–515

FAO (1993) Forest Resources Assessment 1990 – Tropical countries. FAO Forestry paper 112. FAO, Rome

FAO (2006) Global Forest Resources Assessment 2005. FAO Forestry paper 147. FAO, Rome

Fatoux C, Castella JC, Zeiss M, Manh PH (2002) From rice cultivator to agroforester within a decade: the impact of Doi Moi on agricultural diversification in a mountainous commune of Cho Moi district, Bac Kan province, Vietnam. In: Castella JC, Quang DD (eds) Doi Moi in the mountains. Land use changes and farmers' livelihood strategies in Bac Kan province, Vietnam. The Agricultural Publishing House, Ha Noi, pp 73–98

Foster AD, Rosenzweig MR (2003) Economic growth and the rise of forests. Q J Econ 118(2):601–637

Fox J, Dao MT, Rambo AT, Nghiem PT, Le TC, Leisz S (2000) Shifting Cultivation: A New Old Paradigm for Managing Tropical Forests. Bioscience 50(6):521–528

Fritzen S (2002) Growth, inequality and the future of poverty reduction in Vietnam. J Asian Econ 13(5):635–657

General Statistical Office (2001) Results of the 1999 Population and Housing Census. Statistical Publishing House, Ha Noi

Hagen-Zanker A, Straatman B, Uljee I (2005) Further developments of a fuzzy set map comparison approach. Int J Geogr Info Sci 19(7):769–785

Hansen MC, DeFries RS, Townshend JRG, Carroll M, Dimiceli C, Sohlberg RA (2003) Global percent tree cover at a spatial resolution of 500 meters: first results of the MODIS vegetation continuous fields algorithm. Earth Interact 7(10):1–15

Hardy A (2000) Strategies of migration to upland areas in contemporary Vietnam. Asia Pacific Viewpoint 41(1):23–34

Henin B (2002) Agrarian change in vietnam's Northern Upland Region. Journal of Contemporary Asia 32(1):3–28

Houghton RA (2005) Aboveground forest biomass and the global carbon balance. Global Change Biol 11(6):945–958

Hyde WF, Amacher GS, Magrath W (1996) Deforestation and forest land use: theory, evidence and policy implications. World Bank Res Obs 11(2):223–248

Irvin G (1995) Vietnam: assessing the achievements of Doi Moi. J Dev Stud 31(5):725–750

Jakobsen J, Rasmussen K, Leisz S, Folving R, Quang NV (2007) The effects of land tenure policy on rural livelihoods and food sufficiency in the upland village of Que, North Central Vietnam. Agric Syst 94(2):309–319

Jamieson NL, Cuc LT, Rambo TA (1998) The development crisis in Vietnam's mountains. East-West Cent Spec Rep 6:1–32

Jenkins R (2004) Why has employment not grown more quickly in Vietnam? J Asia Pacific Econ 9(2):191–208

Kauppi PE, Ausubel JH, Fang J, Mather AS, Sedjo RA, Waggoner PE (2006) Returning forests analyzed with the forest identity. Proc Natl Acad Sci U S A 103(46):17574–17579

Kerkvliet BJT (1995) Village-state relations in Vietnam: the effect of everyday politics on decollectivization. J Asian Stud 54(2):396–418

Kerkvliet BJ, Porter DJ (eds) (1995) Vietnam's rural transformation. Westview Press, Boulder, CO

Lambin EF, Meyfroidt P (2009) Land use transitions: ecological feedback versus socio-economic change. In revision for Land Use Policy

Laurance WF (1997) Tropical forest remnants: ecology, management and conservation of fragmented communities. The University of Chicago Press, Chicago, IL

Lundberg M (2004) Kinh settlers in Vietnam's northern highlands: natural resources management in a cultural context. Ph.D. Thesis, Linköping University, Linköping
Luu TDH (2003) The organization of the liberalized rice market in Vietnam. Ph.D. Thesis, Rijksuniversiteit Groningen, Groningen
Marland G, Boden T, Andres RJ (2007) National CO2 emissions from fossil-fuel burning, cement manufacture, and gas flaring: 1751–2004. Carbon Dioxide Information Analysis Center, US Department of Energy, Oak Ridge National Laboratory, Oak Ridge, TN
Mather AS (2007) Recent Asian forest transitions in relation to forest-transition theory. Int For Rev 9(1):491–502
Mather AS, Needle CL (1998) The forest transition: a theoretical basis. Area 30:117–124
McElwee P (2004) You say illegal, I say legal: the relationship between 'illegal' logging and land tenure, poverty, and forest use rights in Vietnam. J Sustainable For 19(1–3):97–135
Meyfroidt P, Lambin EF (2008a) The causes of the reforestation in Vietnam. Land Use Policy 25(2):182–197
Meyfroidt P, Lambin EF (2008b) Forest transition in Vietnam and its environmental impacts. Global Change Biol 14(6):1319–1336
Ministry of Agriculture and Rural Development (2001) Five million hectare reforestation program partnership. Synthesis report. Technical Report, Ministry of Agriculture and Rural Development, Ha Noi
Minot N (2003) Income diversification and poverty in the northern uplands of Vietnam—patterns, trends and policy implications. Technical Report, International Food Policy Research Institute (IFPRI), Washington, DC
Minot N, Goletti F (2000) Rice market liberalization and poverty in Vietnam—Research Report 114. Technical Report, International Food Policy Research Institute (IFPRI), Washington, DC
Muller D (2003) Land-use change in the central highlands of Vietnam. Ph.D. Thesis, Georg-August-Universität, Göttingen
NEC (2000) First greenhouse gas inventory. Technical Report, National Environment Commission, Thimphu
Netting RM (1993) Smallholders, householders. Farm families and the ecology of intensive, sustainable agriculture. Stanford University Press, Stanford, CA
Nguygen HD, Ho MT, Soriaga R, Walpole P (2004) The return of limestone forests. Northeastern Vietnam. Research network report no. 11. Asian Forestry Network
Nguyen QT (2006a) Trends in forest ownership, forest resources tenure and institutional arrangements: are they contributing to better forest management and poverty reduction? Case studies from Vietnam. In Understanding forest tenure in South and Southeast Asia. Forestry Policy and Institutions Working Paper 14. FAO, Rome
Nguyen QT (2006b) Forest devolution in Vietnam: differentiation in benefits from forest among local households. For Policy Econ 8(4):409–420
Nguyen TL, Patanothai A, Rambo AT (2004) Recent changes in the composite swidden farming system of a Da Bac Tay ethnic minority community in Vietnam's northern mountain region. Southeast Asian Stud 42(3):273–293
Olson DM, Dinerstein E, Wikramanayake ED et al (2001) Terrestrial ecoregions of the world: a new map of life on Earth. BioScience 51(11):933–938
Pandey S, Dang VM (1998) A socio-economic analysis of rice production systems in the uplands of northern Vietnam. Agric Ecosyst Environ 70(2–3):249–258
Pingali PL, Nguyen TK, Gerpacio RV, Xuan VT (1997) Prospects for sustaining Vietnam's reacquired rice exporter status. Food Policy 22(4):345–358
Poverty Working Group (1999) Vietnam Development Report 2000: attacking poverty–country economic memorandum. World Bank, Washington, DC
Reardon T, Crawford E, Kelly V (1994) Links between nonfarm income and farm investment in African households: adding the capital market perspective. Am J Agric Econ 76(5):1172–1176
Reid RS, Tomich TP, Xu JC, Geist H, Mather AS (2006) Linking land-change science and policy: current lessons and future integration. In: Lambin EF, Geist HJ (eds) Land-use and land-cover change. Local processes and global impacts. Springer, Berlin

Rigg J (2001) More than the soil. Rural changes in Southeast Asia. Prentice-Hall, London
Rudel TK, Coomes OT, Moran E, Achard F, Angelsen A, Xu J, Lambin E (2005) Forest transitions: towards a global understanding of land use change. Global Environ Change Part A 15(1):23–31
Sargent C, Sargent O, Parsell R (1985) The forest of Bhutan: a vital resource for the Himalayas? J Trop Ecol 1:265–286
Satake A, Rudel TK (2007) Modeling the forest transition: forest scarcity and ecosystem service hypotheses. Ecol Appl 17(7):2024–2036
Scott S (2000) Changing rules of the game: local responses to decollectivisation in Thai Nguyen, Vietnam. Asia Pacific Viewpoint 41(1):69–84
Shively G, Martinez E (2001) Deforestation irrigation, employment and cautious optimism in southern Palawan, the Philippines. In: Angelsen A, Kaimowitz D (eds) Agricultural technologies and tropical deforestation. CAB International, Wallingford, UK, pp 335–346
Sikor T (2001) The allocation of forestry land in Vietnam: did it cause the expansion of forests in the northwest? For Policy Econ 2(1):1–11
Sikor T, Dao MT (2002) Agricultural policy and land use changes in a Black Thai commune of northern Vietnam, 1952–1997. Mount Res Dev 22(3):248–255
Sikor T (2006) Politics of rural land registration in post-socialist societies: contested titling in villages of Northwest Vietnam. Land Use Policy 23(4):617–628
Sikor T, Pham TTV (2005) The dynamics of commoditization in a Vietnamese uplands village, 1980–2000. J Agrarian Change 5(3):405–428
Silver WL, Ostertag R, Lugo AE (2000) The potential for carbon sequestration trough reforestation of abandoned tropical agricultural and pasture lands. Restor Ecol 8(4):394–407
Sowerwine JC (2004) Territorialisation and the politics of highland landscapes in Vietnam: negotiating property relations in policy, meaning and practice. Conserv Soc 2(1):97–137
Sowerwine J, Nguyen HD, Poffenberger M (1998) Ba Vi national park and the Dzao. In: Poffenberger M (ed) Stewards of Vietnam's upland forests. Research network report n° 10. Asian Forestry Network
Sterling EJ, Hurley MM, Le Duc Minh (2006) Vietnam: a natural history. Yale University Press, New Haven, CT
Sunderlin WD, Huynh TB (2005) Poverty alleviation and forests in Vietnam. Center for International Forestry Research (CIFOR), Bogor
Tachibana T, Nguyen TM, Otsuka K (2001) Agricultural intensification versus extensification: a case study of deforestation in the northern-hill region of Vietnam. J Environ Econ Manag 41(1):44–69
Tran TU, Kajisa K (2006) The impact of green revolution on rice production in Vietnam. Dev Econ 44(2):167–189
Turner IM (1996) Species loss in fragments of tropical rain forest: a review of the evidence. J Appl Ecol 33(2):200–209
Uddin SN, Taplin R, Xiaojiang Y (2007) Energy, environment and development in Bhutan. Renew Sustainable Energy Rev 11:2083–2103
Wezel A, Luibrand A, Le QT (2002) Temporal changes of resource use, soil fertility and economic situation in upland northwest Vietnam. Land Degrad Dev 13:33–44
Wikramanayake E, Dinerstein E, Loucks C et al (2001) Terrestrial ecoregions of the Indo-Pacific: a conservation assessment. Island Press, Washington, DC
Woodroffe R, Ginsberg JR (1998) Edge effects and the extinction of populations inside protected areas. Science 280(5372):2126–2128
World Bank (2005) Vietnam environment monitor 2005: Biodiversity. World Bank, Washington, DC
Zingerli C, Castella JC, Pham HM, Pham VC (2002) Contesting policies: rural development versus biodiversity in the Ba Be National Park area, Vietnam. In: Castella JC, Quang DD (eds) Doi Moi in the mountains. Land use changes and farmers' livelihood strategies in Bac Kan province, Vietnam. The Agricultural Publishing House, Ha Noi, pp 249–276

# Chapter 15
# Forest Cover in China from 1949 to 2006

**Conghe Song and Yuxing Zhang**

## 15.1 Introduction

Forests are the largest and most important terrestrial ecosystem on Earth, and they provide fundamental goods and services on which the welfare of human society depends (Salim and Ullsten 1999). Man's attitude towards forests is

C. Song (✉)
Department of Geography, University of North Carolina, NC, USA
e-mail: csong@email.unc.edu

Y. Zhang
Academy of Forest Inventory and Planning, State Forestry Administration, Beijing, P. R. China

H. Nagendra and J. Southworth (eds.), *Reforesting Landscapes: Linking Pattern and Process*, Landscape Series 10,
DOI 10.1007/978-1-4020-9656-3_15, 

evolving with human civilization, ranging from venerating them as sacred for their immensity, fearing them due to dangers associated with forests, finding them as impediments to agricultural development, to seeing them as sources of economic value (Vogt et al. 2007). Laarman and Sedjo (1992) summarized the goods and services that forests can provide for human welfare in five categories: (1) protective services and influences, such as soil and water conservation, climate regulation as well as conservation of biodiversity; (2) educational and scientific services, where forests are used in research and teaching to obtain and transmit basic knowledge; (3) psycho-physiological influences, including tourism, recreation, inspiration for art, religion and philosophy, etc.; (4) consumption of plants, animals and derivatives, referring to timber, fuelwood, and other derived products from forests; (5) source of land and living space. Traditional forest practice is primarily aimed at extraction of direct and derivative products from forests to maximize economic gains, while other goods and services that forests offer are frequently not adequately valued. As a result, large areas of forests are cut for timber or fuelwood every year. In addition, due to the rapid growth of human population, large areas of forests are either converted to cropland for food production or to developed area.

About a third of the world's forests have been lost due to human activities. Current forest areas in the world are 3.95 billion hectares (FAO 2006). Deforestation is at the core of most contemporary forestry problems (Hyde et al. 1996). Deforestation contributed approximately a third of the extra $CO_2$ in the atmosphere since the industrial revolution (Watson et al. 2000; Foley et al. 2005). As forests provide habitats for many other plants and animals, deforestation poses a great threat to biodiversity (Dobson et al. 1997). Due to forest decline, about 12.5% of the world 270,000 plant species and 75% of the world's mammals are under threat of disappearing from the Earth for good (Salim and Ullsten 1999).

Forest cover is one of the most important indicators for forest resources in a country. The change in forest cover in a country is closely related to its economic status (Ewers 2006). Forest cover in a country generally decreases as the economy grows when the gross domestic production (GDP) is low, and forest cover increases as the economy continues to grow at a higher GDP. Some characterize the nonlinear development of forest cover with the economic growth as environmental Kuznets curve (Stern et al. 1996; Mather and Needle 1998; Ehrhardt-Martinez et al. 2002). Therefore, countries with the highest rate of deforestation are found among the poorest countries in the world. During 2000–2005, 4.3 million hectares of forest were lost each year in South America, followed by 4.0 million hectares in Africa (FAO 2006). The top 10 countries that lost the most forests during 2000–2005 are all developing countries, including Brazil (3,103,000 ha/year), Indonesia (1,871,000 ha/year), Sudan (589,000 ha/year), Myanmar (466,000 ha/year), Zambia (445,000 ha/year), Tanzania (412,000 ha/year), Nigeria (410,000 ha/year), Democratic Republic of Congo (319,000 ha/year), Zimbabwe (313,000 ha/year), and Venezuela (288,000 ha/year) (FAO 2006). In contrast, many developed countries saw an increase in forest areas in recent years (Kauppi et al. 2006). For example, forest areas increased 296,000 ha/year for Spain and 195,000 ha/year for the United States during 2000–2005. Forest areas in China and Vietnam, two fast growing economies, also increased significantly during the same period (FAO 2006).

As the biggest developing country in the world, changes in the forest ecosystems in China not only impact the environment in the country, but significantly influence the global environment as well. Most of the forests in China were lost before 1949 (Xiong 1989). China's forest cover was merely 8.6% in 1949 (Chinese Government Documents 1977–2005). China managed to increase its forest cover to 18.2% in 2003, and their experience in doing so may be valuable for forestry in other developing countries. In this chapter, we plan to understand the processes that are responsible for China's forest cover change based on data from six national forest inventories (NFI) (Chinese Government Documents 1977–2005), the recent reforestation and afforestation statistics (China State Forestry Administration 2000–2007), along with remotely sensed images (Tucker et al. 2005) and economic growth data (China National Bureau of Statistics 2005). Our goals are to understand how reforestation, afforestation and deforestation contributed to the change in China's forest cover; what the driving forces are for reforestation and afforestation; what the current status of forests in China is; and what potential challenges are for sustainable forest development in the future.

## 15.2 The Natural Environment

China is the biggest country by population in the world, and the third largest country by area. Its land area spans from 3°51′N to 53°31′N in latitude, and from 73°22′E to 135°03′E in longitude, constituting 9.6 million square kilometers with Hong Kong (1,100 $km^2$), Taiwan (36,000 $km^2$) and Macao (29 $km^2$) included. The large land area contains complex landforms, including mountains (33%), high plateaus (26%), basins (19%), plains (12%), and hilly areas (10%). The overall landforms constitute a three-step terrace flowing from the west to the east, as indicated by the dashed lines in Fig. 15.1. The first terrace consists of the Himalayas and the Qinghai-Tibet Plateau with elevation above 4,000 m. The line that separates the second and the third terraces is marked by a series of mountains from the north to the south: the Greater Xingan Mountains, the Yan Mountains, the Taihang Mountains, the Wu Mountains, the Wuling Mountains and the Xuefeng Mountains. The third terrace is to the east of this line with elevation generally below 1,000 m, where most of the fertile agricultural lands are located.

The large land area with complex terrains inevitably sustains complex climatic conditions for China. The country can be divided into three major climatic regions: the eastern monsoon moist region (EMMR), the northwestern dry region (NWDR) and the Qinghai-Tibet high and cold region (QHCR). The EMMR roughly overlaps with the third (lowest) terrace of the country. During the summer the southeast monsoon supplies EMMR with sufficient precipitation, synchronizing the heat and moisture conditions. The southeast monsoon brings moisture from the Pacific

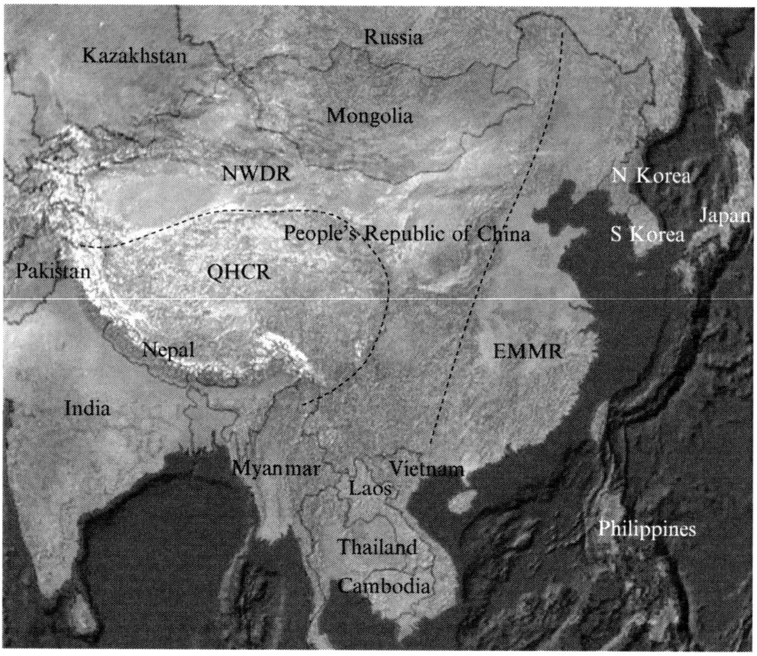


**Fig. 15.1** Topography for People's Republic of China and its major neighboring countries. The *dashed lines* within China mark the separation of the three major terraces in elevation from west to east (background image was created by Google Earth)

Ocean and the Indian Ocean to the entire region with precipitation ranging from 1,600 to 2,000 mm in the coastal southeast China, to 1,200 mm in the Yangtze River basin, and 300–500 mm all the way up to the Greater Xingan Mountains in the northeast. From the south to north, occurring in sequence are rain forests, seasonal rain forests, southern subtropical broadleaf evergreen forests, central subtropical broadleaf evergreen forests, northern subtropical broadleaf evergreen, and deciduous forests, warm temperate broadleaf deciduous forests, mild temperate broad and needle leaf mixed forests, and boreal needle leaf forests.

The NWDR is situated in the middle of the Eurasia landmass overlapping with the second terrace, far from the oceans and surrounded by large mountain ranges. Therefore, moisture can hardly reach the region. The annual precipitation in NWDR is generally below 400 mm/year with some areas having precipitation as low as 15 mm/year. Forests only occur on shaded slopes where sufficient precipitation occurs. Most of the landscapes are dry grass land, semi-desert or desert. The QHCR sits on the highest terrace of the three in China. Due to its high altitude, the Qinghai-Tibet plateau strongly influences global air circulation. During the summer months, the southeast monsoon brings moisture from the Indian Ocean to the region, while the area is controlled by the dry and cold air from the west in the winter. The mean annual temperature is below 0°C for a large portion of the area though there are significant variations over the entire region. Because of the thin air, this region is rich in solar radiation resources. Temperature changes dramatically between day and night. Adapting to the high altitude and low temperature, vegetation in this region appears as short grass or dwarfed woody plants. In the southeast part of the region, the heat and moisture conditions are suitable for the establishment of forests. Therefore, broadleaf evergreen forests, broadleaf and needleleaf mixed forests, and dark needleleaf forests can be found along the elevation gradient.

## 15.3 Changes in Forest Cover from 1949 to 2006

### *15.3.1 Characteristic Development Stages*

Based on the initial natural resources survey and the subsequent six NFI, China's forest cover increased dramatically from 8.6% to 18.2% from 1949 to 2003 (Table 15.1). Though there are some inconsistencies in the NFI data (Zhang and Song 2006), they are the most comprehensive and accurate forest resources data available for the country, and they are frequently used as ground truth for remote sensing (Wang et al. 2007c; Piao et al. 2005). Most of the increase in forest cover

**Table 15.1** Forest cover and areas in China from 1949–2003. Forest cover is calculated based on the country area of 960 million hectares (i.e. 9.6 million square kilometers) for China. Forests in Taiwan, Hong Kong and Macao are not separated into plantation and natural forests in the NFIs. New planted areas in the table are the areas planted during the inventory period. These areas are not forests until the canopy cover reaches the minimum canopy cover required to be classified as forests. The Roman numericals indicate sequence of NFI. There was no forest resource information during 1963–1972. These statistics are different from those in FRA 2005 (FAO 2006) due to difference in forest definition (Zhang and Song 2006)

| Years | New planted areas ($10^6$ ha) | Total plantation forests ($10^6$ ha) | Total forest areas ($10^6$ ha) | Total forest cover (%) |
|---|---|---|---|---|
| 1949 | – | – | – | 8.6 |
| 1950–1962 | 34.12 | 5.11 | 113.36 | 11.81 |
| 1963–1972 | – | – | – | – |
| 1973–1976 (I) | 56.13[a] | 23.69 | 121.86 | 12.70 |
| 1977–1981 (II) | 22.44 | 22.19 | 115.28 | 12.00 |
| 1984–1988 (III) | 43.63[b] | 31.01 | 124.65 | 12.98 |
| 1989–1993 (IV) | 27.76 | 34.25 | 133.7 | 13.92 |
| 1994–1998 (V) | 25.29 | 46.67 | 158.94 | 16.55 |
| 1999–2003 (VI) | 31.85 | 53.26 | 174.91 | 18.21 |

[a] This figure includes new planted areas during 1963–1972.
[b] This figure includes new planted areas during 1982–1988.

was the result of sustained reforestation and afforestation effort. In this chapter, afforestation refers to forest planting or seeding in areas that previously were not classified as forests. Deforestation is the conversion of forests to another land use or long-term reduction of canopy cover below the minimum of that needed to be classified as forests. Reforestation occurs in a deforested area through natural or assisted natural regeneration. From 1949 to 2003, the cumulative afforestation areas reached 241 million hectares, among which 90 million hectares reached forest status. At the last NFI in 2003, there were 53 million hectares of plantation forests remaining, contributing about 80% of the forest cover increase in China from 1962–2003. Forest plantation is the primary form of forest cover increase among the three forms outlined by Rudel (Chapter 3 this volume): natural regeneration, forest plantation and agro-forests. Timber harvesting is the primary process that depletes forest resources in China. The cumulative amount of timber volume harvested from 1949 to 2003 reached 13 billion cubic meters, equivalent to 16% points in forest cover. The change in forest cover can be characterized in three stages: unstable stage (I); recovery stage (II), and expansion stage (III), as shown in Fig. 15.2. The goals of forest management set by the central government were the major factor driving the change.

Unstable Stage (I) 1949–1981: The primary goal of forest management during this stage was extraction of timber resources. Due to the fact that the People's Republic of China was established in 1949 after many years of war, the country was very poor in the beginning, and needed major economic reconstruction. Consequently, demand for timber was high. Forest cover was already as low as 8.6% in 1949. In a typical market economy, the scarcity of forest resources naturally leads to price increase for forest products. The high price of forest products

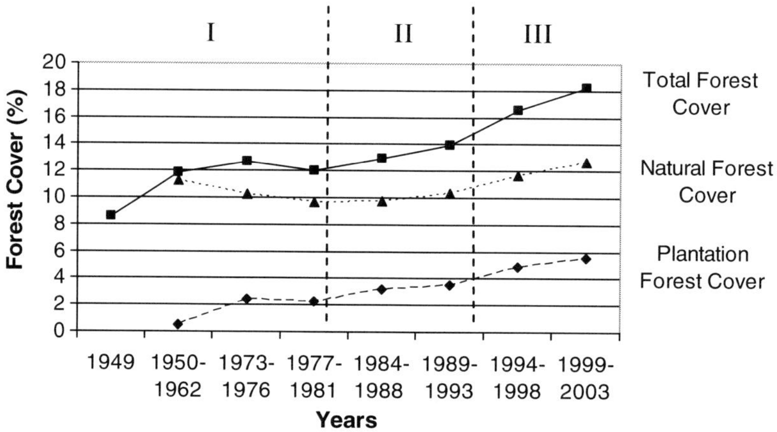


**Fig. 15.2** China forest cover change from 1949 to 2003 separated into natural forests and plantation forests. Plantation forests contributed about 80% to the overall forest cover increase. The *vertical dashed lines* mark the three developmental stages: unstable (I), recovery (II) and expansion (III) stages

**Table 15.2** Timber volume resources in China's forests and timber volume harvested from 1949 to 2003. Note that the total volumes in stock in this table is not equal to the mean volume density times the total forest areas in Table 15.1 because of data inconsistency in NFI at different times. We also see that the mean volume density is higher during 1977–1981 than that during 1973–1976 even though total volume harvested exceeded total volume growth during 1977–1981, indicating comparability problem among data from different NFIs

| | Total volume | Mean volume | Total volume | Total volume |
|---|---|---|---|---|
| Years | In Stock ($10^9$ $m^3$) | Density ($m^3$/ha) | Growth ($10^6$ ha) | Harvested ($10^6$ $m^3$/year) |
| 1949 | – | – | – | – |
| 1950–1962 | 10.2 | 98.9 | 126.0 | 87.8 |
| 1963–1972 | – | – | – | – |
| 1973–1976 (I) | 9.5 | 79.0 | 226.9 | 195.6 |
| 1977–1981 (II) | 10.3 | 83.4 | 275.3 | 194.1 |
| 1984–1988 (III) | 10.6 | 79.2 | 316.0 | 344.0 |
| 1989–1993 (IV) | 11.8 | 83.7 | 420.0 | 320.0 |
| 1994–1998 (V) | 12.5 | 78.1 | 457.5 | 370.0 |
| 1999–2003 (VI) | 13.6 | 84.7 | 497.0 | 365.0 |

provides a positive feedback to afforestation and reforestation, resulting in a transition from contraction to expansion of forest areas (Mather and Needle 1998; Satake and Rudel 2007). However, this theory did not apply to China. After the birth of modern China in 1949, the natural forests were nationalized and the private forests were collectivized (Yin 1998). As a result, 42% of the total forest areas, or 68% of the total timber volume were state-owned (Xu et al. 2004b). The natural forests in China were managed by state-owned enterprises (SOE). The price of timber was not determined by the market, but set by the government. The leadership of SOE was rewarded with promotion for higher profits from timber production, but there was not much incentive for money spent on growing the timber resources (Yin 1998; Rozelle et al. 1998). By the end of this stage, annual timber volume harvested was greater than timber volume growth (Table 15.2). The use of forest resources was not sustainable. As a result, China almost completely depleted its natural forest resources during this stage. The cumulative areas harvested were 75 million hectares, of which 92% were natural forests (Table 15.3). China's forest cover did not change as much because reforestation and afforestation added 70 million hectares of new forests. But the age structure in China's forests was altered significantly. According to Wang et al. (2007c), 80% of China's forests were at or below 40 years old in 2001. Consequently, the ecological services provided by forest ecosystems were seriously compromised.

Recovery Stage (II) 1982–1993: Timber production remained a major forest management goal, but some consideration was given to ecological effects of forests during this period. A major difference in timber harvesting in this stage from the previous stage was that plantation forests contributed 76% of the areas harvested, allowing the natural forests to recover (Table 15.3). In the meantime, the Chinese

**Table 15.3** Change in forest areas caused by afforestation, deforestation and reforestation in China from 1949 to 2003. Units in all columns are in $10^6$ ha. Net Forest Change is the sum of the preceding four columns with discrepancies explained

| Years | New Forests planted | Plantation Forests harvested | Natural Forests regenerated | Natural Forests harvested | Net Forest change |
|---|---|---|---|---|---|
| 1949 | – | – | – | – | – |
| 1950–1962 | 5.1 | 0.0 | 11.3 | −22.8 | +30.8[a] |
| 1963–1972 | – | – | – | – | – |
| 1973–1976 (I) | 18.6 | 0.0 | 24.8 | −34.8 | +8.6 |
| 1977–1981 (II) | 4.3 | −5.8 | 6.6 | −11.8 | −6.7 |
| 1984–1988 (III) | 20.5 | −11.7 | 5.8 | −5.2 | +9.4 |
| 1989–1993(IV) | 12.9 | −9.6 | 7.0 | −1.2 | +9.1 |
| 1994–1998 (V) | 13.6 | −3.4 | 10.4 | −7.1 | +25.2[b] |
| 1999–2003 (VI) | 15.0 | −8.5 | 7.5 | −2.2 | +15.9[c] |

[a] The net increase in forest cover does not match the survey results, indicating that China forest cover in 1949 might have been underestimated.

[b] The net increase includes 11.7 million hectares of forests due to change in forest definition from 30% to 20% in minimum canopy cover in 1994.

[c] This number includes 3.7 million hectares of special purpose scrubs that were recently classified as forests and 0.5 million hectares of forests from Taiwan, Hong Kong and Macao.

government started a series of ecological programs aimed at environmental protection. China achieved increases in both forest area and timber volume stocking by the end of this stage. The reversion of forest area contraction in China was not the result of market forces as seen in other countries in south Asia (Nagendra, Chapter 3 this volume), but more of central government efforts to mitigate increasing adverse ecological impacts due to severe deforestation earlier. The increase in forest cover also partly benefited from a new forest management policy. The Chinese government adopted a higher standard for afforestation in 1985, raising the survival rates from 40% to 85% in the south and to 70% in the north in order to pass afforestation inspection (Chinese Government Documents 1977–2005). Although the timber volume harvested continued to increase with time during this period, it was controlled below the annual timber volume growth (Table 15.2).

Expansion Stage (III) 1994–: Forest cover in China expanded rapidly during this stage. Timber production was no longer the primary goal of forest management. China adopted a new forest policy in 1998 (Zhang et al. 2000), which prioritized environmental protection and sustainable use of forest resources over timber production. By 1998, China's economy had grown at a double-digit rate for two decades, allowing the Chinese government to make a significant investment in forestry. The Chinese government started six key forestry programs with a committed investment of over 700 billion yuan (approximately US$85 billion at the time) by the end of 2010. These programs include (1) natural forest protection, (2) returning agricultural lands to forests/grasses, (3) shelter forest systems for the three-north (northeast, north, and northwest) and middle-to-upper reaches of Yangtze River, (4) controlling dust and sand sources from blowing to Beijing and Tianjin, (5) wildlife protection and natural resources conservation, and (6) the development of fast growth and high yield plantations in key regions. Due to the unprecedented effort in reforestation and afforestation, China's forest cover reached 18.2% according to the sixth NFI in 2003. It is important to note that timber volume harvested during 1999–2003 remained very close to the highest level in history (Table 15.2), indicating timber harvesting remains a threat to forest development.

### *15.3.2 The 1998 Record Flooding and Forest Restoration Efforts*

Years of intensive timber harvesting severely compromised the ecological functions of forest ecosystems in China, particularly for soil and water conservation. The prolonged degradation of the natural environment led to a series of natural disasters. An unprecedented flood occurred in China over almost all the major rivers in 1998. The flooding caused 250 billion Yuan (approximately $31 billion USD in 1998) in damage and a loss of 4,150 human lives (Sun et al. 2002). The alarming phenomenon was that the total precipitation in the Yangtze River basin was lower and lasted longer in 1998 compared to the previous record flooding in 1954, yet the Yangtze River experienced record flooding with eight peaks within 2 months, indicating a severe reduction of soil water retention capacity within the basin. Sun et al. (2002) reported that every year 2.4 billion tons of soil eroded from the Yangtze River basin. As a result, the height of the riverbed in the main stream of Yangtze River increases by a meter every decade. The sediment discharge from erosion in the Yangtze River exceeds the combined discharges from the world's two longest rivers, the Nile and the Amazon (Liu and Diamond 2005). The major lakes along the Yangtze River, which used to provide critical diversions for flood water, have shrunk significantly. The total lake area in the middle to low reaches of Yangtze River declined from 17,000 $km^2$ in 1949 to 6,000 $km^2$ in 1998 (Sun et al. 2002).

After a national reflection on the record flooding in 1998 (e.g. commentary appeared in People's Daily on September 3, 1998), the central government realized the flooding was due to the long-term deforestation and poor agricultural practices that caused serious soil erosion and reduction in soil water retention capacity. The Chinese government issued a series of new policies later in the year, including the ban of logging in the upper and middle reaches of the Yangtze River basin. In the meantime, a new proactive measure was taken to mitigate the problem: returning agricultural lands to forests. These lands included croplands, barren mountains/lands, and mountains with regenerating forests. Among them, returning croplands to forests required the most investments as the livelihoods of the farmers needed to be settled first.

This idea was tested in Sichuan, Shanxi and Gansu provinces with success in 1999 (China State Forestry Administration 2000–2007). Returning cropland to forests was adopted as a national project shortly in 2000. The program was further advanced by being included into the tenth *Five-Year Plan for the Socio-Economic Development of the People's Republic of China* in 2001. The program for returning agricultural lands to forests officially started nationwide in 2002, involving 25 provinces, 32 million households, and 124 million people (China State Forestry Administration 2000–2007). After 2004, the restoration efforts were focused on lands with slopes greater than 25°, areas with serious desertification, riparian zones, or land with low yield potential. The total investment budgeted for the project was 430 billion Yuan (approximately $58 billion USD). By the end of 2006, 9.3 million hectares of croplands were converted to forests (Table 15.4). Given the current moratorium on timber harvesting in the natural forests and the rapid increase in total afforestation areas, forest cover in China is expected to increase significantly in the next NFI.

### 15.3.3 Economic Growth and Forest Cover

The rapid growth of forest cover along with the growth of economy in China led some to argue that forest resources in China are currently in the later stage of the environmental Kuznets curve (Zhang et al. 2006). We examined the relationship between economic growth and increase in vegetation abundance in China based on remotely sensed images. We used increase in GDP as the measure for economic growth, and the change in normalized difference vegetation index (NDVI) as the measure of change in vegetation abundance (Song et al. 2008). NDVI is calculated from remotely sensed data as

$$\mathrm{NDVI} = \frac{\rho_{\mathrm{NIR}} - \rho_{\mathrm{RED}}}{\rho_{\mathrm{NIR}} + \rho_{\mathrm{RED}}} \tag{1}$$

where $\rho_{\mathrm{NIR}}$ and $\rho_{\mathrm{RED}}$ are reflectance values in the near-infrared and red spectrum, respectively. NDVI varies within [−1, 1] with a higher value indicating more abundance of green vegetation.

The NDVI data used in this study is from the Advanced Very High Resolution Radiometer (AVHRR) on board the NOAA-series satellites from 1982 to 2000 with a spatial resolution of 8×8 km (Tucker et al. 2005). AVHRR provides daily NDVI

**Table 15.4** Overall afforestation efforts in China from 1999 to 2006 and areas of croplands converted to forest at the same time (China State Forestry Administration 2000–2007). The effort returning croplands to forests peaked in 2003 and winded down in 2006. Note: Total afforestation also included other types of lands, such as barren mountains or regenerating mountains

| Year | Croplands to forests ($10^3$ ha) | Total afforestation ($10^3$ ha) |
|---|---|---|
| 1999 | 381 | 4,277 |
| 2000 | 405 | 5,105 |
| 2001 | 420 | 4,953 |
| 2002 | 2,586 | 7,771 |
| 2003 | 3,367 | 9,119 |
| 2004 | 667 | 5,598 |
| 2005 | 1,111 | 3,638 |
| 2006 | 267 | 2,718 |
| Total | 9,204 | 43,179 |

images for the entire globe. However, NDVI can be contaminated by aerosols and clouds in the atmosphere. The dataset used in this study is a 15 day composite NDVI, which takes the maximum NDVI from the 15 daily values for each pixel to minimize the contamination from clouds and aerosols in the atmosphere. Therefore, there are 24 AVHRR NDVI images each year. We used the annual total NDVI (ATN) as a measure of vegetation abundance in this study. ATN is calculated as

$$\mathrm{ATN} = \sum_{i=1}^{24} \mathrm{NDVI}_i \tag{2}$$

where $\mathrm{NDVI}_i$ is the 15 day composite NDVI for a pixel, and the subscript indexes the 24 NDVI image series in a year. To minimize the random variation in NDVI in a given year, we used the mean ATN during the first six (1982–1987) and last 6 years (1995–2000) to characterize average vegetation abundance for the corresponding periods. We took the difference between the mean ATN for each province or autonomous region between the two periods as the measure of change in vegetation abundance as seen in Fig. 15.3, where bright areas indicate increase

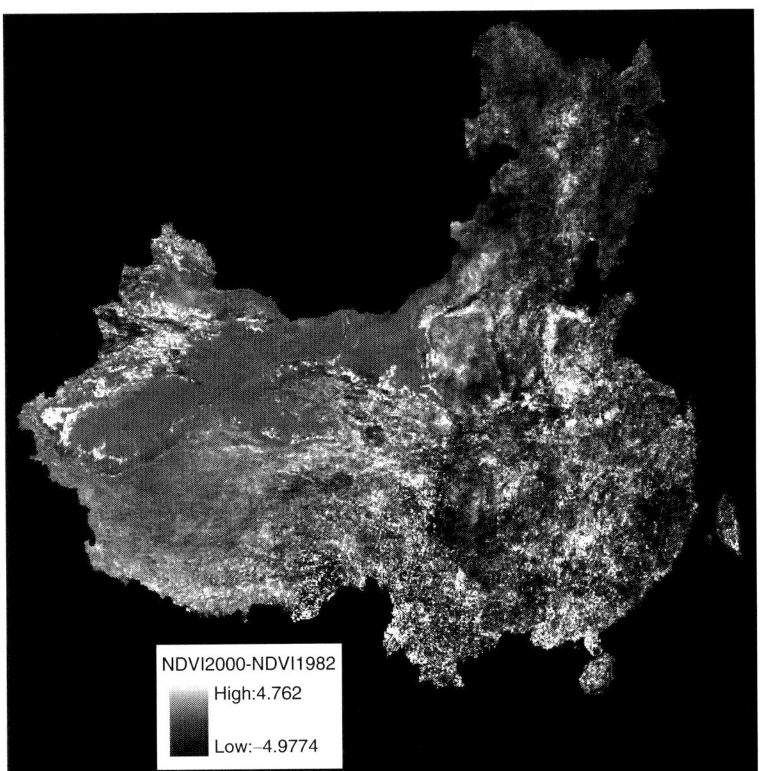


**Fig. 15.3** The difference image between the mean annual total NDVI of 1982–1987 and 1995–2000. The *bright areas* indicate increase in vegetation abundance, and the *dark areas* indicate decrease in vegetation abundance

in vegetation, dark areas indicate decrease in vegetation from 1982–1987 to 1995–2000. We overlaid the provincial boundaries on Fig. 15.3 and calculated the average value for the difference in the mean ATN for each province or autonomous region. We then calculated the difference in the mean annual GDP for each province or autonomous region for 1982–1987 and 1995–2000, respectively (China's National Bureau of Statistics 2005). Assuming an annual inflation rate of 3%, the GDP during 1982–1987 was converted to 1995–2000 value before taking the difference. Figure 15.4 shows how economic growth related to change in vegetation abundance in China from 1982–1987 to 1995–2000. Similar to the conclusion reached by Wang et al. (2007b), our empirical data do not support the argument that forest resources in China are in the later stage of an environmental Kuznets curve. According to the trend in Fig. 15.4, economic growth generally reduces vegetation cover. Contrary to what Zhang et al. (2006) argued, forest resources in China are still in the early stage on the Kuznets curve if we look at the relationship at the provincial scale. The increase in the overall forest cover in China is not the result of feedbacks from the economic development in each province. It is the result of the central government's mandate for afforestation and reforestation. Therefore, the forest cover in China may cease to expand if the government sponsored programs end.

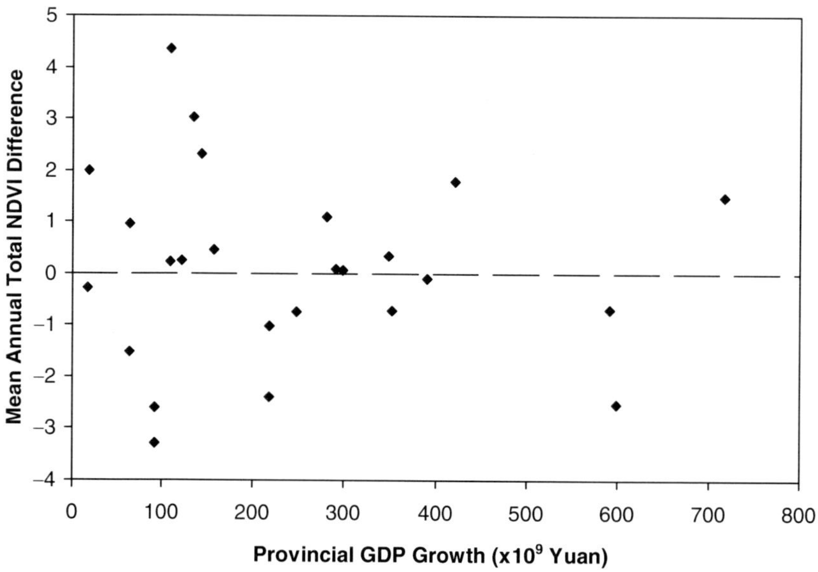


**Fig. 15.4** The relationship between economic growth and vegetation abundance as measured by the provincial average value of the difference in mean annual total NDVI from 1995–2000 to 1982–2987(shown in Fig. 15.3). Note that we excluded Beijing, Shanghai, and Tianjin from the analysis due to their overwhelming urban settings. We also combined Chongqing with Sichuan province and Hainan with Guangdong province as they were part of the corresponding provinces during 1982–1987

## 15.4 Discussion

Increase in forest cover in China was witnessed both on the ground via NFI (Nei 2005) and from space via remote sensing (Fang et al. 2004). However, it remains low compared with the average of 30% forest cover in the world. The per capita forest area is even less with 0.1 ha/person in China, compared with the world average of 0.62 ha/person (FAO 2006). The low forest cover in China is partly due to the fact that the vast northwestern region is not suitable for forest growth, and partly due to the high population that demand large agricultural area for food production. The recent unprecedented investment by the Chinese government in forestry led to significant progress in afforestation (Wang et al. 2007a). The Chinese government set an ambitious goal for the forest cover to reach 26% by 2050. However, several recent studies revealed serious problems for the natural forest protection program and the program of returning agricultural land to forests, two of the six key forestry programs (Trac et al. 2007). Control of logging remains a problem, particularly for large state forests under control of local managers (Xu et al. 2004b). The "top-down" style administration from the central government lacks mechanisms for efficient engagement of local people (Xu et al. 2006). Many participants in the program of returning agricultural lands to forests have not been properly compensated (Xu et al. 2004a). There has also been a lack of long-term planning for how to preserve the newly established forests. Uchida et al. (2005) found that participating farmers may reconvert the forested land back to cultivation after the program ends. The initial success, resulting from the unprecedented investment from the central government, may not last without careful long-term planning.

It is important to note that increase in forest cover does not necessarily signify an increase in forest resources, nor does it bring immediate improvement to the environment (Meyfroidt and Lambin, Chapter 14 this volume). Though the rapid increase in China's forest cover should be applauded, the future of the ecological environments in China remains bleak in the foreseeable future. Improvements of ecological conditions in China require sustained efforts for several generations (Song 2000). China almost completely wiped out its natural old-growth forests. Most of the increase in forest cover is from plantations, which are known for low biodiversity (Rudel, Chapter 3 this volume). The aggregate statistics for China's forest cover may create a false sense of ecological security (Rozelle et al. 1998) as forest successional stages are not accounted for in forest cover. A young plantation forest is counted the same as an old-growth natural forest in forest cover. However, the ecological functions of the two forests cannot be compared. The unique ecological niches that an old-growth forest provides for many other plants and animals cannot be found in any young stands. Soils under an old-growth canopy that have been accumulating organic matter for centuries do not exist under new plantation forests. Banning of logging in the upper and middle reaches of Yangtze River and the new plantations created in the area through the various forestry programs will certainly help reduce soil erosion. However, we cannot expect much alleviation in flooding in the near future should the 1998 rain storms hit China again. The forest ecosystems need many decades, even centuries, of development before they can provide the ecological services we hope for. Some lost habitats may never recover.

Due to both its size and population, improvement of China's environment will affect the most people in the world and contribute significantly to the improvement of the global environment. The reversion of declining forest cover in China offers hope for forestry in developing countries. Though the expansion of forest areas through reforestation and afforestation does little to preserve biodiversity, they do conserve soil and sequester carbon (Rudel et al. 2005). Enhanced vegetation activities were already detected in China from Space (Xiao and Moody 2004; Fang et al. 2004). Several recent studies found that forestry in China has become a significant carbon sink (Zhang and Xu 2003; Piao et al. 2005; Fang et al. 2007). If the ambitious goal of 26% forest cover is achieved, the forest ecosystems in China will play an increasingly bigger role in carbon sequestration in the foreseeable future.

## 15.5 Conclusions

Forest cover in China increased from approximately 8.6% in 1949 to 18.2% in 2003, experiencing three development stages: unstable, recovery, and expansion stages. Government sponsored plantation forests contributed about 80% of the increase from 1962 to 2003. Forest cover in China is expected to increase significantly in the next NFI. In the meantime, China harvested 13 billion cubic meters of timber, equivalent to 16% points in forest cover. Though forest cover in China did not decrease as much, forest age structure shifted dramatically toward the young end. As a result, the ecological functions of forest ecosystems were seriously compromised. The unprecedented flood in 1998 shocked the government into action. The Chinese government dramatically adjusted its forest management policy and started six key forestry programs with unprecedented investment, aiming at improving China's ecological environment. Though China's forest cover increased rapidly, we found China's forest resources are still in the early stage in the environmental Kuznets curve. Therefore, the possibility still exists for China's forest to degrade with economic growth without the mandate from the central government. Though the young forests in China are playing an increasingly greater role in carbon sequestration, the ecological functions of mature and old-growth natural forests cannot be replaced with the young forests in the short-term. Timber harvesting remains a threat to the development of forest ecosystems. Numerous emerging problems in implementing the new forest policy need to be resolved, and long-term planning is urgently needed for the forest ecosystems to reach sustainable development.

**Acknowledgments** This work was partly supported by NASA grant NNX06AE28G. The authors thank Jackson Lord for help collecting provincial GDP data for China, and Limin Zhou for providing the NDVI data.

## References

China National Bureau of Statistics (2005) China compendium of Statistics 1949–2004. China Statistics Press, Beijing

China State Forestry Administration (2000–2007) Annals of Forestry in China. China Forestry Press, Beijing, China

Chinese Government Documents (1977–2005) Chinese Ministry of Forestry, 1977, National Forest Resource Statistics: 1973–1976; Chinese Ministry of Forestry, 1993, National Forest Resource Statistics: 1977–1981; Chinese Ministry of Forestry, 1989, China Forest Resource Statistics: 1984–1988; Chinese Ministry of Forestry, 1994, National Forest Resource Statistics: 1989–1993; Division of Forest Resources, Policy and Management, Chinese Ministry of Forestry, 1996, Forest Resources of Contemporary China: 1949–1993; China State Forestry Administration, 2000, National Forest Resource Statistics: 1994–1998; State Forestry Administration, 2005, Report of China Forest Resources in 2005, China Forestry Press

Dobson AP, Bradshaw AD, Baker JM (1997) Hopes for the future: restoration ecology and conservation biology. Science 277:515–522

Ehrhardt-Martinez K, Crenshaw EM, Jenkins JC (2002) Deforestation and the environmental Kuznets curve: a cross-national investigation of intervening mechanisms. Soc Sci Q 83:226–243

Ewers RM (2006) Interaction effects between economic development and forest cover determine deforestation rates. Global Environ Change 16:161–169

Fang JY, Piao SL, He JS et al (2004) Increasing terrestrial vegetation activity in China, 1982–1999. Sci China (Ser C) 47(3):229–240

Fang JY, Guo ZD, Piao SL et al (2007) Terrestrial vegetation carbon sinks in China, 1981–2000. Sci China Ser D-Earth Sci 50:1341–1350

Food and Agricultural Organization of United Nations (FAO) (2006) Global Forest Resources Assessment 2005: progress toward sustainable forest management. FAO Forestry Paper 147. ISBN 92-5-105481–9.

Foley JA, DeFries R, Asner GP et al (2005) Global consequences of land use. Science 309:570–574

Hyde WF, Amacher GS, Magrath W (1996) Deforestation and forest land use: theory, evidence and policy implications. World Bank Res Obs 11:223–248

Kauppi PE, Ausubel JH, Fang J et al (2006) Returning forests analyzed with the forest identity. PNAS 103:17574–17579

Laarman JG, Sedjo RA (1992) Global forests: issues for six billion people. McGraw-Hill, New York

Liu J, Diamond J (2005) China's environment in a globalizing world. Nature 435:1179–1186

Mather AS, Needle CL (1998) The forest transition: a theoretical basis. Area 30:117–124

Nei J (2005) Forest resources of China. China Forestry Press, Beijing, P. R. China

Piao SL, Fang JY, Zhu B et al (2005) Forest biomass carbon stocks in China over the past 2 decades: estimated based on integrated inventory and satellite data. J Geophys Res 110 (G01006). doi:10.1029/2005JG000014

Rozelle S, Albers H, Li G et al (1998) Forest resources under economic reform in China. China Info 13:106–125

Rudel TK, Coones OT, Moran E et al (2005) Forest transitions: toward a global understanding of land use change. Global Environ Change 15:23–31

Salim E, Ullsten O (1999) Our forests our future. Cambridge University Press, Cambridge, UK

Satake A, Rudel TK (2007) Modeling the forest transition: forest scarcity and ecosystem service hypotheses. Ecol Appl 17:2024–2036

Song C (2000) Comments on China's forest policy for the 21st century. Science. Edebate. http://www.sciencemag.org/cgi/eletters/288/5474/2135

Song C, Lord WJ, Zhou L et al (2008) Empirical evidence for impacts of internal migration on vegetation dynamics in China from 1982 to 2000. Sensors 8:5069–5080

Stern DI, Common MS, Barbier EB (1996) Economic growth and environmental degradation: the environmental kuznets curve and sustainable development. World Dev 24:1151–1160
Sun J, Zhao C, Wang L (2002) The long march of green: the chronicle of returning agricultural land to forests in China. China Modern Economics Press, Beijing
Trac CJ, Harrell S, Hinckley TM et al (2007) Reforestation programs in Southwest China: reported success, observed failure, and the reasons why. J Mount Sci 4:275–292
Tucker CJ, Pinzon JE, Brown ME et al (2005) An extended AVHRR 8-km NDVI dataset compatible with MODIS and SPOT vegetation NDVI data. Int J Remote Sens 26(20):4485–4498
Uchida E, Xu J, Rozelle S (2005) Grains for green: cost-effectiveness and sustainability of China's conservation set-aside program. Land Econ 81:247–264
Vogt KA, Gara RI, Honea JM et al (2007) Historical perceptions and uses of forests. In: Vogt KA, Honea JM, Vogt DJ et al (eds) Forests and society: sustainability and life cycles of forests in human landscapes. Cromwell, Trowbridge, UK
Wang G, Innes JL, Lei J et al (2007a) China's forestry reforms. Science 318:1556–1557
Wang S, Liu C, Wilson B (2007b) Is China in a later stage of a U-shaped forest resource curve? A re-examination of empirical evidence. Forest Policy Econ 10:1–6
Wang S, Chen JM, Ju WM et al (2007c) Carbon sinks and sources in China's forests during 1901–2001. J Environ Manage 85:524–537
Watson RT, Noble IR, Bolin B et al (2000) Land use, land-use change, and forestry. Cambridge University Press, Cambridge, UK
Xiao J, Moody A (2004) Trends in vegetation activity and their climatic correlations: China 1982–1998. Int J Remote Sens 25:5669–5689
Xiong D (1989) Recent history of forestry in China: 1840–1949. China Forestry Press, Beijing
Xu Z, Bennett MT, Tao R et al (2004a) China's sloping land conversion programme four years on: current situation and pending issues. Int Forestry Rev 6(3–4):317–326
Xu J, Tao G, Amacher GS (2004b) An empirical analysis of China's state-owned forests. Forest Policy Econ 6:379–390
Xu J, Yin R, Li Z et al (2006) China's ecological rehabilitation: unprecedented efforts, dramatic impacts, and requisite policies. Ecol Econ 57:595–607
Yin R (1998) Forestry and the environment in China: the current situation and strategic choices. World Dev 26:2153–2167
Zhang and Xu 2003: Zhang XQ, Xu DY (2003) Potential carbon sequestration in China's forests. Environ Sci & Pol 6(5): 421–432
Zhang Y, Song C (2006) Impacts of afforestation, deforestation and reforestation on forest cover in China from 1949 to 2003. J For 104(7):383–387
Zhang P, Shao GF, Zhao G et al (2000) Ecology – China's forest policy for the 21st century. Science 288:2135–2136
Zhang Y, Tachibana S, Nagata S (2006) Impact of socio-economic factors on the changes in forest areas in China. For Policy Econ 9:63–76

# Chapter 16
# Reforestation: Conclusions and Implications

**Harini Nagendra and Jane Southworth**

H. Nagendra (✉)
Indiana University, IN, USA and
Ashoka Trust for Research in Ecology and Environment (ATREE), Bangalore, India
e-mail: nagendra@ indiana.edu

J. Southworth
Department of Geography and Land Use and Environmental Change Institute (LUECI),
University of Florida, FL, USA

H. Nagendra and J. Southworth (eds.), *Reforesting Landscapes: Linking Pattern and Process*, Landscape Series 10,
DOI 10.1007/978-1-4020-9656-3_16, © Springer Science+Business Media B.V. 2010

## 16.1 Major Research Findings from Studies Presented in This Book

This book was written with two major objectives in mind. Our primary goal was to evaluate the drivers and processes associated with reforesting landscapes across a range of contexts and continents, to determine factors which appear to be specific to certain situations, and others which seem to be common across multiple contexts. Since reforestation is an area of study that crosses ecological, biophysical and social boundaries, our second objective was to develop interdisciplinary frameworks that link methods and approaches from different areas of science, to study the patterns and processes associated with reforesting landscapes. We evaluate the major findings of this volume based on these two objectives. First, we compare the research findings from different chapters, to help us discern the common designs and the unique threads that create a variety of reforestation patterns across landscapes (Table 16.1). We then use a combination of FAO data along with the specific case studies presented here for different countries, and develop a typology of forest change (Table 16.2). Finally, the main techniques and approaches used within each research study are presented, and assessed in terms of their effectiveness for reforestation research (Table 16.3). We conclude with an assessment of the implications of these findings, and outline some challenges for future research on reforestation.

## 16.2 Reforesting Landscapes – Drivers, Processes and Frameworks for Study

### *16.2.1 Dominant Drivers of Reforestation*

For each study the main causes or drivers of reforestation are extracted and illustrated in Table 16.1 in an attempt to determine the common drivers of change. This is one of the major goals of the book and can be related to the numerous research examples where dominant drivers of deforestation (Lambin 1994, 1997; Geist and Lambin 2001) and case studies compiling research results have been undertaken. This is one of the first attempts to undertake such a global compilation on drivers of reforestation, regeneration and regrowth studies (Table 16.1).

In summary, we find that the studies presented in this volume confirm many of the ideas linked to the Forest Transition Theory. The 'economic development' and 'forest scarcity' pathways, as proposed by Rudel et al. (2005), are observed in a number of landscapes, including Mexico and Central America as presented in Chapter 5, Eastern Europe in Chapter 6, as one of a number of drivers in Costa Rica from Chapter 10, Poland in Chapter 11 and for Vietnam in Chapter 14. We also find, however, that FTT is inadequate to explain all the instances of reforestation observed here, as also discussed by Rudel in Chapter 3. Institutions of conservation

**Table 16.1** Drivers of reforestation extracted from each case study chapter in the text

| Chapter no. and author | Country/study area | Drivers of reforestation |
|---|---|---|
| 5: Bray | Mexico and Central America (Belize, Guatemala, Honduras, El Salvador, Nicaragua, Costa Rica, and Panama) | Focuses on drivers of forest recovery, maintenance or protection. The four drivers are agricultural abandonment, coffee agroforestry, sustainable forest management and protected areas. |
| 6: Taff et al. | Eastern Europe: with three case studies in Latvia, Romania and Albania | Farm abandonment, population decrease, developing protected areas, sustainable forestry. |
| 7: Nagendra | South Asia: Bangladesh, Bhutan, India, and Nepal | Protection, plantations and agroforestry. |
| 8: Evans and Sweeney | USA – South Central Indiana | Diverse social, economic, institutional and environmental factors. |
| 9: Crews and Moffet | Peruvian Amazon | None. Modeling study. |
| 10: Daniels | Costa Rica | Agriculture intensification in one region facilitated reforestation in less productive lands, protected area establishment, declining beef prices, and revised forest policies. |
| 11: Kozack | Poland | Economic and social transformations leading to declining mountain agriculture, depopulation and land abandonment. |
| 12: Hartter et al. | Uganda | Park establishment and forest protection. |
| 13: Elmqvist et al. | Madagascar | A result of a combination of changes in precipitation, migration and decreased human population and livestock grazing pressure, but only under well defined property rights. |
| 14: Meyfroidt and Lambin | Vietnam and Bhutan | A combination of land scarcity and degradation, agricultural and socio-economic changes and policy in Vietnam. In Bhutan it was a governance and cultural system assigning a high value to nature. |
| 15: Song and Zhang | China | Government sponsored plantation forests. |

are a major driver associated with forest preservation, but also forest regrowth. These can be forest institutions such as government protected national parks (as one of the drivers in Costa Rica from Chapter 10 and in Uganda as illustrated in Chapter 13), but there are also several instances of forests protected by local communities (such

**Table 16.2** Countries studies in chapters in text and their current FAO information with references to change in forest area and forest stocking density, across the two time periods 1990–2000 and 2000–2005

| Country | Percent change in forest areas 1990–2000 | Percent change in forest area 2000–2005 | Dominant process DEF = deforestation REF= Reforestation | Annual change rate in growing stock in 1,000 m³/year 1990–2000 | Annual change rate in growing stock in 1,000 m³/year 2000–2005 | Dominant process DEG = degradation REG = regeneration |
|---|---|---|---|---|---|---|
| China | 1.2 | 2.2 | REF | 186,560 | 181,400 | REG |
| Costa Rica | –0.8 | 0.1 | DEF to recent REF | –1,940 | 200 | Recent REG |
| Guatemala | –1.2 | –1.3 | DEF | –8,806 | –8,806 | DEG |
| Honduras | –3.0 | –3.1 | DEF | –16,800 | –12,800 | DEG |
| India | 0.6 | 0.0 | REF to stable | 29,900 | 7,200 | REG |
| Latvia | 0.4 | 0.4 | REF | 9,500 | 10,600 | REG |
| Madagascar | –0.5 | –0.3 | DEF | –11,400 | –6400 | DEG |
| Mexico | –0.5 | –0.4 | DEF | NA | NA | NA |
| Nepal | –2.1 | –1.4 | DEF | 23,000 | –9,400 | Recent DEG |
| Nicaragua | –1.6 | –1.3 | DEF | –11,400 | –8,000 | DEG |
| Peru | –0.1 | –0.1 | DEF | NA | NA | NA |
| Poland | 0.2 | 0.3 | REF | 25,120 | 25,660 | REG |
| Turkey | 0.4 | 0.2 | REF | 9,890 | 5,679 | REG |
| Uganda | –1.9 | –2.2 | DEF | –3,730 | –3,800 | DEG |
| USA | 0.1 | 0.1 | REF | 189,600 | 210,000 | REG |
| Vietnam | 2.3 | 2.0 | REF | 13,570 | 11,183 | REG |

**Table 16.3** Tools and techniques used to analyze and detect the process of reforestation or forest regrowth

| Chapter no. and author | Remote sensing | GIS | Spatial statistics | Modeling | Forest plots | Household or community surveys | Meta-analysis of published case studies | National and global datasets e.g. FAO |
|---|---|---|---|---|---|---|---|---|
| 2: Grainger | | | | | | | | X |
| 3: Rudel | | | | | | | X | X |
| 5: Bray | | | | | | | X | X |
| 6: Taff et al. | X | | X | | | | | X |
| 7: Nagendra | X | X | X | | | X | X | X |
| 8: Evans et al. | X | X | | X | | X | | X (FIA) |
| 9: Crews and Moffet | X | X | X | X | | | | |
| 10: Daniels | X | X | X | X | | X | | |
| 11: Kozak | X | | | | | | X | X |
| 12: Hartter et al. | X | X | X | | | X | | |
| 13: Elmqvist et al. | X | X | X | | X | X | | |
| 14: Meyfroidt and Lambin | X | X | X | | | | | X |
| 15: Song and Zhang | X | X | X | | | | | X |

as in Nepal, Chapter 7 or Madagascar, Chapter 13, or even as in Chapter 14, Bhutan, where it was a governance and cultural system assigning a high value to nature). The discussions in these chapters clearly indicate that people do not only protect forests from a utilitarian perspective, driven by perceived scarcity of forest products, as proposed by Rudel et al. (2005) – but in addition, there is a basic human tendency towards conservation that makes itself powerfully shown even in cases where extreme poverty and forest dependence coexist with high population densities. Rudel also raises some of these issue is Chapter 3 of this volume.

Plantations and Agroforestry systems are also increasingly common pathways to reforestation and dominate in some cases, such as in China (Chapter 15), where government replanting schemes have created widespread reforestation in the form of plantations. Agroforestry generally occurs on a much smaller scale than government led plantation programs, but can also be a significant reforestation process, as illustrated in Central America and Mexico (Chapter 5 – both coffee agroforestry and sustainable forestry programs) and in South Asia (Chapter 8). With increasing concern about environmental services and the development of payment for environmental services provided, we may well see an increase in these smaller scale reforestation and agroforestry systems and so a better understanding of these systems is warranted. Ultimately, as stated Section 10.6, this volume: in

> context-specific case studies must be linked to broader global forest cover trends and the supply/demand of forest goods in order to appropriately contextualize regional or national-level forest recovery, its driving forces, and its degree of permanence. ….As of now, the issues of land development pressure, the timber trade and deforestation displacement are not explicitly addressed in the forest transition literature for developing countries. (Daniels, page 249)

This is an important addition to this body of literature and these forces of plantation forests, forestry and agroforestry will lead an increasingly important role if payment for environmental services programs continue (see Chapter 10 for more discussion) and also larger programs such as REDD (Reducing emissions from deforestation and degradation) which are policy programs linked to the post-Kyoto climate change agreements (see later for more discussion) which may link payments to reforestation.

### *16.2.2 Typologies of Forest Change*

Before 1990 there was much debate on the quality of the FAO database. However, for the 1990–2000, and 2005 datasets recalculations were undertaken, in which a single forest definition was used (10% canopy closure), and along with the advent of the use of coarse scale remote sensing as a means of collecting the data to supplement expert opinion for each country, these data are now more useable and more accurate (Grainger 2008; FAO 2006; Rudel et al. 2005). By using FAO databases some basic national-level information on the various countries can be discerned for the same time period (FAO 2006).

FAO data and location of the research countries within the FAO database and trends of change are given below for the main countries discussed within the text. For each country the dominant processes of reforestation or deforestation at the national level are given based on the change in forest area values, followed by the change in stocking density values, which is used as an indication of within cover processes of forest degradation or regeneration processes. Such processes may also be a precursor to future changes or shifts in forest area, with continued degradation and thinning of forest areas often occurring as a precursor to deforestation, and forest regeneration and increasing biomass within existing stands as a possible precursor to continued forest expansion and reforestation processes. Such an approach allows us to categorize different countries and regions into broad types, and develop a typology of forest change (Table 16.2).

If we review the trends illustrated in Table 16.2 we can see that there are four basic types of groupings occurring here at these national levels. (For a larger discussion of these types see Kauppi et al. (2006), which discusses the top 50 nations in terms of growing stock.) Here we are looking only at the countries highlighted within this volume and placing these countries within this larger discussion. As such we can see the following eight groups are possible outcomes here: (1) reforestation via the addition of new land in forest cover, shown by an increase in the percent area but not in growing stock or density; (2) deforestation via the loss of land in forest cover, but no real change in growing stock on the remaining land; (3) regeneration and increased density of existing forest lands, without additional area in forest cover; (4) degradation and thinning of existing forest areas, without the loss of actual forest cover extent; (5) processes of both reforestation and regeneration occurring together on the landscape; (6) deforestation and degradation occurring on the landscape; (7) reforestation occurring but with degradation within those, shown by decreasing stocking numbers; and (8) deforestation occurring but regeneration on remaining forested areas.

From the table we can see in reality we only have two main groups as the processes of reforestation and regeneration go together and deforestation and degradation. As such, we have China, Costa-Rica from 2000 to 2005, India, Latvia, Poland, Turkey, USA and Vietnam where reforestation is the dominant process occurring on the landscape as evidenced by an increasing area of land in forest cover, but hand in hand is the process of regeneration with increased stocking densities being found on the already forested lands. On the other hand we have Guatemala, Honduras, Madagascar, Nepal, Nicaragua and Uganda where deforestation or loss of land area in forest cover is occurring, and also a decrease in stocking or density on remaining forest lands. The two remaining countries of Peru and Mexico have deforestation as the dominant process, but data on stocking number changes are unavailable and so we do not know if these would also show the degradation trend. It is of interest to note then, that for those countries covered in this book, only half have reforestation and regeneration as the dominant national process occurring on the landscape. The remaining half of the locations has deforestation and most probably degradation occurring nationally, and yet here in this volume, we have highlighted regions or areas where the reverse process is occurring, in terms of reforestation or regeneration.

### *16.2.3 Tools for Reforestation Studies*

The tools used in each of the research sites are highlighted in Table 16.3 and discussed below.

The use of interdisciplinary tools and techniques, as illustrated in Table 16.3, highlights how reforestation crosses social and natural boundaries both in topic and in the range of tools and techniques used in its study. Even with this plethora of information and case studies as illustrated here though, it still seems there are gaps in some areas, as more studies need to directly measure and add an ecological dimension (Chapter 13 provides an example of this integration), and more studies need to add an economic dimension (Chapter 14 provides an example of this). The available case studies in the literature also provide a fruitful base for data mining, and this is growing rapidly with more people waking up to the idea that reforestation is a land cover trend in need of more research. As such this current volume represents timely information and serves to fill a void in the literature.

## 16.3 Continued and Future Challenges for Reforestation Research

Through this book, our endeavor is to map our current state of knowledge on reforestation, to outline the gaps in our understanding, and to identify the major challenges for reforestation research. Based on discussions with all authors contributing to this volume, we have identified major challenges critical to reforestation studies, which are not addressed in this volume, but which we feel are critical to address in future research. These are listed and discussed below.

### *16.3.1 Definitions of Reforestation*

A common language needs to be developed and agreed upon across the different scientific communities involved in this work, to consolidate the different terms used within this field. Generally speaking we have found that regeneration implies a natural process often occurring after land abandonment, and if left alone, usually leads to reforestation on the landscape. The point at which a regenerating landscape becomes reforested may be uncertain and is usually a function of the ecosystem in question. This process can be hard to 'see' or 'measure' with many of the interdisciplinary methods currently used (such as remote sensing, see Table 16.3) but links well to the need for more ecological measures and data collection as part of these studies. This implies a natural process and no additional role of humans, beyond the abandonment of the land. Additional forms of reforestation are those which come about as a process of planting trees, such as for agroforestry or plantations, also termed afforestation within the forestry community. Given the very different processes and pathways which can be followed a consolidation of the terminology currently in use, to create 'set definitions' would help studies crossing such disciplinary boundaries to communicate much more easily, and facilitate data exchange and comparison of studies.

### *16.3.2 Need Long-Term Assessments of Change*

Just like the long term ecological research sites (LTER) and other long term assessments, we also need these longer term research sites to study the issues of land cover change, including those related to reforestation. These studies should link the social and ecological processes together, and be at multiple temporal and spatial scales. They should incorporate a multitude of available methodologies (Table 16.3) and also emphasize repeat measures and the potential role of remote sensing, not just in terms of land cover classifications and conversions but also to emphasize the 'within-class' changes and modifications which are so key for us to evaluate and understand (Southworth et al. 2004). Such studies could be tied in with some major, key issues which we will be addressing in the coming years, specifically those of Carbon/REDD and Climate Change. This is important as issues of reforestation are long term processes that may be difficult to study in single date snapshots or shorter duration field studies (unlike the abrupt changes brought about immediately upon deforestation) and so to gain spotlight or the attention deserved we should link the issues associated with reforestation and the research needs, to those larger and more critical topics such as 'Carbon' and 'Climate Change'.

### *16.3.3 Drivers of Future Change*

There are a few drivers which we pick up as currently only influencing a small part of our study landscapes, but which are likely to play an increasingly significant role in the future. These drivers link to future research directions, and specifically are linkages between the environment and our existing social, political, and economic systems. As such we will mention these areas briefly in this section as these may well provide future mechanisms by which reforestation will occur. These areas are: climate change; carbon sequestration and REDD; and payment for environmental services.

Climate change has the potential to drive land cover change as people seek to sequester more carbon in our ecosystems (such as via reforesting landscapes) and so decrease the atmospheric carbon dioxide concentrations. While there is still much current debate on the role of forests and potential gains of reforestation schemes (Clark 2004; IPCC 2007), this debate seems likely to continue and the potential for reforestation schemes to increase globally (though there are projected regional differences in the potential impacts of reforestation). This also links into the topic of carbon sequestration and REDD (reducing emissions from deforestation and degradation). This program was created by the United Nations as collaboration between FAO, UNDP and UNEP and is a financial pooling of resources to provide funding of activities which will lead to a reduction in deforestation and forest degradation in developing countries. This program came about as a direct result of the intergovernment panel on climate change (IPCC 2007) report which estimated that deforestation was contributing over 20% of the overall greenhouse gas emissions and that forest degradation added even more to this. As such they stated:

> there is an immediate need to make significant progress in reducing deforestation, forest degradation, and associated emission of greenhouse gases. The UN-REDD Programme is aimed at tipping the economic balance in favour of sustainable management of forests so that their formidable economic, environmental and social goods and services benefit countries, communities and forest users while also contributing to important reductions in greenhouse gas emissions. The aim is to generate the requisite transfer flow of resources to significantly reduce global emissions from deforestation and forest degradation. The immediate goal is to assess whether carefully structured payment structures and capacity support can create the incentives to ensure actual, lasting, achievable, reliable and measurable emission reductions while maintaining and improving the other ecosystem services forests provide (from the UN-REDD Program website http://www.undp.org/mdtf/un-redd/overview.shtml, accessed Feb 2009).

If indeed such an influx of resources, aimed specifically at forest recovery and reforestation in developing countries comes to fruition, then new pathways of reforestation seem likely, driven by this global policy and effort.

While the REDD program has very specific goals there are more general types of trading and payment programs for environmental service, such as the carbon market, and individual country programs (see Chapter 10, Costa Rica for an example). These also seem likely to not only continue but also to expand in the future, as more developing countries realize the financial gain possibilities related to their forests or the processes of reforestation. While not all such ecosystem services may result in reforestation of landscapes, it seems likely that linked in to the carbon markets, some reforestation will occur, although the longevity of these programs and their sustainability over time may lead to future challenges.

Overall though, through all three of the topics discussed here, it seems future trajectories of land cover change may increasingly relate to reforestation and our understanding of these pathways and their associated processes is therefore of paramount importance.

## 16.4 Final Thoughts

To this point there has been a lot less emphasis on reforestation, both in the scientific literature and also within the general public arena (Table 16.4). This book was developed as an attempt to address some of these limitations and also to provide a single summary of a suite of theoretical, methodological and substantive examples

**Table 16.4** The numbers of web based search results of the terminology used in this book, highlighting the dominance of deforestation over reforestation as a subject of study

| Search term | Google | Scholar Google | Web of science |
|---|---|---|---|
| Deforestation | 3,710,000 | 107,000 | 5,653 |
| Reforestation | 2,690,000 | 52,800 | 1,853 |
| Forest degradation | 400,000 | 195,000 | 2,678 |
| Forest regeneration | 462,000 | 153,000 | 5,707 |
| Forest regrowth | 517,000 | 32,700 | 742 |

of reforestation research underway across the globe. It was our hope to attempt to provide researchers with an ability to determine common drivers of reforestation processes (Table 16.1), to illustrate the commonly used tools and techniques (Table 16.2) and also to highlight the gaps in the current research where future work must be directed (this chapter). We hope to have satisfied these criteria and that this volume both initiates new discussions and serves as a basis for continued future research on the development of improved understanding of reforestation pathways, processes and drivers.

## References

Clark D (2004) Tropical forests and global warming: slowing it down or speeding it up? Front Ecol Environ 2(2):73–80

Food and Agriculture Organization (2006) Global Forest Resources Assessment 2005. Food and Agriculture Organization of the United Nations, Rome

Grainger A (2008) Difficulties in tracking the long-term global trend in tropical forest area. Proc Natl Acad Sci U S A 105:818–823

Geist HJ, Lambin EF (2001) What drives tropical deforestation? A meta-analysis of proximate and underlying causes of deforestation based on subnational case study evidence. Louvain-la-Neuve (Belgium): LUCC International Project Office, LUCC Report Series no. 4

IPCC (2007) In: Solomon S, Qin D, Manning M, Chen Z, Marquis M, Averyt KB, Tignor M, Miller HL (eds) Climate change 2007: the physical science basis. Contribution of working group I to the fourth assessment report of the intergovernmental panel on climate change. Cambridge University Press, Cambridge/United Kingdom/New York, NY, 996 pp

Kauppi PE, Ausubel JH, Fang J, Mather AS, Sedjo RA, Waggoner PE (2006) Returning forests analyzed with the forest identity. Proc Natl Acad Sci U S A 103(43):17574–17579

Lambin EF (1994) Modelling deforestation processes: a review. Luxembourg: European Commission, Directorate-General XIII. Report no. EUR-15744-EN. (Available from Office of Official Publications of the European Community, Luxembourg.)

Lambin EF (1997) Modelling and monitoring land-cover change processes in tropical regions. Prog Phys Geogr 21:375–393

Rudel TK, Coomes OT, Moran E, Achard F, Angelsen A, Xu J, Lambin EF (2005) Forest transitions: toward a global understanding of land-use change. Global Environ Change 15:23–31

Southworth J, Munroe DK, Nagendra H (2004) Land cover change and landscape fragmentation - comparing the utility of continuous and discrete analyses for a Western Honduras region. Agric Ecosyst Environ 101(2):185–205

# Index

**D**

**G**



**H**



**I**

## N



## O



## P

**Q**



**R**

**T**



**U**



**V**



**W**



**Y**



**Z**

# Color Plates

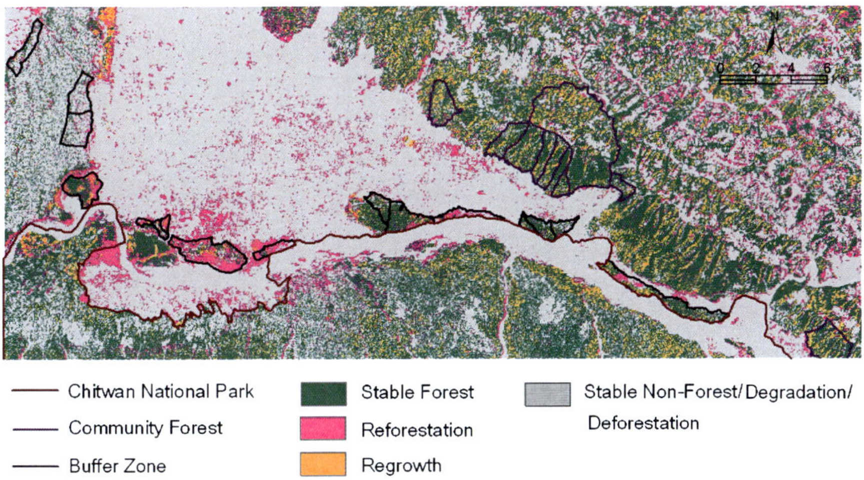


**Fig. 7.2** Map of forest change in Landscape 1 – Chitwan district, Nepal *Terai* plains

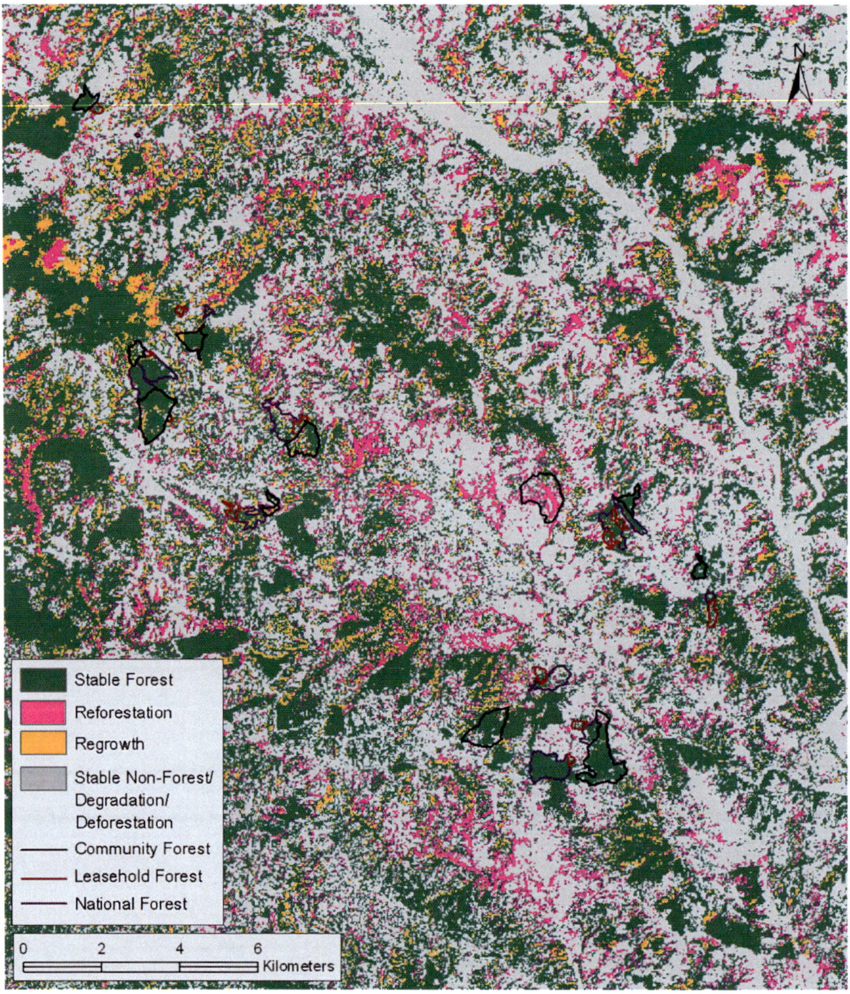


**Fig. 7.3** Map of forest change in Landscape 2 – Kabrepalanchowk district, Nepal middle hills

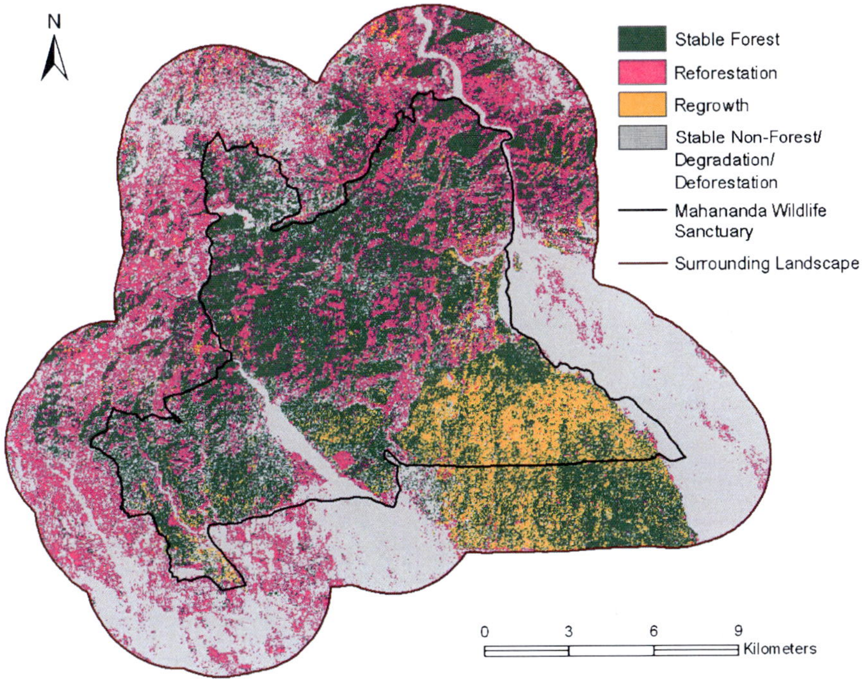


**Fig. 7.4** Map of forest change in Landscape 3 – Mahananda Wildlife Sanctuary, eastern Indian foothills

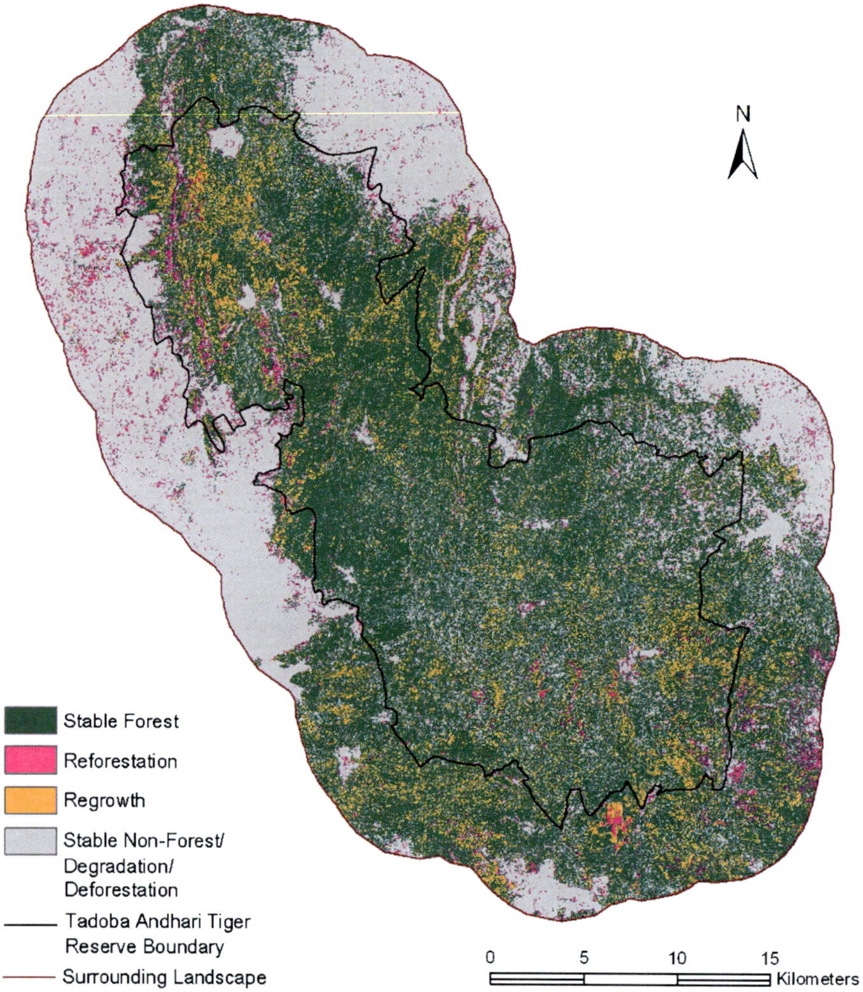


**Fig. 7.5** Map of forest change in Landscape 4 – Tadoba Andhari Tiger Reserve, central Indian plains

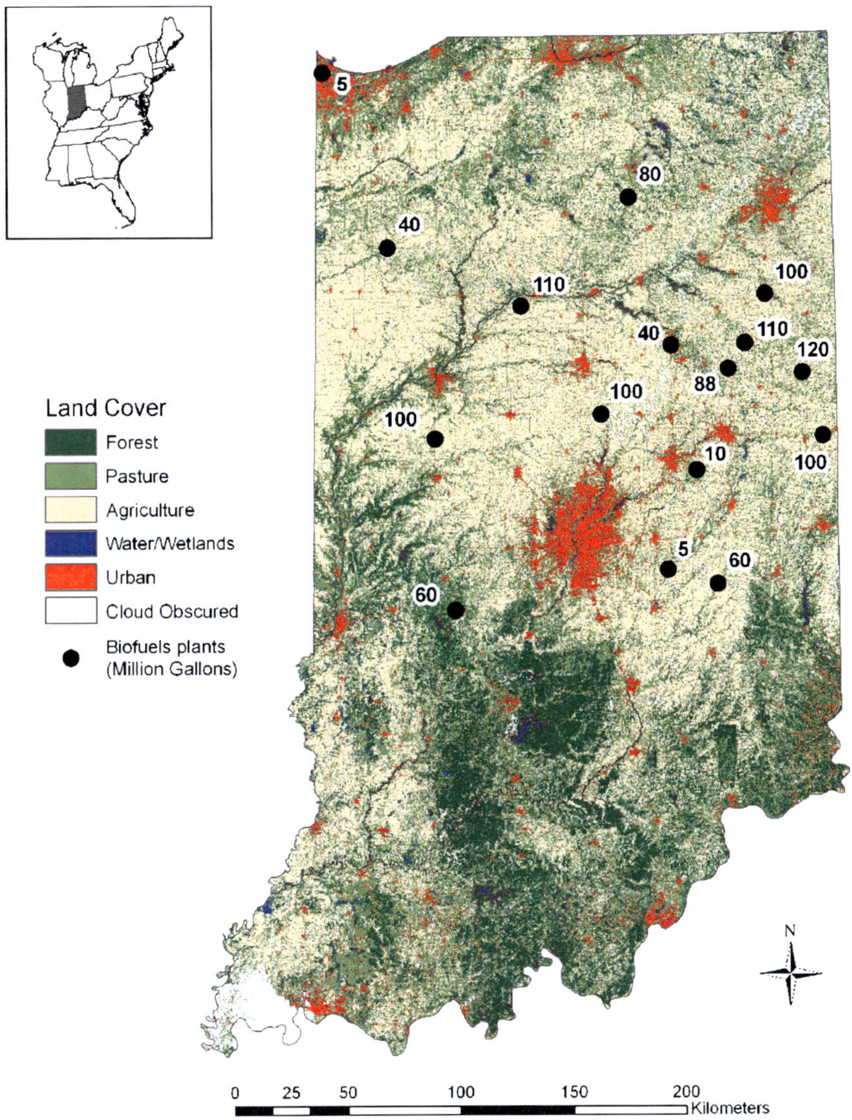


**Fig. 8.6** Land cover in Indiana, 2003, and biofuel production facilities (ISDA 2008, NRCS 2008)

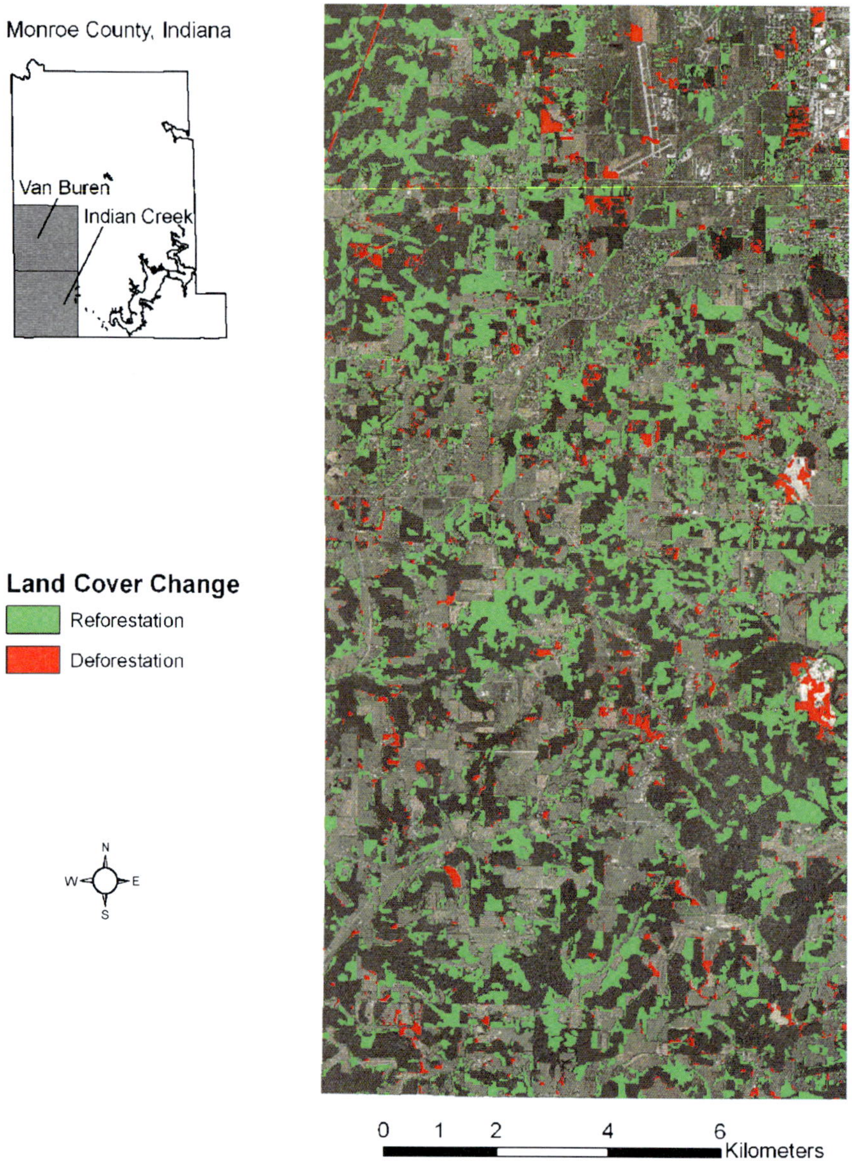


**Fig. 8.8** Land-cover change, 1939–2003, Indian Creek and Van Buren townships, Monroe County, Indiana

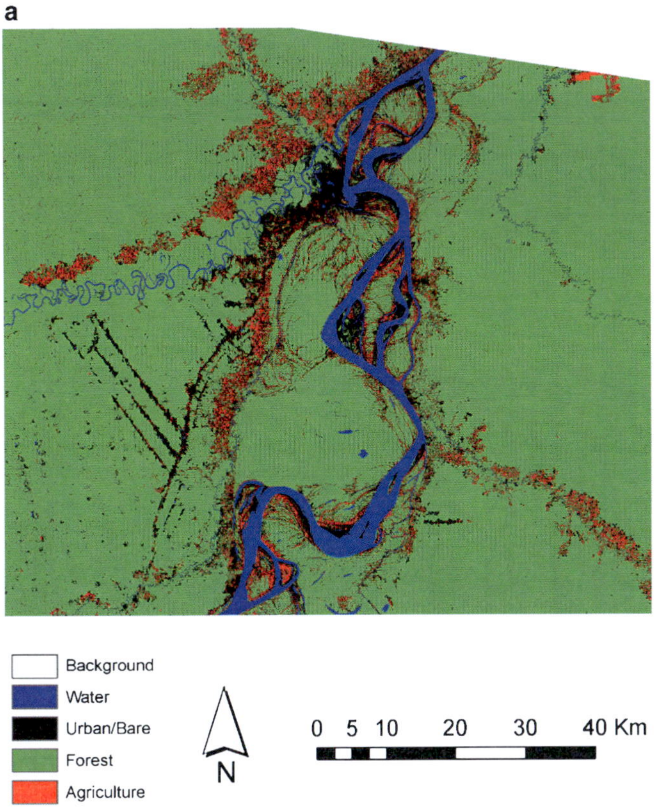


**Fig. 9.3** Landscapes used to initiate simulations: (**a**) the unaltered landscape

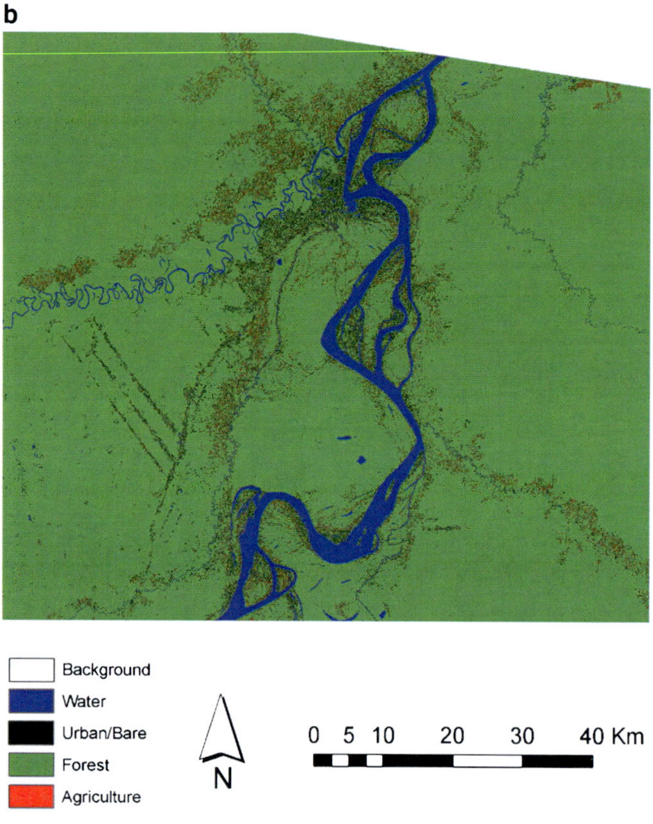


**Fig. 9.3b** Landscapes used to initiate simulations: the landscape altered so as to increase forest cover by 10%

c

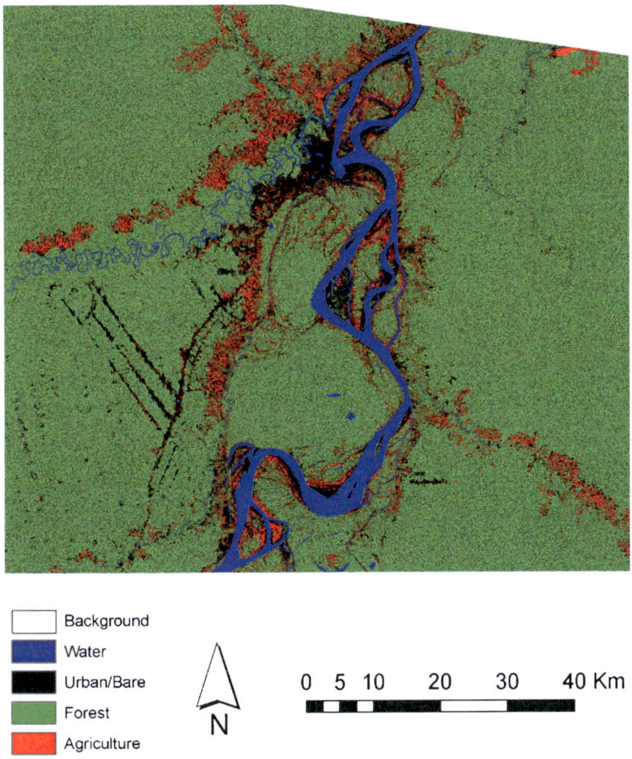


**Fig. 9.3c** Landscapes used to initiate simulations: the landscape altered so as to decrease forest cover by 10%

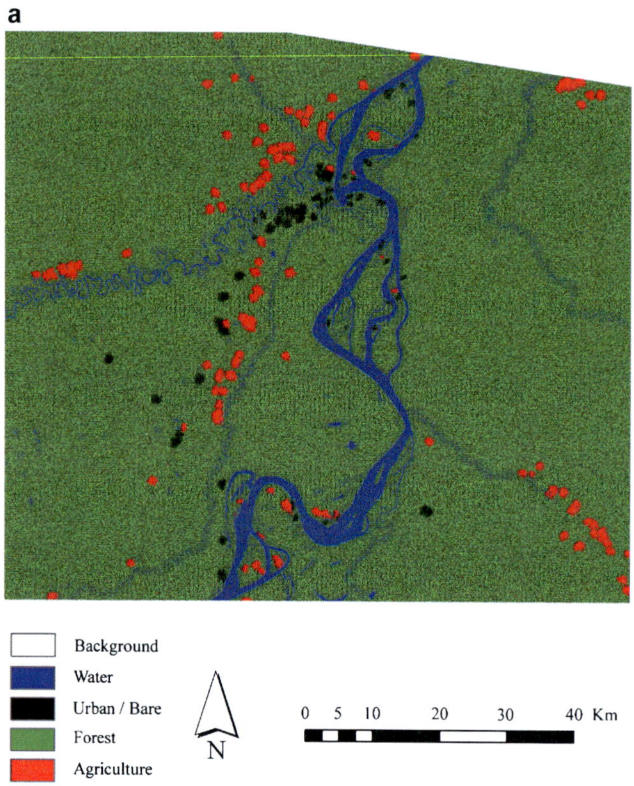


**Fig. 9.4** Simulation results: (**a**) a randomly selected result starting with a 10% increase in forest cover

b

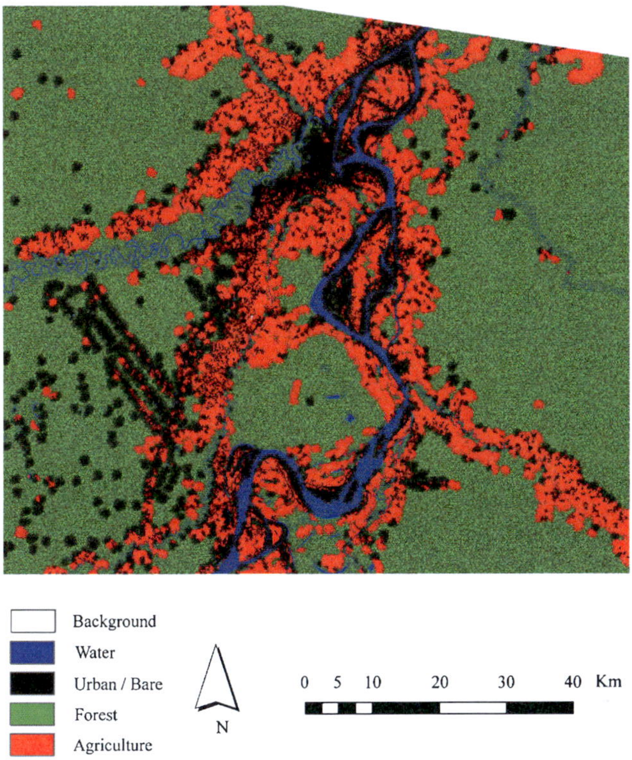


**Fig. 9.4b** Simulation results: a randomly selected result starting with a 10% decrease in forest cover

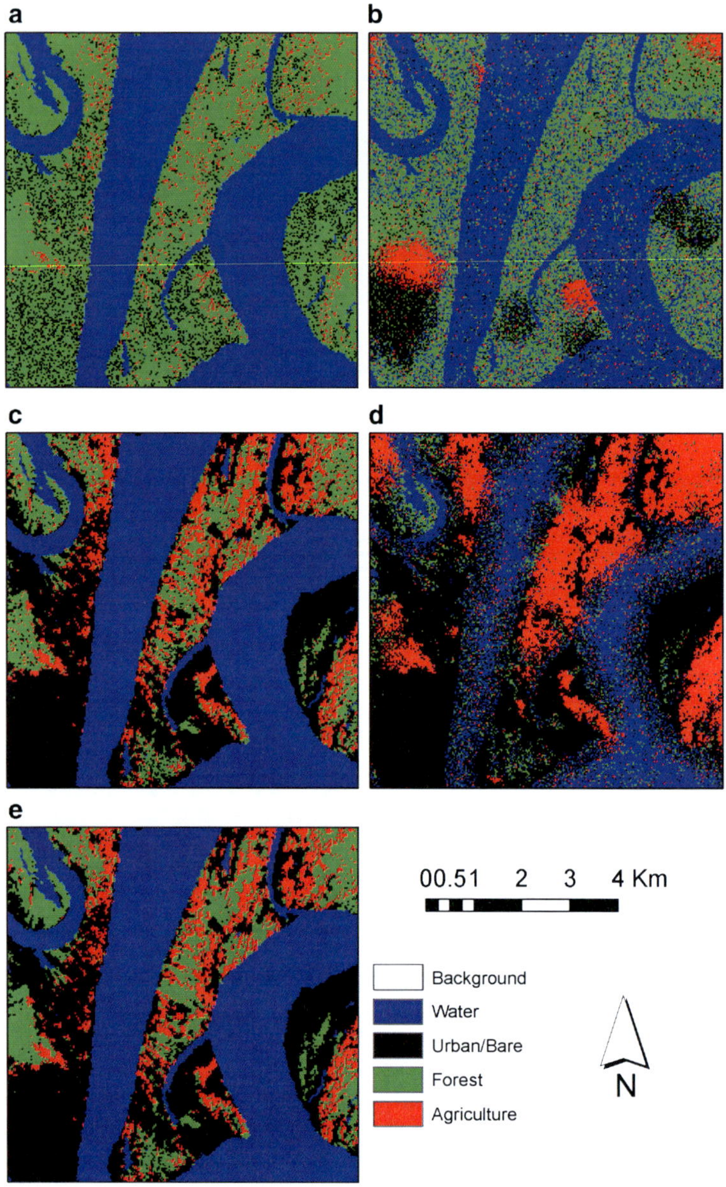


**Fig. 9.5** Simulation results, subset area: the effects of initial forest cover on landscape dynamics when viewed at a finer scale: (**a**) the initial conditions associated with an increase in forest cover; (**b**) a randomly selected landscape produced from the initial conditions in (**a**); (**c**) the initial conditions associated with a decrease in forest cover; (**d**) a randomly selected landscape produced from the initial conditions in (**c**); (**e**) the actual initial conditions, unaltered to reflect an increase or decrease in forest cover

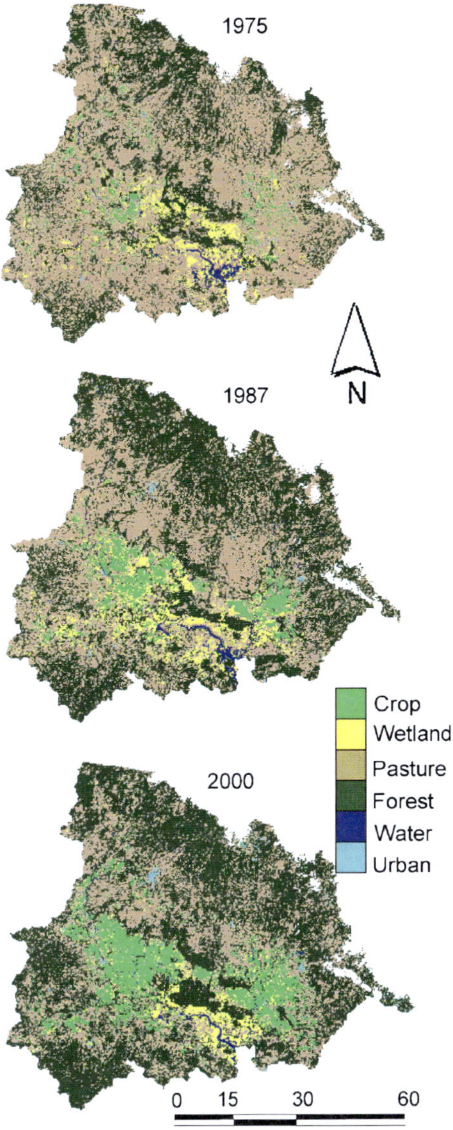


**Fig. 10.2** Land cover maps (plus water) for 1975, 1987, and 2000. Note water and urban classes were not employed in trajectory analysis

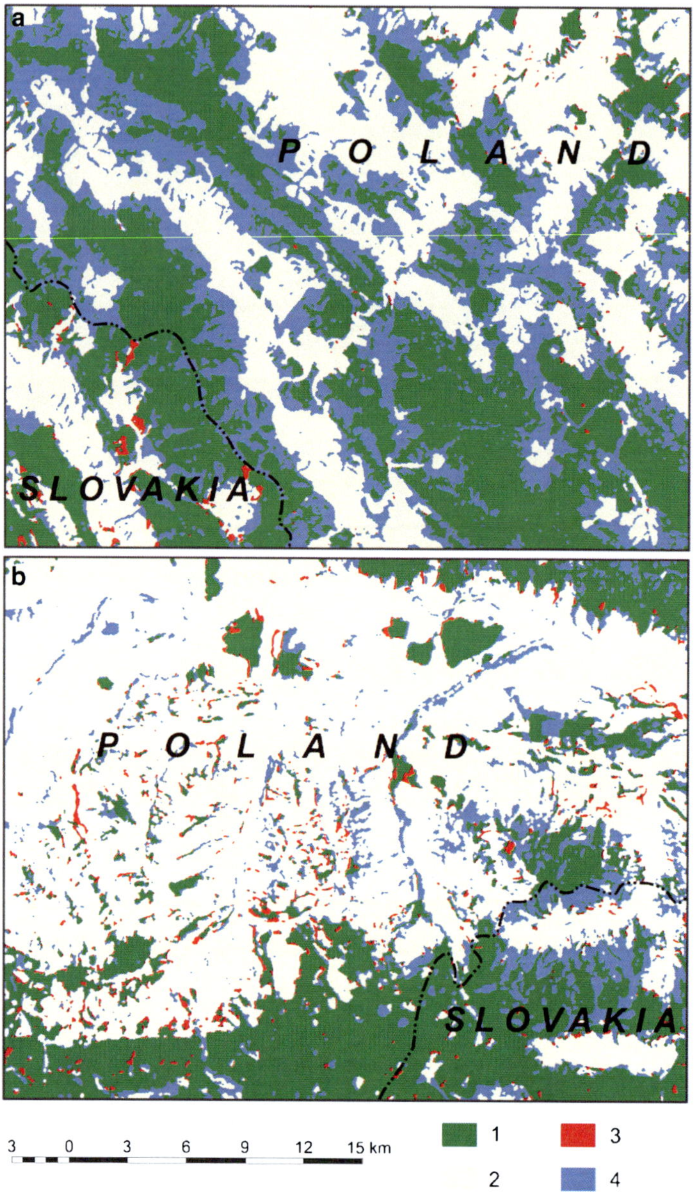


**Fig. 11.2** Forest cover changes in the depopulated areas (Beskid Niski and Bieszczady Mountains), 1936–1992 (**a**) and outside the depopulated areas (Podhale Region), 1931–2000 (**b**), after Kozak et al. (2007a). 1 – forests, no change in the analysed period; 2 – open land, no change in the analysed period; 3 – deforestation; 4 – forest expansion

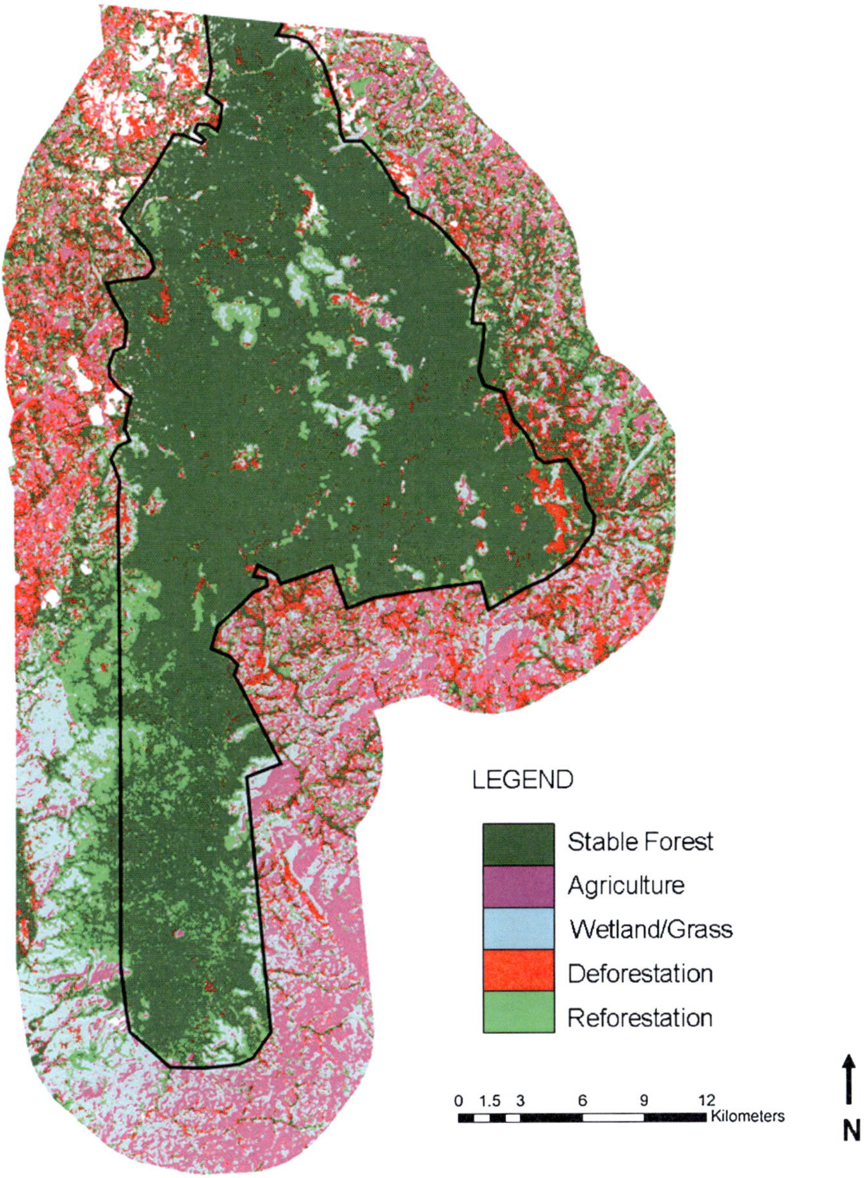


**Fig. 12.2** Trajectories of land cover change occurring within the park boundaries and surrounding landscape from 1984 to 1995 to 2003

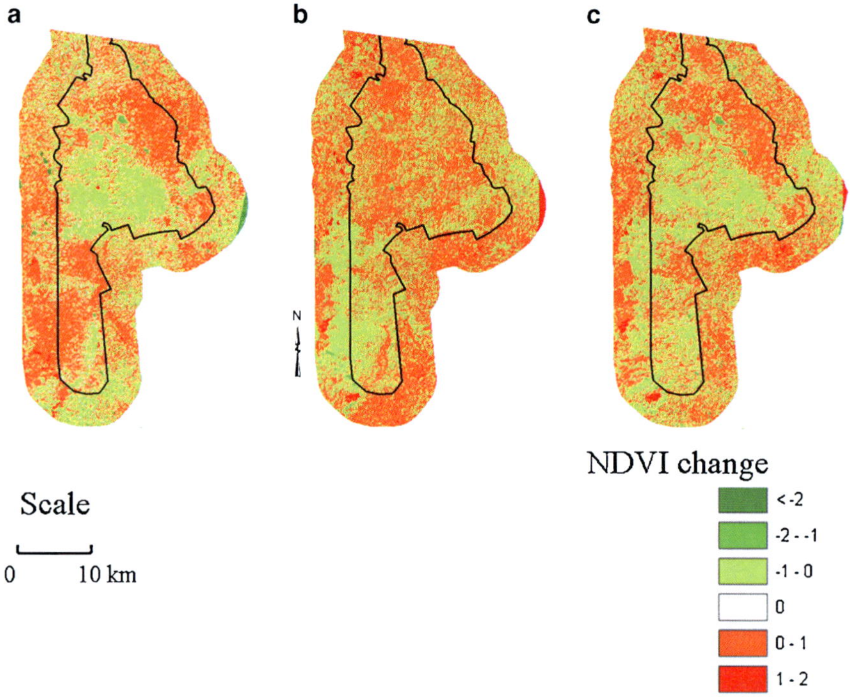


**Fig. 12.3** NDVI change trajectories showing the NDVI difference based on the individual image date NDVI standard normal deviates, for the three different time-steps, (**a**) 1984–1995, (**b**) 1995–2003, and (**c**) 1984–2003, where negative values indicates an increase in NDVI over time

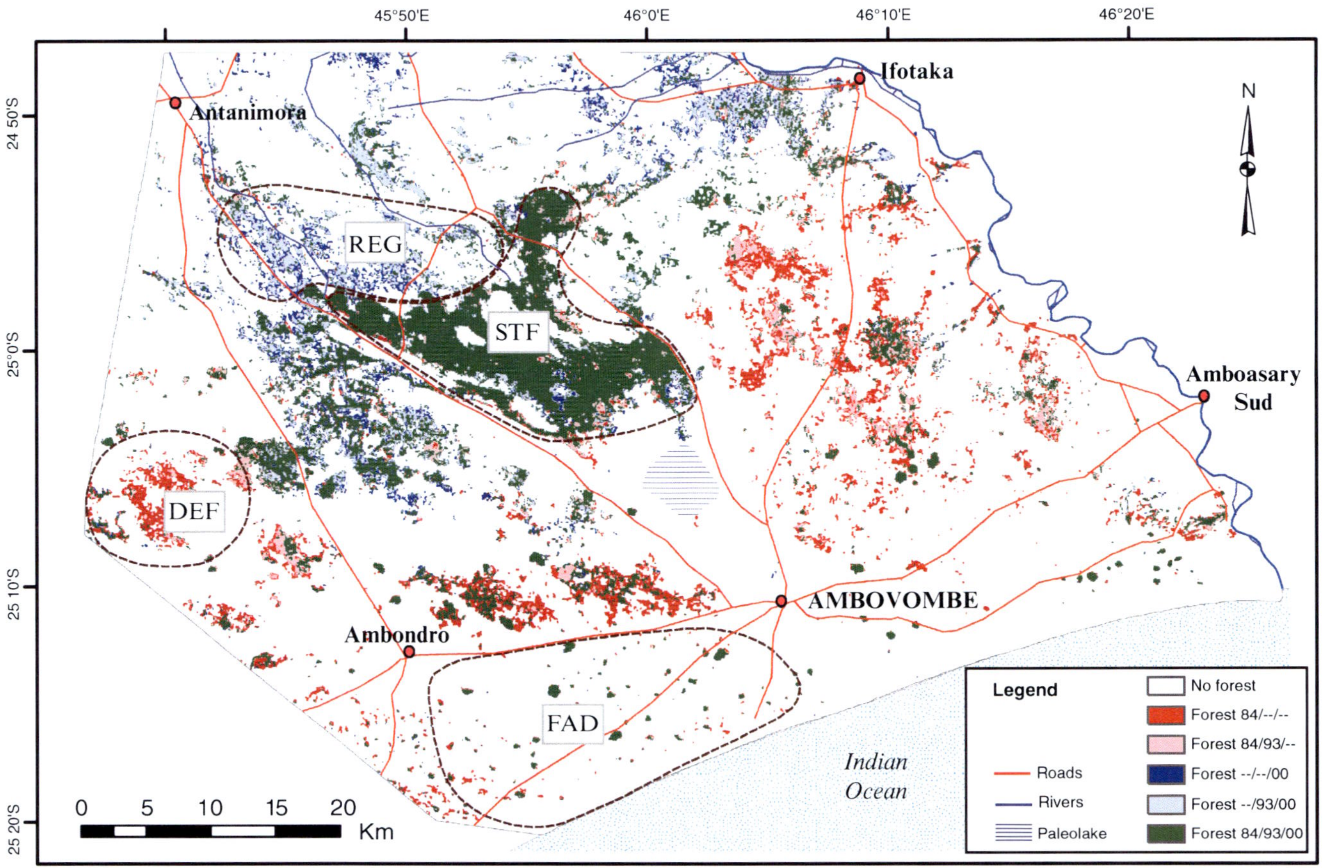


**Fig. 13.1** Time-series analysis of changes in forest cover based on satellite images (dry season synoptic views from 25 June 1984 (Landsat 5 TM), 15 April 1993 (Landsat TM) and 28 May 2000 (Landsat 7 ETM+) from Androy, southern Madagascar. Four areas were used for ecological and social surveys REG=regenerating forest, STF=stable forest, FAD=forest patches protected through taboos, DEF=deforestation (modified after Elmqvist et al. 2007)

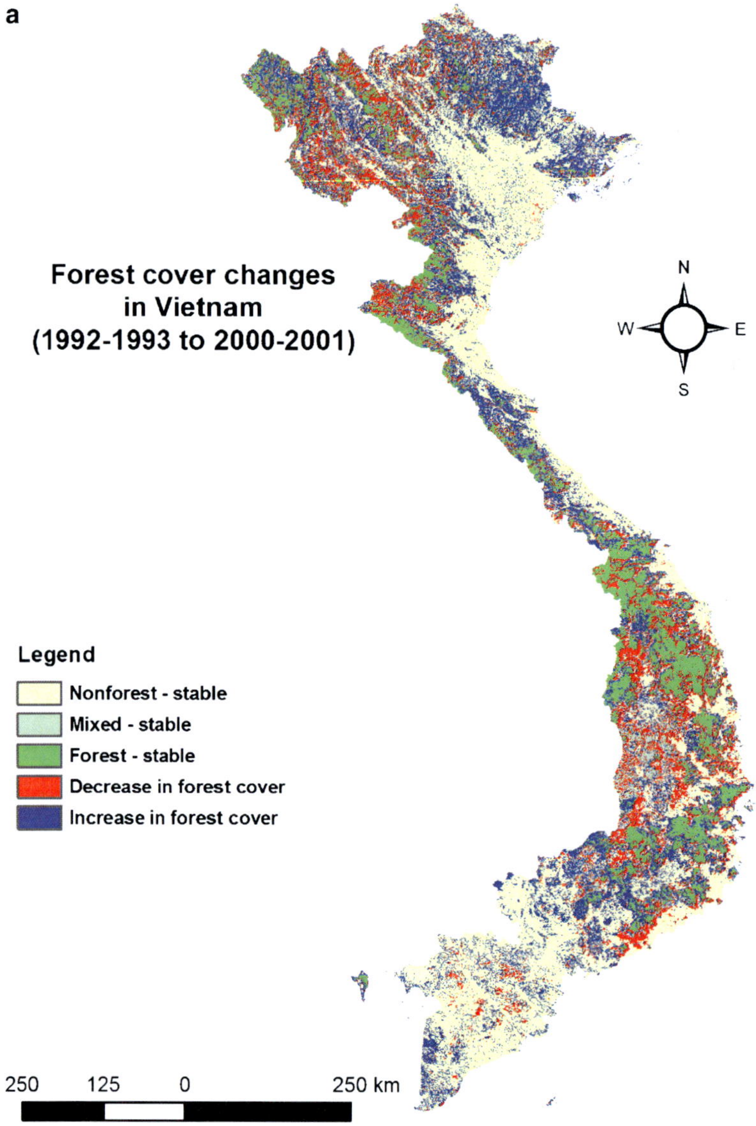


**Fig. 14.2** Forest cover changes and ecoregions in Vietnam (adapted from Meyfroidt and Lambin 2008b). (**a**) Forest cover changes were derived from the University of Maryland VCF maps